Quaternary Climate Change over the Indian Subcontinent

Quaternary Climate Change over the Indian Subcontinent

Edited by
Neloy Khare

CRC Press
Taylor & Francis Group
Boca Raton London New York

CRC Press is an imprint of the
Taylor & Francis Group, an **informa** business

Library of Congress Cataloging-in-Publication Data
Names: Khare, Neloy, editor.
Title: Quaternary climate change over the Indian Subcontinent / edited by
Neloy Khare.
Description: First edition. | Boca Raton : CRC Press, 2021. | Includes
bibliographical references and index.
Identifiers: LCCN 2020055183 (print) | LCCN 2020055184 (ebook) |
ISBN 9780367537579 (hardback) | ISBN 9781003083238 (ebook)
Subjects: LCSH: Climatic changes—Research—South Asia. |
South Asia—Environmental conditions. | Climatic geomorphology—South Asia.
Classification: LCC QC903.2.S64 Q38 2021 (print) | LCC QC903.2.S64
(ebook) | DDC 551.6954—dc23
LC record available at https://lccn.loc.gov/2020055183
LC ebook record available at https://lccn.loc.gov/2020055184

ISBN: 978-0-367-53757-9 (hbk)
ISBN: 978-0-367-53758-6 (pbk)
ISBN: 978-1-003-08323-8 (ebk)

Typeset in Times
by codeMantra

Dedicated to My revered Guru
Late Prof Indra Bir Singh

Late Prof Indra Bir Singh

(08.07.1943 – 11.02.2021)

Prof Indra Bir Singh received his BSc (Hons) and MSc (Geology) from Lucknow University in the early 1960s, joined as a lecturer in 1972 and served for approximately 40 years. He was Head of the Geology Department, Lucknow University. He published many books and over 200 research papers. Prof. Singh was famous among students, teachers, and all science fraternity members for his simplicity and over five decades of scientific contribution.

Prof. Singh has also taught at Louisiana State University (US) and Friedrich-Alexander-Universität Erlangen-Nürnberg (FAU) (Germany). He was the Alexander Humboldt Fellow in Germany (1978-79) and the recipient of the prestigious L. Rama Rao Centenary Award and National Geoscience Award for Excellence.

He was a teacher par excellence who has not only taught geology but life lessons as well. He nurtured and trained a band of dedicated students, perpetuating his legacy in different institutions in India and abroad. He developed the Centre of Sedimentological Research at Lucknow University. His contribution was not confined to Lucknow University, but at the national and international level as well. He was associated with the International Union of Geological Sciences (IGCP) and contributed to Dwarka Excavation. He did a comprehensive study of the Ganga Plains, providing an in-depth analysis of landform development.

He was a forceful free thinker who nurtured further in Wilhelmshaven on the North Sea where he was trained to see sedimentary structures forming live on the tidal flats. Along with his mentor Prof. H.E Reineck, Prof. Singh wrote a remarkable book Depositional Sedimentary Environment

*published by Springer Verlag, a book
that received more than 15 thousand
citations in the world literature. With
an exceptionally inquisitive and
analytical mind, Prof. Singh challenged
many age-old geological beliefs.
He picked up loopholes in the existing
literature by asking simple questions
like "Why we do not get prolific biota
in the tidal flat Himalayan sediments
assigned to Mesozoic ages?" It led to
a complete overhaul of the Himalayan
stratigraphy, making the entire
Himalayan succession as Proterozoic.
His geological contributions included
reformatting the sedimentary
architecture of the Vindhyan sequence
in Central India. Prof Singh's novel
interpretations of the sedimentary fill in
the Ganga Plains opened a new window
to the study of Kutch successions. His
remarkable contributions include the
innovative understanding of the Karewa
sediments in the Kashmir Himalaya,
archaeological investigations on
the Ganga alluvium, and a novel
interpretation of the sedimentation
of the Ladakh region. His scientific
contributions spanned between
Precambrian and Anthropocene,
the entire stratigraphic spectrum.*

*****

Contents

Foreword

The Indian subcontinent has unique geographic position having "the Himalayas" in the north and embraced by vast seas in South India. The tropical seas are the generation zone of low cumulus clouds. Thus, the Indian subcontinent preserves a mosaic of ecosystems subject to the Indian summer monsoon's influence and strength. The region comprises diverse ecological zones with complex geological and climatic histories, including a biannual monsoon prevalence. A stratigraphic evaluation of the Indian subcontinent during Quaternary reveals a strong correction among biotic (paleo-life), abiotic components, and significant Tectonic and climate-induced sea levels and vegetation changes. Indian summer monsoon (ISM) is a prominent component of atmospheric circulation and significantly influences the global climate; indubitably, the Quaternary is characterized by the large-scale environment and associate sea-level changes on different time scale, including the Younger Drays event (12,700–11,500 years ago) at the end of the Pleistocene.

Nevertheless, many important events remained unexplored despite many studies from this region dedicated to Quaternary climates, such as the 74-ka Toba volcanic super eruption's possible effect. Similarly, the impact of abrupt shifts in the Indian Summer Monsoon (ISM) on Indian socio-economy, associated with the rise and fall of ancient human civilization, remained an enigma. Such signature of environmental/climatic condition is well preserved in many natural archives. Therefore, the Indian subcontinent presents a critical, but inadequately documented, region for gleaning clues and investigating significant facets of Quaternary climates.

The present book *Quaternary climate change over Indian Subcontinent* aptly assimilates and collates various recent findings related to the reconstruction of Quaternary climate change through diversified geological basins/proxies like lakes, rivers, estuaries, speleothem, shallow/deeper ocean, besides other in and around the Indian subcontinent. I am sure this book will be useful reference for climate changes research and motivate young scholars to embark upon climate science's most challenging world.

Prof. Talat Ahmad
FNA, FASc, FNA Sc, JC Bose National fellow
Vice chancellor
University of Kashmir
Hazratbal Srinagar – 190006
Jammu and Kashmir (India)
February 13, 2021

Preface

The Quaternary Period is unique among all divisions of geologic time of the 4.5 billion years of Earth history that most preoccupies humans. The modern environments of the Indian subcontinent are diverse with unique distinctive characteristics such as deserts, tropical rainforests, alpine tundra and savannahs, and subtropical woodland and mangroves. Over a sixth of the world's current human population (United Nations Department of Economic and Social Affairs and Population Division 2015) is sustained in this region. Owing to its geographical situation, the Indian subcontinent is subject to warm temperatures with a limited seasonal change. Increases in solar insolation, predominately due to 19,000–23,000-year precessional cycles, result in the northward movement of the Intertropical Convergence Zone (ITCZ). The northward movement of the ITCZ results in an increased monsoonal circulation, leading to a greater level of precipitation.

The Indian subcontinent is known to have new discoveries regarding Lower Paleolithic hominins and their biological and behavioral evolution. This region comprises diverse ecological zones with complex geological and climatic histories, including a biannual monsoon prevalent since the Miocene. The Neogene–Quaternary mammal genera at 1-Ma intervals have been correlated with the faunal turnovers (first and last appearances and immigrations), diversity, and major tectonic and climate-induced sea-level and vegetation changes.

The signatures of environmental and/or climatic conditions during Quaternary Period are well preserved in many natural archives available over land and ocean. Many efforts are made to unravel such mysteries of climate during Quaternary Period through land-based geological archives coupled with marine counterparts during this important period. Although Quaternary scientists have recognized the close linkage between Earth's atmosphere and its oceans play an essential role in redistributing energy on the planet. Climate change and sea-level variation have been rapid during the Pleistocene. Major restructurings in climate that happened on time scales of decades, such as during the Younger Dryas interval (12,700–11,500 years ago) at the end of the Pleistocene, are well documented.

Nevertheless, we have little sense or appreciation of recent history, and even less so of geologic history. As we look back in time, humans become increasingly disconnected from it. Moreover, they have considerable interest in events of the past several years to several decades, less interest in events hundreds to thousands of years ago, and, with some exceptions, almost no interest in events millions or billions of years old. However, the Quaternary is characterized by a large-scale climate and associated sea-level changes. In order to gain a better understanding of the relative contribution and functioning of natural processes on climate and landscape changes, understanding this particular time period is of paramount importance.

Undoubtedly in the wake of ongoing climate change worldwide, the Quaternary Period has gained global significance owing to the burgeoning interest to understand the climate change science and its related issues.

The Quaternary geoscience has evolved through several specializations such as paleoclimate, paleoenvironment, and past global changes. A large number of new data, insights, and discoveries have been made focusing on the Quaternary climate change over the Indian subcontinent through developments of new trends, concepts, and techniques. The progress made in this field during the last few decades is remarkable. Valuable climate data from the Himalaya, Indo-Gangetic plains, lakes, speleothems, Arabian Sea, Bay of Bengal, Indian Ocean, coastal regions, rivers etc. has been generated and added to our understanding of the climate change during Quaternary Period over the Indian subcontinent.

It is appropriate to evaluate the current status of patterns of climatic change over the Quaternary Period in and around the subcontinent by combining all possible archives of paleoenvironmental variability at different time scales. It is therefore an attempt that needs to be made to synthesize the climate change data generated in Indian context. Such an attempt becomes important in the present scenario of perceptible climatic changes and the generation of predictive models and scenarios based on available records which are richly expressed by a variety of archives in a wide geographic extent.

In order to assimilate and collate various recent findings related to the reconstruction of Quaternary climate change, gleaning clues from various natural archives, the present book titled *Quaternary Climate Change over Indian Subcontinent* is prepared, which deals with many such aspects of Quaternary Period. It covers various depositional regimes (ocean, estuaries, coastal rivers, delta etc.) and proxies (planktonic, benthic, pollens and spores, invertebrates, geochemistry, sedimentology). It also explores a possible teleconnection on regional- and global-scale besides sealevel changes and neotectonics.

The book begins with an exhaustive review of the Late Quaternary climatic history of the Indian subcontinent by Rajan. The climatic and anthropogenic influences on Himalayan glacial and nonglacial lakes have been reviewed by Meena et al. However, monsoon variability during Holocene Period over Peninsular India has been studied in detail by Achyuthan using lake and marine sediment records. On the other hand, an exhaustive account of speleothems-based monsoonal assessment and Holocene–Quaternary climate understanding from the Indian subcontinent has been provided by Tewari. A 45-ka record of productivity in the western Bay of Bengal with an aim to understand their implications on the Indian monsoon and Atlantic climate has been demonstrated by Mir. The interplay of Mid- to Late Holocene tectonism and northeast monsoon on the landscape evolution has been ably studied by Reshmi and Achyuthan. On the other hand, the importance of the geochemistry of Northern Indian Ocean sediments for assessing the Quaternary climate change has been highlighted by Manjunatha et al. Similarly, pollen analysis and paleoenvironmental studies of archeological deposits from the Konkan Coast of India have been attempted by Naik. In-depth evaluation has been done by Panchang and Sen on Late Quaternary sea levels along the Indian subcontinent vis-à-vis an ongoing climate change. Recent and paleo-tsunami deposits over the Indian subcontinents have been evaluated by Chaturvedi and Khare. With regard to the response of climate change to modern microlife, the Ostracod diversity from continental slope sediments of Gulf of Mannar, India, has been studied by Rajkumar et al. However, ecological assessment of recent benthic Ostracoda, off Kurusadai Island, Gulf of Mannar, India, has been made by Maniyarasan et al. Benthic foraminifera, texture, and sediment geochemistry along the depositional environment of Palk Strait, East Coast of India, have been studied by Suresh Gandhi and Kasilingam. Additionally, the Quaternary faunal records from Upper Reaches of Sina Basin focusing on Math Pimpri – an early historic site – have been studied by Sabale et al.

Neloy Khare
February 2021
New Delhi

Acknowledgments

Kind support and help received from all contributing authors are gratefully acknowledged; without their valuable inputs on various facets of the Quaternary climate over the Indian subcontinent, this book would not have been possible. Various learned experts who have reviewed different chapters are sincerely acknowledged for their timely important and critical reviews.

I express my sincere thanks to the Ministry of Earth Sciences, Government of India, New Delhi (India), for various inputs, support, and encouragements. Dr. M.N. Rajeevan, Secretary, Ministry of Earth Sciences, Government of India, and Dr. Arbind Mitra, Secretary, Office of the Principal Scientific Advisor to Prime Minister, Government of India, New Delhi, have always been the source of inspiration and acknowledged for their kind support. Dr. Vipin Chandra, Joint Secretary to Government of India at Ministry of Earth Sciences, New Delhi (India); Dr. Prem Chan Pandey, former Director, National Centre for Polar and Ocean Research, Goa (India); and Dr. K.J. Ramesh, former Director General, India Meteorological Department (IMD), New Delhi (India), have always been supportive as true well-wisher. Dr. Om Prakash Mishra, the National Centre for Seismology (NCS), New Delhi (India); Prof. Rajesh Kumar Dubey, M.L. Sukhadia University (MLSU), Jodhpur (India); and Dr. Akhilesh Kumar Gupta, Advisor, Department of Science and Technology, Government of India, New Delhi, are deeply acknowledged for providing many valuable suggestions at various stages of the preparation of this book. This book has received blessings of eminent Quaternary Climate experts of the country, namely, Prof. Anil Kumar Gupta, from the Indian Institute of Technology, Kharagpur, India, and Dr. Rajiv Nigam, former Deputy Director, National Institute of Oceanography, Goa, India.

Akshat Khare and Ashmit Khare have unconditionally supported enormously during various stages of this book. Dr. Shabnam Chaudhury and Shri. Haridas Sharma from Ministry of Earth Sciences, New Delhi (India), have helped immensely in formatting the text and figures of this book, and bring this book to its present form. Publishers (Taylor and Francis) have done commendable job and are sincerely acknowledged.

Neloy Khare
Date: February 2021
Place: New Delhi

Editors

Dr. Neloy Khare, presently Adviser/Scientist G to the Government of India at MoES, has a very distinctive acumen not only of administration but also of quality science and research in his areas of expertise covering a large spectrum of geographically distinct locations like Antarctic, Arctic, Southern Ocean, Bay of Bengal, Arabian Sea, and Indian Ocean. Dr. Khare has almost 30 years of experience in the field of paleoclimate research using paleobiology (paleontology)/teaching/science management/administration/coordination for scientific programs (including Indian Polar Program) etc. Having completed his doctorate (PhD) on tropical marine region and Doctor of Science (DSc) on southern high-latitude marine regions towards environmental/climatic implications using various proxies, including foraminifera (microfossil). He has made significant contributions in the field of paleoclimatology of southern high-latitude regions (Antarctic Ocean and Southern Ocean) using micropaleontology as a tool. These studies coupled with his paleoclimatic reconstructions from tropical regions helped understand causal linkages and teleconnections between the processes taking place in southern high latitudes with that of climate variability occurring in tropical regions. Dr. Khare has been conferred Honorary Professor and Adjunct Professor by many Indian universities. He has a very impressive list of publications to his credit (121 research articles in National and International Scientific journals; 3 Special Issues of National Scientific Journals as Guest Editor; Edited Special Issue of Polar Science as its Managing Editor). He has authored/edited many books and contributed 130 abstracts to various seminars (23 popular science articles; 5 technical reports). Government of India and many professional bodies have bestowed him with many prestigious awards for his humble scientific contributions to past climate changes/oceanography/polar science, and southern oceanography. The most coveted award is Rajiv Gandhi National Award 2013 conferred by Honorable President of India. Other awards include ISCA Young Scientist Award, BOYSCAST Fellowship, CIES French Fellowship, Krishnan Gold Medal, Best Scientist Award, Eminent Scientist Award, ISCA Platinum Jubilee Lecture, IGU Fellowship, besides many. Dr. Khare has made tremendous efforts to popularize ocean science and polar science across the country by way of delivering many invited lectures, radio talks, and publishing popular science articles.

Dr. Khare has sailed in Arctic Ocean as a part of "Science Pub" in 2008 during the International Polar Year campaign for scientific exploration in Arctic Ocean and became the first Indian to sail in the Arctic Ocean.

Contributors

Hema Achyuthan
Institute for Ocean Management
Anna University
Chennai, India

G. L. Badam
Deccan College Post-Graduate and
    Research Institute
Deemed University
Pune, India

Keshava Balakrishna
Department of Civil Engineering
Manipal Institute of Technology,
    Manipal Academy of Higher
    Education
Manipal, India

Subodh Kumar Chaturvedi
Department of Geology
Arba Minch University
Arba Minch, Ethiopia

Kumari Deepali
Department of Applied Geology
School of Earth and Atmospheric
    Sciences
University of Madras
Chennai, India

M. Suresh Gandhi
Department of Geology
School of Earth and Atmospheric
    Sciences
University of Madras
Chennai, India

Pawan Kumar Gaury
Department of Environmental Sciences
Central University of Himachal Pradesh
Dharamshala, India

S. M. Hussain
Department of Geology
School of Earth and Atmospheric
    Sciences
University of Madras
Chennai, India

Jithin Jose
Department of Marine Geology
Mangalore University
Mangalagangothri, India

K. Kasilingam
Department of Geology
School of Earth and Atmospheric
    Sciences
University of Madras
Chennai, India

Neloy Khare
Ministry of Earth Sciences
Prithvi Bhawan
New Delhi, India

S. D. Kshirsagar
Deccan College Post-Graduate and
    Research Institute
Deemed University
Pune, India

A. Naveen Kumar
Department of Marine Geology
Mangalore University
Mangalagangothri, India

A. K. Mahajan
Department of Environmental Sciences
Central University of Himachal Pradesh
Dharamshala, India

S. Maniyarasan
Department of Applied Geology
School of Earth and Atmospheric
    Sciences
University of Madras
Chennai, India

Busnur Rachotappa Manjunatha
Department of Marine Geology
Mangalore University
Mangalagangothri, India

Narendra Kumar Meena
Department of Climate Change
Wadia Institute of Himalayan Geology
Dehradun, India

Ishfaq Ahmad Mir
State Unit: Karnataka and Goa
Geological Survey of India
Bengaluru, India

Satish S. Naik
Palaeobotany and Palynology Lab
Department of A.I.H.C and
    Archaeology
Deccan College Post-Graduate
    and Research Institute, Deemed
    University
Pune, India

Mohammed Noohu Nazeer
Department of Geology
School of Earth and Atmospheric
    Sciences
University of Madras
Chennai, India

Rajani Panchang
Department of Environmental Science
Savitribai Phule Pune University
Pune, India

K. Radhakrishnan
Department of Geology
School of Earth and Atmospheric
    Sciences
University of Madras
Chennai, India

S. Rajan
National Centre for Polar and Ocean
    Research (Formerly)
Goa, India

A. Rajkumar
Department of Geology
School of Earth and Atmospheric
    Sciences
University of Madras
Chennai, India

M. R. Resmi
School of Earth Sciences
Banasthali Vidyapith
Vanasthali, Rajasthan, India

P. D. Sabale
Deccan College Post-Graduate and
    Research Institute
Deemed University
Pune, India

Abhilash Sen
Department of Environmental Science
Savitribai Phule Pune University
Pune, India

S. G. D. Sridhar
Department of Applied Geology
School of Earth and Atmospheric
    Sciences
University of Madras
Chennai, India

V. C. Tewari
Department of Geology
Sikkim Central University, School of
    Physical Sciences
Sikkim, India

1 Late Quaternary Climatic History of the Indian Subcontinent
A Review of the Data Sources on the Land and the Sea

S. Rajan

National Centre for Polar and Ocean Research (Formerly)

CONTENTS

INTRODUCTION: BACKGROUND

Natural climate change at all temporal scales, with or without human interference, represents one of the most fundamental issues of scientific and social concern today. With the perceived influence of climate change on all forms of ecosystem stresses including extreme natural events, there has been a subtle but definite transformation in the ongoing studies on climate change, from why it happens to how it has changed in the past and how it will change in the future. At first instance, this might sound paradoxical considering that the earth's climate system especially at the centennial to the millennial scale (10^2–10^3 years) tends to be abrupt and rapid and hence difficult to predict. Furthermore, these changes do not follow a simple pattern and their impacts vary from region to region making it difficult to find analogs of a certain period in the geological record. In addition, these abrupt changes are superimposed upon the more cyclic multimillennial (10^4–10^5-year) responses of the earth system

1

to variations in the orbital parameters or to the still longer time scales (>10^6 years) related to plate tectonics.

Despite the challenges that complicate the picture of a changing climate through time, documentation of past climate remains an important element of earth system studies. A rationale for these studies stems from the observation that the processes related to the interactions among the ocean, atmosphere, geosphere, and the cryosphere, which drive/modulate the climate changes, tend to be cyclic. Therefore, it is only logical that with increased and better documentation of past climates, such cyclic recurrences and the driving forces behind them can be better understood and the climate models predicting future climates can be evaluated and fine-tuned where necessary. Increased knowledge is also the best way to improve the effectiveness of our response and, maybe, even to increase our ability to adapt to a changing climate.

This chapter is an attempt to present the late Quaternary climatic history of the Indian subcontinent through a review of some of the recently published studies carried out utilizing proxy data from the terrestrial and offshore areas as well as model simulations. The emphasis is on the last ~100 kyr (=100,000 years) spanning the Holocene and the last glaciation for which relatively detailed information is available. The choice of the late Quaternary timeframe was dictated by three considerations: (1) Late Quaternary climate during much of the last glacial period has been documented to be highly unstable marked by abrupt and rapid quasi-periodic shifts in temperature that occurred on decadal time scales and lasted for a few centuries (described further, below). (2) Correlative multi-proxy records of such climatic perturbations at temporal resolutions ranging from decadal-to-centennial and beyond are available from the offshore and terrestrial environments of the Indian sub-continent. (3) At a shorter interannual scale, instrument and proxy records as well as model results indicate marked variability in Indian summer monsoon (ISM) precipitation over the past 1–2 millennia (e.g. Berkelhammer et al., 2010; Feng and Hu, 2008; Goswami et al., 2006; Sinha et al., 2007). The forcing functions behind these precipitation variations remain topics of academic debate and call for careful consideration to understand the full spectrum of monsoon behavior on all time scales.

There have been many review papers of late on the paleoclimate studies in India focused on both terrestrial and marine realms. One of the early reviews of the Asian and Indian paleomonsoon variability at different time scales, from tectonic to centennial, has been by Clemens (2006). Singhvi and Kale (2010) published a comprehensive status report on the paleoclimate of India and adjacent regions as a part of the IGBP-WCRP-SCOPE-Report series, detailing the key issues in climate change studies and some of the salient results from marine and terrestrial proxy records. Tiwari et al. (2009, 2011) provided reviews on the spatial and temporal variability in multiproxy paleomonsoon records from the Indian region during the past 30 kyr and since the last glacial maximum (LGM) respectively. Gupta et al. (2012) described the paleoceanographic studies carried out by Indian scientists in the Arabian Sea and the Bay of Bengal sectors of the Indian Ocean between 2006 and 2012. Some of the other notable reviews of the Quaternary monsoon history of India based on terrestrial and/or marine proxy records from the Indian subcontinent and the adjoining seas are by Saraswat et al. (2014), Achyuthan et al. (2016), and Ramesh et al. (2017).

Most recently, Krishnan et al. (2020) have brought out an edited volume providing an assessment of the Indian climate system and its short-term variability based on observational data and analyses, as a report of the Ministry of Earth Sciences, Government of India. The present chapter is an update to these reviews with consideration of all recent contributions published up to June 2020.

QUATERNARY CLIMATE SYSTEM

The Quaternary period encompassing the last ~2.6 million years of earth history has essentially been one of frigidity with the climate being cold enough over 90% of the timespan to support major ice sheets (Holmgren and Karlen, 1998). Proxy records of late Quaternary climate derived from deep-sea sediments and Antarctic ice cores show a repeating pattern of glacial (cold and arid)/interglacial (warm and humid) cycles characterized by a 100-kyr periodicity, with global temperature differences between the cycles averaging 9°C–12°C (Petit et al., 1999; Shackleton, 2000). Such transitions between the glacial and the interglacial conditions have been suggested to have been paced by cyclical variations in the summer insolation at high latitudes in the northern hemisphere. Three major cycles of earth's orbital variability around the sun ("Milankovitch cycles") recur over time associated with eccentricity (100-kyr cycles), obliquity (41-kyr cycle), and precession (23-kyr cycle; e.g. Hays et al., 1976; Imbrie and Imbrie, 1979; Sharaf and Boudnikova, 1967). The insolation variability is also strongly modulated by such global boundary conditions as atmospheric CO_2, sea level, ice-sheet extent, and sea surface temperature (Labeyrie et al., 2003).

Despite a consensus on the role of orbital-scale forcing on earth's climate, the mechanism(s) by which these changes in insolation pace the timing of the climate cyclicity remain matters of debate. There is also an element of uncertainty with regard to which of the two orbital elements – obliquity and precession – paced the dominant 100 kyr glacial–interglacial cycles of the Quaternary, considering that the 100 kyr eccentricity band is small in the insolation spectrum (e.g. Feng and Bailer-Jones, 2015; Imbrie et al., 1989, 1993). Furthermore, the extent to which the feedbacks from processes internal to the climate system such as, for example, from the continental ice sheet and other climate components, including the atmospheric concentration of water vapor, CO_2, and other gases, as well as the atmospheric concentration of volcanic dust and variations in the cloud cover impact the external orbital forcing is also not well known (Labeyrie et al., 2003; Shackleton, 2000).

Superimposed on the orbital-scale variations are centennial to millennial-scale events marked by large, abrupt, and rapid alternations between stadial (cold) and interstadial (warm) conditions in time scales of a decade or so and which persist over hundreds to one thousand years ago or longer. First recognized in the Greenland ice cores from within the ~30–80 kyr BP interval (Baldini et al., 2015; Bond et al., 1993; Dansgaard et al., 1984; Oeschger et al., 1984), correlative evidences of such abrupt climate perturbations referred to as "Dansgaard–Oeschger cycles or D–O cycles" have been documented from several parts of the world, from the Greenland ice cores to the marine and terrestrial sedimentary records as well as from the Antarctic peninsula (e.g. Behl and Kennett, 1996; Bender et al., 1994; Blunier et al., 1998; Charles et al., 1996; Clark and Bartlein, 1995; Cowley, 1992; Curry and Oppo, 1997;

Deplazes et al., 2013; EPICA Community Members, 2006; Genty et al., 2003; Greenland Ice-core Project (GRIP) Members, 1993; Kanner et al., 2012; Keigwin et al., 1994; Kennett and Ingram, 1995; Leduc et al., 2007; Leuschner and Sirocko, 2000; Mayewski et al., 1996; Oeschger et al., 1984; Oppo and Lehman, 1995; Porter and Ann, 1995; Schmidt and Hertzberg, 2011; Schulz et al., 1998; Stocker et al., 1992; Street-Perrot and Perrot, 1990; Wang et al., 2001; also Rajan and Khare, 2002, for an overview). Similar millennial-scale changes probably occurred during previous ice ages as well (Franco et al., 2012; Thouveny et al., 1994; Voelker, 2002).

The synchronous global signature of these abrupt climate changes has been suggested to be indicative of the potential for a global response related to an initial forcing linked to the instabilities of the Northern Hemisphere ice sheets and consequent variations in freshwater flux to the North Atlantic, or feedback related to Atlantic thermohaline circulation, the effects in both cases being transmitted and amplified elsewhere by way of oceans and/or atmosphere (e.g. Alley, 1995; Broecker, 1994, 1995; Rahmstorf, 2002). Other hypotheses for the origin of these millennial-scale cycles involve rhythmic solar forcing (Bond et al., 2001) or internal oscillations of the coupled ocean-atmosphere system (e.g. Alley et al., 2001).

Nested within the millennial-scale climate cycles are abrupt submillennial to decadal events, during which the climate has been documented to change from one state to another. High-resolution observational records from terrestrial, marine, and ice cores and model simulations, show that paleoclimate trends over the late Quaternary and especially during the Holocene period have been punctuated by significant fluctuations at century to multidecadal time scales (Steffensen et al., 2008). These fluctuations are mostly marked from the LGM (21 ± 2 calendar ka BP) to the present day. At least some of these climatic fluctuations have been proposed to be the result of melt-water-driven changes in the Atlantic meridional overturning circulation – the large-scale oceanic overturning circulation that transports heat from the South Atlantic to the North Atlantic high latitudes. For example, in some regions of Europe, monthly temperatures increased by >15°C over a century during the Bølling–Allerød (B–A) warming event (14,700–12,900 BP) and cooled by >10°C during the Younger Dryas cooling event (12,900–11,600 BP) (Fordham et al., 2017). A similar pattern is characteristic of the Asian monsoon system as well (described further, below).

Besides oceanic circulation, various other hypotheses have been proposed to explain this decadal-to-centennial scale variability, such as solar forcing (Agnihotri et al., 2002; Hoyt and Schatten, 1997), the El Niño-Southern Oscillation (ENSO), Indo-Pacific climate variability (Cobb et al., 2003; Newton et al., 2006; Prasad et al., 2014), movement of the mean position of the Intertropical Convergence Zone, ITCZ (Sachs et al., 2009), changes in the Indian Ocean Dipole, and so on (Ding et al., 2010). However, like the millennial cycles, the specific mechanisms by which the climate responds to these forcings are not known.

Occupying the shorter end of the climate spectrum is the variability at interannual (2–8 years) to decadal and interdecadal scales, which have been documented in instrumental records coupled with modeling results. The quasi-periodic ENSO marked by temperature fluctuations between the ocean and atmosphere in the Equatorial Pacific is an example of changes at the interannual scale. El Niño (the "warm phase" of ENSO) and La Niña ("the cold phase") episodes typically last for 9–12 months and occur on average every 2–7 years (https://oceanservice.noaa. gov/facts/ninonina.html). The ENSO signal is transmitted from its Pacific locus to other tropical oceans through global tropospheric anomalies affecting the climate of much of the tropics and subtropics (e.g., Enfield and Mayer, 1997; Hastenrath et al., 1987). In contrast, the "Pacific Decadal Oscillation" is a longer-lived (20–30 years) El Niño-like pattern of coupled ocean-atmosphere climate variability centered over the Pacific basin (Zhang et al., 1997). The less-than a decade to multidecadal (>25 years) North Atlantic Oscillation (NAO) is another irregular fluctuation of atmospheric pressure over the North Atlantic Ocean that has a strong effect on winter weather in the Northern Hemisphere (https://www.britannica.com/science/North-Atlantic-Oscillation). The NAO can occur on a yearly basis, or the fluctuations can take place decades apart (Enfield and Mestas-Nunez, 1999).

LATE QUATERNARY CLIMATE OF INDIA

The dominantly agrarian economy of Asia and, in particular, of the Indian region is overwhelmingly dependent on the seasonal reversal of the regional patterns of monsoon winds and attendant precipitation. For instance, the Indian agriculture, which accounts for 25% of the GDP and employs 70% of the population, relies heavily on the monsoon rains, over 75% of which arrives during the summer months (the southwest or ISM; Gadgil, 2007). A delay of a few days in the arrival of the monsoon can badly affect the economy, as evidenced in the numerous droughts in India in the 1990s. It is therefore hardly surprising that in the Indian context, any consideration of climate change in the long term or climate trends on a shorter scale basically translates to an examination of monsoon rainfall variability through time.

The Indian Monsoon (or South Asian monsoon, as referred to, at times) is a major dynamic component of the regional Asian monsoon system. The strong inherent seasonality of the Indian monsoon circulation results in warm, wet summers (summer monsoon period) and cool, dry winters (winter monsoon) over the Indian landmass and contiguous oceanic areas. In its present form, the monsoon circulation has been suggested to have initiated sometime between 9 and 7 million years ago (Clemens, 2006). Historical and instrumental records, modeling studies, and high-resolution multiproxy reconstructions suggest that large precipitation anomalies in the ISM across a spectrum of time scales had been a characteristic trait of the monsoon since inception and more so during the late Quaternary period. Paleoclimate records indicate the persistence of a weaker monsoon system over much of Asia during the last glacial period. Model simulations also show a reduction in ISM precipitation by 1.7 mm/day, due to the LGM ice sheets and reduced atmospheric CO_2 (Braconnot et al., 2007).

Paleo reconstructions of ISM have come from a variety of archives on land and the marine realm – deep-sea and terrestrial sediments (e.g. Agnihotri et al., 2002; An et al., 2011; Banakar et al., 2010; Gupta et al., 2003, 2005, 2013; Gupta and Thomas, 2003; Kudrass et al., 2001; Prasad and Enzel, 2006; Schulz et al., 1998; Sharma et al., 2004), corals (Ahmad et al., 2011), lakes and wetland systems (Achyuthan et al., 2016 and references therein), lacustrine and paleolake sections (Kumar et al., 2017; Sanwal et al., 2019), peat (Sukumar et al., 1993), loess-paleosol sequences (Singhvi et al., 2001), speleothems (e.g. Fleitmann et al., 2007; Yadava and Ramesh, 2005;

also Kaushal et al., 2018 and references therein for a comprehensive review of stalagmite $\delta^{18}O$ proxy records of ISM variability), tree rings (Yadav, 2013), and glaciers and ice-cores (Thompson et al., 2006).A wide variety of physical, chemical, and biological proxies have been utilized to extract the paleomonsoon signals from these archives. Singhvi and Kale (2010), Tiwari et al. (2011), and Wang et al. (2005) provided an overview of the various proxy indicators of monsoon variability together with the inferences made from the records; Singhvi and Kale (2010) have also briefly discussed the different techniques of establishing an accurate chronology of the proxy events, which is critical for evaluating the synchroneity or leads and lags of the climate event in a regional/global framework.

In the following pages, I describe some of the very recent studies carried out on the ISM variability on a variety of temporal scales, from orbital to millennial, centennial, multidecadal, and interannual. The purpose of the review is twofold: (1) to highlight the continuing interest, significant developments, and recent advances in this dynamic field of paleoclimatology and (2) to stress on the fact that despite a plethora of observations, models, and inferences, the textured pattern of the Indian monsoon climate through time still remains an enigma.

One of the earliest studies on the long-term late Quaternary evolution of the Indian monsoon was carried out by Prell et al. (1980), in which they observed that SW monsoon was weaker during the glacial periods and stronger during the interglacials. In general, the last interstadial period (58–24 ka) was characterized by a moderate climate, may be only slightly cooler and drier than the present (Goodbred, 2003). Proxy data from the Arabian Sea point to a phase of relatively strong monsoon winds during a greater part of this interstadial period (Prell, 1984; Prell and Kutzbach, 1987). The succeeding LGM saw low insolation and strong glacial boundary conditions, which weakened the summer monsoon with consequent widespread aridity (Goodbred, 2003), as documented in the $\delta^{18}O$ and pollen data from the sediments off the SW Indian coast (e.g. Van Campo, 1986).

Multiproxy studies from Arabian Sea sediment cores by several workers (e.g. Gupta et al., 2003; Naidu and Malmgren, 1995; Overpeck et al., 1996; Sirocko et al., 1993; Tiwari et al., 2009) show that the monsoonal variations since the LGM were however, not gradual but occurred in a series of abrupt steps, with at least two peak intensities in monsoon during the late stages of the LGM and early part of the Holocene (at ~15.5 and ~8.5 ka, Sirocko et al., 1993; between 13 and 12.5 ka and 10 and 9.5 ka, Overpeck et al., 1996; at ~13 ka and between 10 and 5 ka, Naidu and Malmgren, 1995; at ~8.5 ka and declining since then to ~1.5 ka, Gupta et al., 2003; at 9.5 and 9.1 ka, weakening gradually thereafter, Meloth et al., 2007; major precipitation increase occurred only after ~9 ka, Tiwari et al., 2009, among others. See Tiwari et al. (2009) for more comprehensive coverage of marine data).

Compared to the studies on the Arabian Sea basin, studies on the monsoon variability from proxies in the Bay of Bengal sediments are much less. Partly, the reason has been the presence of one of the largest turbiditic system in the world (the Bengal fan) over much of the basin and the consequent difficulties in retrieving undisturbed high-resolution cores. Nonetheless, there have been some classical studies on monsoon variability focused on the Bay of Bengal. Duplessy (1982), for instance, reconstructed the Holocene and LGM sea surface characteristics of the Bay of Bengal from $\delta^{18}O$ of *Globigerinoides ruber* in the core samples. He concluded that the SW monsoon was weaker during the LGM and the NE monsoon was stronger than the present with more precipitation south of 10°N. Kudrass et al. (2001) reconstructed the ISM history for the last 80 ka based on salinity fluctuations from proxy data in a core from the northernmost part of the Bay of Bengal.

Goodbred (2003) carried out a novel reconstruction of the late Quaternary changes in the Ganges sediment dispersal system from its source in the Himalayas to its sink in the Bay of Bengal in an attempt to understand the dispersals in the context of major shifts in monsoon strength and precipitation. Basing his observations on all available data, Goodbred (2003) contends that the Ganges dispersal system can be seen to have responded well to much of the multimillennial-scale climate changes during late Quaternary.

Temporal and spatial variations in the textural and mineralogical characteristics of sediments across a transect in the Bengal fan were utilized by Chauhan and Vogelsang (2006) to assess the sediment dispersal pattern in terms of climate-induced variations in the Bay. The studies bring out the dominant role of winter hydrography in dispersing the sediments prior to 12.5 ka BP. The initiation of a stronger summer monsoon regime at 12.5 ka BP resulted in a change in the dispersal pattern.

High-resolution terrestrial records of climate change during the late Quaternary have come principally from two sources – lacustrine sections and speleothems. Lake records of climate change have been fairly well reported (e.g. see the review by Achyuthan et al. (2016) and references therein). Kumar et al. (2017) carried out a compilation of high-resolution multiproxy records from 30 lacustrine archives covering the major geographical regions of peninsular India. A comparative evaluation of the results indicates a period of intense monsoon during 13–8 ka straddling the late Glacial–Holocene, conforming to the marine records. A short phase of weak monsoon precipitation characterizes the mid-Holocene period during ~5–4 ka. Interestingly, none of the major lake records bear strong imprints of Younger Dryas cooling (12.9–11.5 ka).

Sanwal et al. (2019) adopted a different approach to characterize a 320-m thick fluvial-lacustrine sedimentary profile in the Kumaun Central Himalaya in terms of late Quaternary climatic fluctuations. They tracked the variations in carbon isotopic composition of organic matter in the sediments along the profile in terms of the dominant vegetation, C3 or C4. Sanwal et al. (2019) contend that segments in the profile marked by $\delta^{13}C$ values characteristic of C3 vegetation are indicative of warm and wet conditions coinciding with the intensification of the ISM, while the intervening segments dominated by C4 vegetation point to cold and arid conditions and strengthening of winter westerlies. The studies also relate a short warm spike identified in the profile to the oscillation of B–A interstadial at ~15,000 years BP, as also a phase of reduced monsoon precipitation toward the top of the profile as probably corresponding to Older Dryas.

Dixit (2020) in a very recent paper describes an interesting study highlighting the spatial heterogeneity of Holocene climatic information from paleolake sections in northwest India. Age-constrained proxy information from five lake sections across an E-W transect representing a precipitation gradient from arid Thar desert in the west to semi-arid and humid conditions to the east reveals their divergent

Holocene monsoon history. All these lakes appeared when the monsoon intensified during the early Holocene. However, the coherent response of these lakes was short-lived. Proxy records from only the subhumid to semi-arid lakes mimic the mid-to-late Holocene reconstructions of ISM variability from correlated marine sediment and speleothem archives. The arid lakes in contrast carry largely signals of local climate history than global monsoon history. These observations highlight the need to exercise caution while translating the lacustrine records to the global monsoon climate.

Speleothems provide a good source of high-resolution paleoclimatic information particularly over the late Quaternary period. Much progress has been made in this regard on the evolutionary history of Asian monsoon in general and the East Asian monsoon (EAM) in particular, from studies of caves in China (e.g. see Cai et al., 2006; Kathayat et al., 2016; Wang et al., 2005; Wang et al., 2001, 2008, 2017; Zhao et al., 2015). As regards the Indian speleothem studies, there have been good contributions by Berkelhammer et al. (2010), Cai et al. (2006), Kathayat et al. (2016), Ramesh et al. (2010), Tiwari et al. (2009), and so on. To date, the oldest and longest published stalagmite record from India is from Bittoo cave covering the last 240 ka, bearing the imprints of orbital, millennial, and Holocene-scale monsoon variability (Kathayat et al., 2016; Kaushal et al., 2018).

A recent noteworthy contribution has been the comprehensive review of ISM variability from Indian stalagmite records collated in the Speleothem Isotope Synthesis and AnaLysis version 1 (SISAL_v1) database and other published accounts/data by Kaushal et al. (2018). As indicated by these authors, the available stalagmite $\delta^{18}O$ data in SISAL show that regional ISM responses differ in terms of sensitivity and timing. Global-scale climate events such as the Heinrich, B–A, Younger Dryas, 4.2 ka, and the little ice age are not discernible in all the cave records studied. Where detectable, the amplitude of the $\delta^{18}O$ excursions of these events tends to be low compared to variability over orbital time scales (Kaushal et al., 2018). A caveat on utilizing the cave isotope records for paleoclimate interpretations is that the $\delta^{18}O$ signatures can be influenced by multiple factors including local meteorology. This, therefore, calls for exercising caution when translating the $\delta^{18}O$ values in terms of regional/global variability and necessitates local calibration with measurements of rainfall, temperature, and so on.

MULTIMILLENNIAL SCALE VARIABILITY OF THE INDIAN SUMMER MONSOON

As mentioned earlier, insolation change related to the earth's orbital forcing plays a primary role in driving quasi-periodic variations in climate at geological time scales. Many high-resolution speleothem oxygen isotope ($\delta^{18}O$) records and proxy records from the Arabian Sea document the sensitivity of the late Quaternary ISM to orbital forcing at the obliquity and precession bands, in conjunction with the high-latitude changes in global ice volume (e.g. Clemens, 2006; Clemens and Prell, 2003; Clemens et al., 2010; Kathayat et al., 2016; Kutzbach, 1981; Kutzbach et al., 2008; Leuschner and Sirocko, 2003; Wang et al., 2017). Modeling studies also strongly emphasize on the direct link between increased NH summer insolation and strengthened Indian and Asian monsoons (e.g. de Noblet et al., 1996; Liu et al., 2020).

Despite the above, significant debates remain on the primary drivers of the ISM on orbital time scales (e.g. Clemens, 2006; Clemens et al., 2010; DiNezio et al., 2018; Kathayat et al., 2016; Kaushal et al., 2018; Liu et al., 2020; Overpeck et al., 1996; Tabor et al., 2018; Tiwari et al., 2011). One of the major issues concerning the response of the ISM to orbital forcing relates to the physical mechanism responsible for the observed phase relationships between NH summer insolation maxima and high latitude changes in global ice. For instance, climate simulations suggest that the intensity of the "global monsoon" (Caley et al., 2011), of which the ISM can be considered a dynamic component, varies nearly in-phase with changes in precession-dominated NH summer insolation (Kutzbach, 1981; Kutzbach et al., 2008). This is supported by speleothem records from China, which point to a coherence between EAM variability at precession bands and the NH Summer insolation during peak Asian monsoon intensity (Wang et al., 2008). In contrast, proxy records from the Arabian Sea indicate prominent lags between the phase estimates of changes in NH Summer insolation and variability in ISM (Clemens and Prell, 2003). This lag has been attributed to the influence of latent heat transport from the southern Indian Ocean in conjunction with the global ice volume and the summer insolation (Clemens, 2006). Recent studies of speleothem oxygen isotope data from the Bittoo cave, India, however, bring out a strong coherence between the Bittoo and Chinese speleothem $\delta^{18}O$ records, indicative of a coupled response of the Indian and East Asian monsoons to changes in NH summer insolation without much lags (Kathayat et al., 2016). This would also tend to support the view that the ISM variability is driven directly by precession-induced changes in NH summer insolation.

MILLENNIAL-SCALE VARIABILITY OF THE INDIAN SUMMER MONSOON

The Asian monsoon system, which is sensitive to orbitally controlled changes of insolation as discussed above, also varied on millennial time scales as shown by terrestrial and marine sediment records. The Indian monsoon has also been suggested to vary on these time scales (Rajan, 2018). High-resolution proxy records of climate change from the northern Indian Ocean (the Arabian Sea and the Bay of Bengal) and speleothem in China and India suggest that the abrupt and rapid D–O and Heinrich-style events known from Greenland and the northern North Atlantic characterized the late Quaternary climate variability of the Indian subcontinent (including the contiguous oceanic areas).

Monsoon variations on D–O time scales (~1,500 years) have been reported from the varved sediment sections off Pakistan (Schulz et al., 1998), from bioturbated high-resolution sediment sections in the Arabian Sea (Altabet et al., 1995, 2002; Sirocko et al., 1993) and the Bengal Fan (Kudrass et al., 2001). These marine-based results are also consistent with the terrestrial records of the Indian peninsula and the Himalayas, as discussed below.

Multiproxy studies carried out by Schulz et al. (1998) on sediment cores from the OMZ off Pakistan show sequences of laminated organic carbon-rich and bioturbated carbon-poor sediment units, reflective of strong monsoon-induced biological productivity and weak monsoon respectively. Constraining the sediment sequences by AMS-radiocarbon ages and geochemical and micropaleontological data, Schulz et al. (1998)

contend that these alternate sequences indicate millennial to centennial rapid fluctuations in the intensity of monsoonal circulation during the past 110 kyr that might correspond to the D–O and Heinrich-style events of Greenland and North Atlantic.

Sirocko et al. (1999) by considering the eolian dust in the sediment record of a western Arabian Sea core corroborated the findings of Schulz et al. (1998) of an arid climate (weak monsoon periods) in the proxy record. The studies by Sirocko et al. (1999) demonstrated that these arid phases were associated with drought in the Arabian Desert, resulting in stronger dust flux to the ocean. Constraining the core sequences by age dates and based on the findings of Schulz et al. (1998), Sirocko et al. contended that the periods of drought corresponded to the stadials of the D–O oscillations.

Altabet et al. (2002) noted the strong correspondence between the D–O events recorded in the Greenland ice core (GISP2) and the denitrification $\delta^{15}N$ records in two cores from the Oman margin within the present-day oxygen-minimum zone and relatively high sediment accumulation rates. $\delta^{15}N$ and thus denitrification in the Arabian Sea cores were found to be close to near-modern, high values during the warm interstadial phases corresponding of the D–O events, and at near-minimal glacial values during the cold stadials. Altabet et al. (2002) contend that such a coupling of Northern Hemisphere climate and the Arabian Sea denitrification points to their linkage through meteorological forcing.

As mentioned earlier, Kudrass et al. (2001) carried out a reconstruction of the SW monsoon history for the last 80 ka based on salinity fluctuations obtained from $\delta^{18}O$ of G. ruber and alkenone SST in a sediment core from the Bay of Bengal. They observed a strong correlation between the Greenland temperature record and ISM variability, with the warmer and the colder phases corresponding respectively to stronger and weaker monsoon.

Based on interpreted salinity variations from proxy records of surface water $\delta^{18}O$ in a core from the northern Andaman Sea, Marzin et al. (2013) documented millennial-scale oscillations in salinity from 40k to 11k BP. The timing and sequence of the variations indicate that events of high salinity correspond to weak Indian monsoon while low salinity represents periods of intense monsoon. In turn, Marzin et al. (2013) related these periods respectively to the cold and warm events recorded in the Greenland (GISP2) ice cores.

A good record of D–O oscillations from a stalagmite from Socotra Island has been provided by Burns et al. (2003). The oxygen-isotope ratios of this stalagmite show a record of changes in monsoon precipitation and climate for the time period from 55,000 to 42,000 years BP. The precipitation pattern has been shown to correspond to the oxygen-isotope record from Greenland ice cores, with increased precipitation associated with warm periods in the high northern latitudes.

Premathilake and Gunatilaka (2013) provide an interesting perspective on millennial-scale monsoon climate variability in Sri Lanka over the past 24 kyr as reconstructed from age-dated botanical proxy records from a peat sequence in the Horton rainforest plains of Central Sri Lanka. Commencing ~17,600 years BP, the proxy record of the vegetation history of rainforest expansion, diversification, fluctuation, and decline indicates a succession of stages of strong monsoonal activity with intervening phases of weaker monsoon. Broadly, these events correspond to the millennial records of south-west monsoon variability in the Northern Indian Ocean.

CENTENNIAL TO MULTIDECADAL-SCALE VARIABILITY
OF THE INDIAN SUMMER MONSOON

The summer monsoon has been suggested to be declining with variability at the multidecadal-to-centennial scales (Chao and Chen, 2001). Abrupt changes in the Indian monsoon precipitation on decadal and centennial time scales superimposed on an overall declining trend are evident in the Holocene monsoon proxy records from the speleothems and marine data, in addition to model results (e.g. Berkelhammer et al., 2012; Sanyal and Sinha, 2010). As described earlier, marine proxy records, especially from the Arabian Sea, show that the monsoonal variations since the LGM were not gradual but occurred in a series of abrupt steps, with at least two peak intensities in monsoon during the late stages of the LGM and early part of the Holocene. Proxy terrrestrial records from the Indian subcontinent supports a substantial weakening of the ISM around 7,000 years BP (Gupta et al., 2006). In addition, prolonged periods of weak monsoon/drought at multidecadal and century scales have occurred during the last 4,000 years (Sinha et al., 2011) with notable century-scale long declining trends 1,550–2,200 years BP and 100–550 years BP.

Many recent studies have sought to establish a centennial-scale teleconnection between the North Atlantic climate variations and the ISM during the Holocene. For instance, a broad correspondence between the dry (weak) phases of the ISM and the cold intervals of the North Atlantic climate has been documented (Bond et al., 2001; Gupta et al., 2003; Schulz et al., 1998). Based on a comparison of the variations in the abundance of *Globigerina bulloides* in the ocean drilling program (ODP) samples from site 723 in the OMZ on the Oman margin with the 65°N July solar insolation and data on the percentage of hematite-stained grains from the North Atlantic (indicative of cooling phases), Gupta et al. (2003) concluded that the ISM exhibits good correlation with the high latitude climate on centennial time scales (weaker monsoon during colder periods). Monsoon maxima occur in several pulses while the North Atlantic was warmest (marked by low hematite abundance).

Although it has been suggested that short-term variations in solar input might be responsible for the observed decadal-to-centennial variability in the Holocene monsoon record (e.g. Agnihotri et al., 2002; Bond et al., 2001; Gupta et al., 2005; Sinha et al., 2018; Tiwari and Ramesh, 2007), the specific mechanisms by which these small solar irradiance variations would have forced the monsoon variability have not yet been convincingly established. A recent contribution in this regard has been made by Gupta et al. (2013), who, on the basis of studies of the decadal-to-centennial scale record of summer monsoon proxy *G. bulloides* in the ODP site 723A (described above, earlier), identified two weak summer monsoon intervals coinciding with the cold phases within the Ållerød inerstadial as well as a correlated strong, centennial-scale solar cycle during the Ållerød period (13.6–13.1 kyr). These observations point to the influence of solar signals in forcing the summer monsoon on centennial scales.

DECADAL-TO-INTERANNUAL SCALE VARIABILITY OF THE INDIAN SUMMER MONSOON

As described earlier, phase studies show that the ISM is sensitive to global forcings and atmospheric dynamics, including insolation, ENSO, Pacific Decadal Oscillation dynamics, and changes triggered in the North Atlantic realm. Several studies also indicate the influence of North Atlantic SST anomalies on ISM precipitation on interannual and multidecadal time scales (e.g. Goswami et al., 2006; Rajeevan and Sridhar, 2008; Zhang and Delworth, 2006). Observation and modeling studies show that the short-term multidecadal variations of the North Atlantic SST, i.e. the Atlantic Multidecadal Oscillation (AMO), have corresponding variations in the ISM (Feng and Hu, 2008; Goswami et al., 2006, Lu et al., 2006). These authors have shown that the warm phase of AMO is linked to increased precipitation over the Indian subcontinent. Berkelhammer et al. (2010) for the first time corroborated the model studies of multidecadal variability with results from an annually resolved speleothem oxygen isotope ($\delta^{18}O$) record from the Dandak Cave in Chhattisgarh, India. The persistence of multidecadal cycles in the millennial-long (AD 600–1550) speleothem records has been suggested to demonstrate the presence of a recurrent multidecadal scale pattern of climatic variability. This record also corresponds well with similar speleothem-based Asian monsoon reconstruction from the Wanxiang Cave of China testifying to the regional significance of such short-term climate variability in the monsoon record.

SUMMARY

Despite the great strides that have been made during the past two odd decades in unraveling the late Quaternary climate history of the Indian subcontinent across a spectrum of time scales, a comprehensive picture of the monsoon variability still remains elusive. We certainly know much more about the Indian monsoon today than what we presumed a decade back. However, our predictive skills have not kept pace with our knowledge base. The reasons range from the non-availability of high-resolution ground-truth data without spurious discontinuities and trends to the spatial heterogeneity of monsoon precipitation and our limited success in understanding the feedback mechanism(s) by which the monsoon responds to the three major forcings – orbital, greenhouse gases, and ice sheets – at different timescales.

An interesting contribution on this latter aspect has recently been made by Jalihal et al. (2019) who, based on their results from a climate simulation model, showed that feedbacks amplify the effect of insolation changes on the ISM. Jalihal et al. (2019) demonstrated that during the deglacial times, the ISM was predominantly influenced by rising water vapor due to increasing sea surface temperature while during the Holocene, cloud feedback was more important. Considering the pace of growth of our ideas and skills, it is only to be expected that a holistic picture of the ISM evolutionary history is not far away.

ACKNOWLEDGMENTS

This chapter represents a panorama of sterling contributions made by several scientists from India and across the world in advancing our understanding of climate change and variability focused on the ISM. I would like to sincerely acknowledge them all. Many worthy names and ideas might not have figured in this paper, certainly not intentional, but due to my ignorance. To my friend Neloy Khare, I owe a special debt of gratitude. But for his perseverance in goading me to sit, read, and write, this chapter would not have become a reality.

REFERENCES

Achyuthan, H., A. Farooqui, V. Gopal, B. Phartiyal, and A. Lone. 2016. Late Quaternary to Holocene South West monsoon reconstruction: A review based on lake and wetland systems (studies carried out during 2011-2016). *Proceedings of the Indian National Science Academy* 82: 847–868. doi: 10.16943/ptinsa/2016/48489.

Agnihotri, R., K. Dutta, R. Bhushan, and B. L. K. Somayajulu. 2002. Evidence for solar forcing on the Indian monsoon during the last millennium. *Earth and Planetary Science Letters* 198: 521–527.

Ahmad, S. M., V. M. Padmakumari, W. Raza, et al. 2011. High-resolution carbon and oxygen isotope records from a scleractinian (Porites) coral of Lakshadweep Archipelago. *Quaternary International* 238: 107–114.

Alley, R. B. 1995. Resolved: The Antarctic controls Global climate change. In *Arctic Oceanography: Marginal Ice Zones and Continental Shelves*, eds. W. O. Smith Jr, and J. M. Grebmeir, pp. 263–283.Coastal and Estuarine Studies, Vol. 49, Washington, DC: American Geophysical Union. doi: 10.1029/CE049.

Alley, R. B., S. Anandakrishnan, and P. Jung. 2001. Stochastic resonance in the North Atlantic. *Paleoceanography* 16: 190–198.

Altabet, M. A., M. J. Higginson, and D. W. Murray. 2002. The effect of millennial-scale changes in Arabian Sea denitrification on atmospheric CO_2. *Nature* 415: 159–162.

Altabet, M. A., R. Francois, D. W. Murray, and W. L. Prell. 1995. Climate related variations in denitrification in the Arabian Sea from sediment $^{15}N/^{14}N$ ratios. *Nature* 373: 506–509.

An, Z., S. C. Clemens, J. Shen, et al. 2011. Glacial-interglacial Indian summer monsoon dynamics. *Science* 333: 719–723.

Baldini, J. U. L., R. J. Brown, and J. N. McElwaine. 2015. Was millennial scale climate change during the Last Glacial triggered by explosive volcanism? *Scientific Reports* 5: 17442. doi: 10.1038/srep17442.

Banakar, V. K., B. S. Mahesh, G. Burr, and A. R. Chodankar. 2010. Climatology of the Eastern Arabian Sea during the last glacial cycle reconstructed from paired measurement of foraminiferal $\delta^{18}O$ and Mg/Ca. *Quaternary Research* 73: 535–540.

Behl, R. J., and J. P. Kennett. 1996. Brief interstadial events in the Santa Barbara Basin, NE Pacific, during the last 60 kyr. *Nature* 379: 243–246.

Bender, M., T. Sowers, M. L. Dickson, J. Orchardo, P. Grootes, P. A. Mayewski, and D. A. Messe. 1994. Climate correlations between Greenland and Antarctica during the past 100,000 years. *Nature* 372: 663–666.

Berkelhammer, M., A. Sinha, L. Stott, H. Cheng, F. S. R. Pausata, and K. Yoshimura. 2012. An abrupt shift in the Indian monsoon 4000 years ago. *Geophysical Monograph Series* 198: 75–87.

Berkelhammer, M., A. Sinha, M. Mudelsee, H. Cheng, R. Edwards, and K. Cannariato. 2010. Persistent multidecadal power of the Indian Summer Monsoon. *Earth and Planetary Science Letters*, 290: 166–172.

Blunier, T., J. Chappellaz, J. Schwander, et al. 1998. Asynchrony of Antarctic and Greenland climate change during the last glacial period. *Nature* 394: 739–743.

Bond, G., B. Kromer, J. Beer, et al. 2001. Persistent solar influence on North Atlantic climate during the Holocene. *Science* 294: 2130–2136. doi: 10.1126/science.1065680.

Bond, G., W. S. Broecker, J. Johnsen, et al. 1993. Correlations between climate records from North Atlantic sediments and Greenland ice. *Nature* 365: 143–147.

Braconnot, P., B. Otto-Bliesner, S. Harrison, et al. 2007. Results of PMIP2 coupled simulations of the Mid-Holocene and last glacial maximum –Part 1: experiments and large-scale features. *Climate of the Past* 3: 261–277. doi: 10.5194/cp-3-261-2007.

Broecker, W. S. 1994. Massive iceberg discharges as triggers for global climate change. *Nature* 372: 421–424.

Broecker, W. S. 1995. *The Glacial World according to Wally*, 2nd edition. Palisades, NY: Eldigio Press, Lamont-Doherty Earth Observatory of Columbia University.

Burns, S. J., D. Fleitmann, A. Matter, J. Kramers, and A. A. Al-Subbary. 2003. Indian Ocean climate and an absolute chronology over Dansgaard/Oeschger Events 9 to 13. *Science* 301: 1365–1367.

Cai, Y., Z. An, C. Hai, et al. 2006. High-resolution absolute-dated Indian Monsoon record between 53 and 36 ka from Xiaobailong Cave, southwestern China. *Geology* 34: 621–624. doi: 10.1130/G22567.1.

Caley, T., B. Malaizé, M. Revel, et al. 2011. Orbital timing of the Indian, East Asian and African boreal monsoons and the concept of a 'global monsoon'. *Quaternary Science Reviews* 30: 3705–3715.

Chao, W. C., and B. Chen. 2001. The origin of monsoons. *Journal of the Atmospheric Sciences* 58: 3497–3507.

Charles, C. D., J. Lynch-Stieglitz, U. S. Ninnemann, and R. G. Fairbanks. 1996. Climate connections between the hemisphere revealed by deep sea sediment core/ice core correlations. *Earth and Planetary Science Letters* 142: 19–27.

Chauhan, O. S., and E. Vogelsang. 2006. Climate induced changes in the circulation and dispersal patterns of the fluvial sources during late Quaternary in the middle Bengal. *Journal of Earth System Science* 115: 379–386.

Clark, P. U., and P. J. Bartlein. 1995. Correlation of the late-Pleistocene glaciation in the western United States with North Atlantic Heinrich events. *Geology* 23: 483–486.

Clemens, S. 2006. Extending the historical record by proxy. In *The Asian Monsoon*, ed. B. Wang, pp. 615–630. Berlin, Heidelberg: Springer Praxis Books. doi: 10.1007/3-540-37722-0_16.

Clemens, S., and W. L. Prell, 2003. A 350 000 year summer-monsoon multi-proxy stack from the Owen Ridge, Northern Arabian Sea. *Marine Geology* 201: 35–51.

Clemens, S. C., W. L. Prell, and Y. Sun. 2010. Orbital-scale timing and mechanisms driving Late Pleistocene Indo-Asian summer monsoons: Reinterpreting cave speleothem $\delta^{18}O$. *Paleoceanography* 25:PA4207. doi: 10.1029/2010PA001926.

Cobb, K. M., C. D. Charles, H. Cheng, and R. L. Edwards. 2003. El Niño/Southern Oscillation and tropical Pacific climate during the last millennium. *Nature* 424: 271–276. doi: 10.1038/nature01779.

Cowley, T. J. 1992. North Atlantic water cools the Southern Hemisphere. *Paleoceanography* 7: 489–497.

Curry, W. B., and D. W. Oppo. 1997. Synchronous, high-frequency oscillations in tropical sea surface temperatures and North Atlantic Deep-Water production during the last glacial cycle. *Paleoceanography* 12: 1–14.

Dansgaard, W., S. J. Johnsen, H. G. Clausen, et al. 1984. North Atlantic climate oscillations revealed by deep Greenland ice cores. In *Climate Processes and Climate Sensitivity*, eds. J. E. Hansen and T. Takahashi, pp. 288–298. Washington D. C: AGU. Geophysical Monograph Series, 29.

de Noblet, N., P. Braconnot, S. Joussaume, and V. Masson. 1996. Sensitivity of simulated Asian and African summer monsoons to orbitally induced variations in insolation 126, 115 and 6 kBP. *Climate Dynamics.* 12: 589–603.

Deplazes, G., A. Lückge, L. C. Peterson, et al. 2013. Links between tropical rainfall and North Atlantic climate during the last glacial period. *Nature Geoscience* 6: 213–217. doi: 10.1038/ngeo1712.

DiNezio, P. N., J. E. Tierney, J. B. L. Otto-Bliesner, et al. 2018. Glacial changes in tropical climate amplified by the Indian Ocean. *Science Advances* 4:eaat9658.

Ding, R., K-J. Ha, and J. Li. 2010. Interdecadal shift in the relationship between the East Asian summer monsoon and the tropical Indian Ocean. *Climate Dynamics* 34: 1059–1071. doi: 10.1007/s00382-009-0555-2.

Dixit, Y. 2020. Regional Character of the "Global Monsoon"; Paleoclimate insights from Northwest Indian lacustrine sediments. *Oceanography* June 2020 Early Online Release.

Duplessy, J. C. 1982. Glacial to interglacial contrasts in the northern Indian Ocean. *Nature* 295: 494–498.

Enfield, D. B., and A. M. Mestas-Nuñez. 1999. Interannual-to-multidecadal climate variability and its relationship to global sea surface temperatures. In *Present and Past Inter-Hemispheric Climate Linkages in the Americas and their Societal Effects*, ed. V. Markgraf. Cambridge, UK: Cambridge University Press.

Enfield, D. B., and D. A. Mayer. 1997. Tropical Atlantic SST variability and its relation to El Niño-Southern Oscillation. *Journal of Geophysical Research* 102: 929–945.

EPICA Community Members. 2006. One-to-one coupling of glacial climate variability in Greenland and Antarctica. *Nature* 444: 195–198.

Feng, F., and C. A. L. Bailer-Jones. 2015. Obliquity and precession as pacemakers of Pleistocene deglaciations. *Quaternary Science Reviews* 122: 166–179.

Feng, S., and Q. Hu. 2008. How the North Atlantic multidecadal oscillation may have influenced the indian summer monsoon during the past two millennia. *Geophysical Research Letters* 35: L01707. doi: 10.1029/2007GL032484

Fleitmann, D., S. J. Burns, A. Mangini, et al. 2007. Holocene ITCZ and Indian monsoon dynamics recorded in stalagmites from Oman and Yemen (Socotra). *Quaternary Science Reviews* 26: 170–188.

Fordham, D. A., F. Saltré, S. Haythorne, et al. 2017. PaleoView: a tool for generating continuous climate projections spanning the last 21, 000 years at regional and global scales. *Ecography* 40: 1348–1358. doi: 10.1111/ecog.03031.

Franco, D. R., L. A. Hinnov, and M. Ernesto. 2012. Millennial-scale climate cycles in Permian–Carboniferous rhythmites: Permanent feature throughout geologic time? *Geology* 40: 19–22. doi: 10.1130/G32338.1.

Gadgil, S. 2007. The Indian monsoon: 3. Physics of the monsoon. *Resonance* 12: 4–20. doi: 10.1007/s12045-007-0045-y.

Genty, D., D. Blamart, R. Ouahdi, et al. 2003. Precise dating of Dansgaard–Oeschger climate oscillations in Western Europe from stalagmite data. *Nature* 421: 833–837.

Goodbred, S. 2003. Response of the Ganges dispersal system to climate change: A source-to-sink view since the last interstade. *Sedimentary Geology* 162: 83–104. doi: 10.1016/S0037-0738(03)00217-3.

Goswami, B. N., M. S. Madhusoodanan, C. P. Neema, and D. Sengupta. 2006. A physical mechanism for North Atlantic SST influence on the Indian summer monsoon. *Geophysical Research Letters* 33: L02706. doi: 10.1029/2005GL024803.

Greenland Ice-Core Project (GRIP) Members. 1993. Climate instability during the last interglacial period recorded in the GRIP ice core. *Nature* 364: 203–207.

Gupta, A. K., D. K. Sinha, A. K. Singh, P. D. Naidu, R. Saraswat, and A. K. Rai. 2012. Indian contributions in the field of Paleoceanography (2006-2012). *Proceedings of the Indian National Science Academy* 78: 313–319.

Gupta, A. K., D. M. Anderson, D. N. Pandey, and A. K. Singhvi. 2006. Adaptation and human migration, and evidence of agriculture coincident with changes in the Indian summer monsoon during the Holocene. *Current Science* 90: 1082–1090.

Gupta, A.K., D. M. Anderson, and J. T. Overpeck. 2003. Abrupt changes in the Asian southwest monsoon during the Holocene and their links to the North Atlantic Ocean. *Nature* 421: 354–357.

Gupta, A. K., and E. Thomas. 2003. Initiation of Northern hemisphere glaciation and strengthening of the northeast Indian monsoon: ocean drilling program site 758, eastern equatorial Indian Ocean. *Geology* 31: 47–50.

Gupta, A. K., K. Mohan, M. Das, and R. K. Singh. 2013. Solar forcing of the Indian summer monsoon variability during the Ållerød period. *Scientific Reports* 3: 2753. doi: 10.1038/srep02753.

Gupta, A. K., M. Das, and D. M. Anderson. 2005. Solar influence on the Indian summer monsoon during the Holocene. *Geophysical Research Letters* 32: L17703. doi: 10.1029/2005GL022685.

Hastenrath, S., L.C. de Castro, and P. Aceituno. 1987. The Southern Oscillation in the Atlantic sector. *Contributions in Atmospheric Physics* 60: 447–463.

Hays, J. D., J. Imbrie, and N. J. Shackleton. 1976. Variations in the Earth's orbit: Pacemaker of the ice ages. *Science* 194: 1121–1132.

Holmgren, K., and W. Karlén. 1998. *Late Quaternary changes in climate.* Stockholm: Swedish Nuclear Fuel and Waste Management Co. Technical Report 98-13.

Hoyt, D. V., and K. H. Schatten. 1997. *The Role of the Sun in Climate Change.* New York: Oxford University Press.

Imbrie, J., A. Berger, E. A. Boyle, et al. 1993. On the structure and origin of major glaciation cycles 2. The 100,000-year cycle. *Paleoceanography* 8: 699–735. doi: 10.1029/93PA02751.

Imbrie, J., A. Mcintyre, and A. Mix. 1989. Oceanic response to orbital forcing in the late Quaternary: Observational and experimental strategies. In *Climate and Geosciences*, eds. A.L. Berger, S. Schneider, and J. Cl. Duplessy, pp. 121–164. Boston, MA: Kluwer Academic.

Imbrie, J., and K. P. Imbrie. 1979. *Ice Ages: Solving the Mystery.* Berkeley Heights, NJ: Enslow.

Jalihal, C., J. Srinivasan, and A. Chakraborty. 2019. Modulation of Indian monsoon by water vapor and cloud feedback over the past 22,000 years: *Nature Communications* 10: 5701. doi: 10.1038/s41467-019-13754-6.

Kanner, L. C., S. J. Burns, H. Cheng, and R. L. Edwards. 2012. High-latitude forcing of the South American summer monsoon during the last glacial. *Science* 335: 570–573. doi: 10.1126/science.1213397.

Kathayat, G., H. Cheng, A. Sinha, et al. 2016. Indian monsoon variability on millennial-orbital timescales. *Scientific Reports* 6: 24374. doi: 10.1038/srep24374.

Kaushal, N., S. F. M. Breitenbach, F. A. Lechleitner, et al. 2018. The Indian summer monsoon from a Speleothem $\delta^{18}O$ perspective—A review. *Quaternary* 1: 29. doi: 10.3390/quat1030029.

Keigwin, L. D., W. B. Curry, S. J. Lehman, and S. Johnsen. 1994. The role of the deep ocean in North Atlantic climate change between 70 and 130 kyr ago. *Nature* 371: 323–326.

Kennett, J. P., and B. L. Ingram. 1995. A 20,000-year record of ocean circulation and climate change from the Santa Barbara basin. *Nature* 377: 510–514.

Krishnan, R., J. Sanjay, C. Gnanaseelan, M. Majumdar, A. Kulkarni, and S. Chakraborty, ed. 2020. *Assessment of Climate Change over the Indian Region.* Singapore: Springer. doi: 10.1007/978-981-15-4327-2.

Kudrass, H. R., A. Hofmann, H. Doose, K. Emeis, and H. Erlenkeuser. 2001. Modulation and amplification of climatic changes in the Northern Hemisphere by the Indian summer monsoon during the past 80 k.y. *Geology* 29: 63–66.

Kumar, O., R. Devrani, and A. L. Ramanathan. 2017. Deciphering the past climate and monsoon variability from lake sediment archives of India: A review. *Journal of Climate Change* 3: 11–23. doi: 10.3233/JCC–170011.

Kutzbach, J. E. 1981. Monsoon climate of the early Holocene: Climate experiment with the Earth's orbital parameters for 9000 years ago. *Science* 214: 59–61.

Kutzbach, J. E., X. D. Liu, Z. Y. Liu, and G. S. Chen. 2008. Simulating the evolution response of global monsoons to orbital forcing over the past 280,000 years. *Climate Dynamics* 30: 567–579.

Labeyrie, L., J. Cole, K. Alverson, and T. Stocker. 2003. The history of climate dynamics in the late Quaternary. In *Paleoclimate, Global Change and the Future*. Global Change — The IGBP Series, ed. K. D. Alverson, T. F. Pedersen, and R. S. Bradley, pp. 33–62. Berlin, Heidelberg: Springer. doi: 10.1007/978-3-642-55828-3_3.

Leduc, G., L. Vidal, K. Tachikawa, et al. 2007. Moisture transport across Central America as a positive feedback on abrupt climatic changes. *Nature* 445: 908–911.

Leuschner, D. C., and F. Sirocko. 2000. The low-latitude monsoon climate Dansgaard-Oeschger cycles and Heinrich Events. *Quaternary Science Reviews* 19: 243–254.

Leuschner, D. C., and F. Sirocko. 2003. Orbital insolation forcing of the Indian Monsoon- a motor for global climate changes? *Palaeogeography, Palaeoclimatology, Palaeoecology* 197: 83–95.

Li, S. L., J. Perlwitz, X. W. Quan, and M. P. Hoerling. 2008. Modelling the influence of North Atlantic multidecadal warmth on the Indian summer rainfall. *Geophysical Research Letters* 35: 33–61. doi: 10.1029/2007GL032901.

Liu, G., X. Li, H.-W. Chiang, et al. 2020. On the glacial-interglacial variability of the Asian monsoon in speleothem $\delta18O$ records. *Science Advances* 6: eaay8189. doi: 10.1126/sciadv.aay8189.

Lu, R., B. Dong, and H. Ding. 2006. Impact of the Atlantic multidecadal oscillation on the Asian summer monsoon. *Geophysical Research Letters* 33: L24701. doi: 10.1029/2006GL027655.

Marzin, C., N. Kallel, M. Kageyama, J. C. Duplessy, and P. Braconnot. 2013. Glacial fluctuations of the Indian monsoon and their relationship with North Atlantic climate: New data and modelling experiments. *Climate of the Past* 9: 2135–2151.

Mayewski, P. A., M. S. Twickler, S. I. Whitlow, et al. 1996. Climate change during the last deglaciation in Antarctica. *Science* 272: 1636–1638.

Meloth, T., H. Kawahata, and V. P. Rao. 2007. Indian summer monsoon variability during the Holocene as recorded in sediments of the Arabian Sea: Timing and implications. *Journal of Oceanography* 63: 1009–1020.

Naidu, P. D., and B. A. Malmgren. 1995. A 2,200 years periodicity in the Asian Monsoon System. *Geophysical Research Letters* 22: 2361–2364.

Newton, A., R. Thunell, and L. Stott. 2006. Climate and hydrographic variability in the Indo-Pacific Warm Pool during the last millennium. *Geophysical Research Letters* 33: L19710. doi: 10.1029/2006GL027234.

Oeschger, H., J. Beer, U. Siegenthaler, B. Stauffer, W. Dastard, and Langway, C. C. 1984. Late glacial climate history from ice core. In *Climate Processes and Climate Sensitivity*, ed. J. E Hansen, and T. Takahashi, pp. 299–306. Washington, DC: American Geophysical Union. Monograph Series 29.

Oppo, D., and S. J. Lehman. 1995. Suborbital timescale variability of North Atlantic Deep Water during the past 200,000 years. *Paleoceanography* 10: 901–910.

Overpeck, J., D. Rind, A. Lacis, and R. Healy. 1996. Possible role of dust-induced regional warming in abrupt climate change during the last glacial period. *Nature* 384: 447–449.

Petit, J-P., J. Jouzel, D. Raynaud, et al. 1999. Climate and atmospheric history of the past 420,000 years from the Vostok Ice Core, Antarctica. *Nature* 399: 429–436.

Porter, S.C., and Z. Ann. 1995. Correlation between climate events in the North Atlantic and China during the last glaciations. *Nature* 375: 305–308.

Prasad, S., A. Anoop, N. Riedel, et al. 2014. Prolonged monsoon droughts and links to Indo-Pacific warm pool: A Holocene record from Lonar Lake, central India. *Earth and Planetary Science Letters* 391: 171–182. doi: 10.1016/j.epsl.2014.01.043.

Prasad, S., and Y. Enzel. 2006. Holocene paleoclimates of India. *Quaternary Research* 66: 442–453.

Prell, W. L. 1984. Variations of monsoon upwelling: A response to changing solar radiation. In *Climate Processes and Climate Sensitivity*, ed. J. E Hansen, and T. Takahashi, pp. 48–57. Washington, DC: American Geophysical Union. Geophysical Monograph Series 29.

Prell, W. L., and J. E. Kutzbach. 1987. Monsoon variability over the past 150,000 years. *Journal of Geophysical Research* 92: 8411–8425.

Prell, W. L., W. H., Hutson, D. F. Williams, A. W. H. Bé, K. Geitzenauer, and B. Molfino. 1980. Surface circulation of the Indian Ocean during the last Glacial Maximum, approximately 18,000 yr BP. *Quaternary Research* 14: 309–336.

Premathilake, R., and A. Gunatilaka. 2013. Chronological framework of Asian Southwest Monsoon events and variations over the past 24,000 years in Sri Lanka and regional correlations. *Journal of the National Science Foundation of Sri Lanka* 41: 219–228. doi: 10.4038/jnsfsr.v41i3.6057.

Rahmstorf, S. 2002. Ocean circulation and climate during the past 120,000 years. *Nature* 419: 207–214. doi: 10.1038/nature01090.

Rajan, S. 2018. Abrupt climate shifts over the past 10,000+ years: An Arctic-Antarctic-Asian imbroglio? In *Science and Geopolitics of the White World*, ed. P. S. Goel, R. Ravindra and S. Chattopadhyay, pp. 83–91. Switzerland: Springer International. doi: 10.1007/978-3-319-57765-4_1.

Rajan, S., and N. Khare. 2002. Emerging Visions in Indian marine Geosciences- the Southern Oceans. In Four Decades of Marine Geosciences in India- A Retrospective, Vol. 74, pp. 70–76. Kolkata: Geological Survey of India Special Publication.

Rajeevan, M., and L. Sridhar. 2008. Inter-annual relationship between Atlantic sea surface temperature anomalies and Indian summer monsoon. *Geophysical Research Letters* 35: L21704. doi: 10.1029/2008GL036025

Ramesh, R., H. Boragaonkar, S. Band, and M. G. Yadava. 2017. Proxy climatic records of past monsoons. In *Observed Climate Variability and Change over the Indian Region*, eds. M. Rajeevan, and S. Nayak, pp. 271–284. Singapore: Springer Geology. doi: 10.1007/978-981-10-2531-0_15.

Ramesh, R., M. Tiwari, S. Chakraborty, S. R. Managave, M. G. Yadava, and D. K. Sinha. 2010. Retrieval of south Asian monsoon variation during the Holocene from natural climate archives. *Current Science* 99: 1770–1786.

Sachs, J. P., D. Sachse, R. H. Smittenberg, Z. Zhang, D. S. Battisti, and S. Golubic. 2009. Southward movement of the Pacific Intertropical Convergence Zone AD 1400–1850. *Nature Geoscience* 2: 519–525. doi: 10.1038/ngeo554.

Sanwal, J., C. P. Rajendran, and M. S. Sheshshayee. 2019. Reconstruction of late quaternary climate from a paleo-lacustrine profile in the Central (Kumaun) Himalaya: Viewing the results in a regional context. *Frontiers in Earth Science* 7: 2. doi: 10.3389/feart.2019.00002.

Sanyal, P., and R. Sinha. 2010. Evolution of the Indian summer monsoon: synthesis of continental records. *Geological Society, London, Special Publications* 342: 153–183 doi: 10.1144/SP342.11.

Saraswat, R., R. Nigam, and T. Correge. 2014. A glimpse of the Quaternary monsoon history from India and adjoining seas. *Palaeogeography, Palaeoclimatology, Palaeoecology* 397: 1–6. doi: 10.1016/j.palaeo.2013.11.001.

Schmidt, M. W., and J. E. Hertzberg. 2011. Abrupt climate change during the last ice age. *Nature Education Knowledge* 3: 11.

Schulz, H., U. von Rad, and H. Erlenkeuser. 1998. Correlation between Arabian Sea and Greenland climate oscillations of the past 110 000 years. *Nature* 393: 54–57.

Shackleton, N. J. 2000. The 100,000-year ice-age cycle identified and found to lag temperature, carbon dioxide, and orbital eccentricity. *Science* 289: 1897–1902.

Sharaf, S. G., and N. A. Boudnikova. 1967. Secular variations of elements of the Earth's orbit which influences the climates of the geological past. *Trudy Institute; Theoretical Astronomy at Leningrad* 11: 231–261.

Sharma, S., M. Joachimski, M. Sharma, et al. 2004. Late glacial and Holocene environmental changes in Ganga plain, Northern India. *Quaternary Science Reviews* 23: 145–159.

Singhvi, A. K. and V. S. Kale. 2010. *Paleoclimate studies in India; Last Ice Age to the Present.* IGBP-WCRP-SCOPE-Report Series: 4. New Delhi: Indian National Science Academy.

Singhvi, A.K., A. Bluszcz, M. D. Bateman, and M. S. Rao. 2001. Luminescence dating of loess-palaeosol sequences and coversands: Methodological aspects and palaeoclimatic implications. *Earth-Science Reviews* 54: 193–211.

Sinha, A., K. G. Cannariato, L. D. Stott, H. Cheng, R. L. Edwards, M. G. Yadava, R. Ramesh, and I. B. Singh. 2007. A 900-year (600 to 1500 A.D.) record of the Indian summer monsoon precipitation from the core monsoon zone of India. Geophysical Research Letters 34: L16707. doi: 10.1029/2007GL030431.

Sinha, A., L. D. Stott, M. Berkelhammer, et al. 2011. A global context for monsoon mega droughts during the past millennium. *Quaternary Science Reviews* 30: 47–62. doi: 10.1016/j.quascirev.2010.10.005.

Sinha, N., N. Gandhi, S. Chakraborty, R. Krishnan, M. G. Yadava, and R. Ramesh. 2018. Abrupt climate change at ~2800-year BP evidenced by a stalagmite record from the peninsular India. *Holocene* 28: 1720–1730. doi: 10.1177/0959683618788647.

Sirocko, F., D. Leuschner, M. Staubwasser, J. Maley, and L.Heusser. 1999. High-frequency oscillations of the last 70,000 years in the tropical/subtropical and polar climates. In *Mechanisms of Global Climate Change at Millennial Time Scales*, eds. P. U. Clark, R. S. Webb, and L. D. Keigwin, pp. 113–126. Washington, DC: AGU. Geophysical Monograph 12.

Steffensen, J. P., K. K. Andersen, M. Bigler, et al. 2008. High-resolution Greenland ice core 536 data show abrupt climate change happens in few years. *Science* 321: 680–684.

Stocker, T. F., D. G. Wright, and W. S. Broecker. 1992. The influence of high – Latitude surface forcing on the global thermohaline circulation. *Paleoceanography* 7: 529–541.

Street-Perrot, F. A., and R. A. Perrott. 1990. Abrupt climate fluctuations in the tropics: The influence of Atlantic Ocean circulation. *Nature* 343: 607–612.

Sukumar, R., R. Ramesh, R. K. Pant, and G. Rajagopalan. 1993. A $\delta^{13}C$ record of late Quaternary climate change from tropical peats in southern India. *Nature* 364: 703–706.

Tabor, C. R., B. L. Otto-Bliesner, E. C. Brady, et al. 2018. Interpreting precession driven $\delta^{18}O$ variability in the South Asian monsoon region. *Journal of Geophysical Research: Atmospheres* 123: 5927–5946. doi: 10.1029/2018JD02842.

Thompson, L. G., E. M. Thompson, M. E. Davis, et al. 2006. Ice core evidence for asynchronous glaciation on the Tibetan plateau. *Quaternary International* 154–155: 3–10.

Thouveny, N., J. L. Beaulieu, E. Bonifay, et al. 1994. Climate variations in Europe over the past 140 kyr deduced from rock magnetism. *Nature* 371: 503–506.

Tiwari, M., A. K. Singh, and R. Ramesh. 2011. High-resolution monsoon records since last glacial maximum: A comparison of marine and terrestrial paleoarchives from South Asia. *Journal of Geological Research* 2011. doi: 10.1155/2011/765248.

Tiwari, M., S. Managave, M. G. Yadava, and R. Ramesh. 2009. Spatial and temporal coherence of paleo-monsoon records from marine and land proxies in the Indian region during the past 30 ka. In *Current Trends in Science: Platinum Jubilee Special*, ed. N. Mukunda, pp. 517–535. Bangalore: Indian Academy of Sciences.

Tiwari. M., and R. Ramesh. 2007. Indian monsoon-solar connection. *Proceedings of 12th ISMAS Symposium cum IRP-7 Workshop on Mass Spectrometry*. Goa, India.

Van Campo, E. 1986. Monsoon fluctuations in two 20,000-yr B.P. oxygen-isotope/pollen records off southwest India. *Quaternary Research* 26: 376–388.

Voelker, A. 2002. Global distribution of centennial-scale records for Marine Isotope Stage (MIS) 3: A database. *Quaternary Science Reviews* 21: 1185–1212.

Wang, P. X., B. Wang, H. Cheng, et al. 2017. The global monsoon across time scales: Mechanisms and outstanding issues. *Earth-Science Reviews* 174: 84–121.

Wang, P., S. Clemens, L. Beaufort, et al. 2005. Evolution and variability of the Asian monsoon system: State of the art and outstanding issues. *Quaternary Science Reviews* 24: 595–629.

Wang, Y. J., H. Cheng, R. L. Edwards, et al. 2001. A High-Resolution Absolute-Dated late Pleistocene monsoon record from Hulu Cave, China. *Science* 294: 2345–2348. doi: 10.1126/science.1064618.

Wang, Y., H. Cheng, R. L. Edwards, et al. 2008. Millennial-and orbital-scale changes in the East Asian monsoon over the past 224,000 years. *Nature* 451: 1090–1093.

Yadav, R. R. 2013. Tree ring-based seven-century drought records for the Western Himalaya, India. *Journal of Geophysical Research: Atmosphere* 118: 4318–4325. doi: 10.1002/jgrd.50265.

Yadava, M.G., and R. Ramesh. 2005. Monsoon reconstruction from radiocarbon dated tropical Indian speleothems. *Holocene* 15: 48–59.

Zhang, R., and T. L. Delworth. 2006. Impact of Atlantic multidecadal oscillations on India/Sahel rainfall and Atlantic hurricanes. *Geophysical Research Letters* 33: L17712. doi: 10.1029/2006GL026267.

Zhang, Y., J. M. Wallace, and D. S. Battisti. 1997. ENSO-like interdecadal variability: 1900–93. *Journal of Climate* 10: 1004–1020. doi: 10.1175/1520-0442(1997)010,1004:ELIV.2.0.CO;2.

Zhao, K., Y. Wang, R. L. Edwards, H. Cheng, D. Liu, and X. Kong. 2015. A high-resolved record of the Asian Summer Monsoon from Dongge Cave, China for the past 1200 years. *Quaternary Science Reviews* 122: 250–257.

2 The Climatic and Anthropogenic Influences on Himalayan Glacial and Non-Glacial Lakes
A Review

Narendra Kumar Meena
Wadia Institute of Himalayan Geology

Pawan Kumar Gaury and A. K. Mahajan
Central University of Himachal Pradesh

CONTENTS

INTRODUCTION

The Himalaya is a mountain belt of 2,400 km from east to west and an abode of several hundred glacial and non-glacial lakes, which are formed due to glacier melting and various types of earth processes, respectively. The water of these lakes is being used for drinking, irrigation, fishing, recreation, and other basic needs. The Himalayan lakes can be divided into two groups viz. (1) the glacial lakes, situated at a higher

21

altitude of greater Himalaya and (2) non-glacial lakes, situated at lesser and outer Himalaya and its frontal part. There are ~15,000 glaciers and ~9,000 glacial lakes over the Himalaya (Mool, 2005; Lohani et al., 2015), which are considered as a major water source for the Asian river system, upon which the lives of more than 1.3 billion people depend (Bajracharya et al., 2006). For decades, the glacial environment has been influenced by the changing climate (Bajracharya et al., 2008; Kang et al., 2010; Wang et al., 2012), and as a result, the glaciers are melting at a fast speed and the numbers as well as size of glacial lakes are also increasing at an unexpected rate (Lama and Devkota, 2009; Chen et al., 2013; Riaz et al., 2014).

Glacial lakes have a free surface in front of glaciers and receive water from their melting and snow melting (Xu et al., 2009; Jain et al., 2015). Glacial lakes can be ice-dammed, moraine-dammed, supra-glacial, sub-glacial, or en-glacial (Benn and Evans, 2010; Veettil et al., 2016). These lakes are very sensitive to precipitation, temperature, and soil frost. (Liu et al., 2009). The ice-dammed lakes generally become extinct with failure of dam and sediments deposition (Ghimire, 2005) and infrequently survive more than one summer (Hewitt and Liu, 2010). The moraine-dammed lakes are structurally weak and unstable and always show risks of outburst (Dahal, 2008). The stability of moraine-dammed lakes also depends on the porosity of the wall as well as the height-to-width ratio (Hambrey et al., 2008).

The glacial lakes can cause disastrous effect downstream due to glacial lake outburst floods (GLOFs) (Mal and Singh, 2014), which include a heavy load of debris and mud (Rasul et al., 2011), and the water from these lakes is released due to destruction of moraine or ice wall at the end of lakes (Rana et al., 2000). The lake outburst generally depends on the depth and size of lakes, geomorphology of the valley, and nature of outburst (Fujita et al., 2009). The unstable glacial lakes easily burst out due to certain driving forces like compressive stress and strain produced by tectonic activities, an overload of water mass on becoming more than the material strength (Massey et al., 2010), earthquake, massive landslide, cloud burst, and so on (Post and Mayo, 1971). The water of these lakes can overflow due to debris flow and displaced waves produced by snow avalanche, heavy rain, slumping, and slope failure (Shrestha et al., 2010; Worni et al., 2012a; Banshtu and Prakash, 2013). The overflow of water generally accelerates the coastal erosion, which results in the failure of the moraine dam or glacial lakes followed by GLOFs (Carey et al., 2012). As a rough estimate, the GLOFs have caused death of thousands of people all over the world (Clague and Evan, 2000) and the risk is still prevalent everywhere.

The non-glacial lakes are situated at lesser and outer Himalaya and its frontal parts (Figure 2.1). These lakes generally contain freshwater (Dal (J & K), Wular, Renuka, Nainital, Bhimtal, Dal (Mcleodganj) and so on), but may have brackish (Tso Moriri and Chang Chenmo) and saline water (Pongong Tso, Namtso, and Nam Co). These lakes are formed by tectonic activities as well as extension and compression of the surface of earth. Freshwater lakes are either oligotrophic, mesotrophic, and eutrophic or hyper-eutrophic (Kumar et al., 2019a). The oligotrophic lakes are deep, having clear water with less species density and diversity. Eutrophic lakes are highly nutritious with a rich biodiversity of flora and fauna. Besides, the lakes that fall between the eutrophic and oligotrophic stages are termed as mesotrophic. The hyper-eutrophic lakes are those lakes that are extremely polluted. The eutrophication is the main

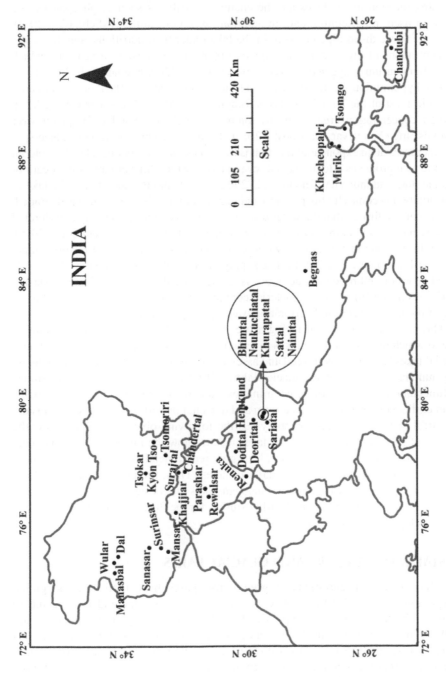

FIGURE 2.1 The map showing some of the famous lakes in the Indian Himalayan regions.

problem (Dixit and Tiwari, 2005) and is highly responsible for changing the natural scenario of Himalayan lakes (Litke, 1999; Guyuan et al., 2011). The major causes of eutrophication in the lakes are the chemical fertilizers such as phosphorus and nitrogen used in agriculture and, in addition, excreta of animals (Schindler, 2012), domestic waste disposal, and so on. In the lake under high nutritious conditions, the concentrations of algae and other aquatic organisms increase at a fast rate, which thereafter consume oxygen dissolved in water (Leng, 2009; Kumar et al., 2020a).

The hypolimnion, i.e. the deepest water layer, shows anoxic (lack of oxygen) condition in most of the Himalayan lakes (Singh et al., 2008). The water with reduced oxygen concentration creates an environment somehow favorable for plants over animals' life (Yannawar and Bhosle, 2013). Thence, the concentration of algae species increases at an exponential rate that is known as an algal bloom. The algal bloom is toxic as it produces a toxin named 'cyanotoxin', which further causes the death of aquatic flora and fauna (Anderson et al., 2002; Pathak and Pathak, 2012). The lake in that situation becomes highly polluted and thereafter the natural scenario is degraded. It has been confirmed that the water quality of non-glacial lakes is being deteriorated both by natural and anthropogenic factors (Sheikh et al., 2014; Saleem et al., 2015), but anthropogenic factors are known to contribute more than natural factors (Gupta et al., 2012; Kumar et al., 2019b). The supply of nutrients inside the lakes due to different types of anthropogenic activities is termed as cultural eutrophication (Hasler, 1947; Guyer and Ilhan, 2011). Bacterial contamination also has been observed to be one of the major causes of deteriorating water quality of the lakes at present.

The non-glacial lakes are facing acute problems of sedimentation and siltation due to which the size of these lakes is reducing at a fast speed. The glacial and non-glacial lakes have been monitored and assessed so far in the past to know the increasing numbers and size of the glacial lakes, GLOFs, water quality, trophic status, sedimentation rate, bacterial contamination, and so on. The information available from earlier studies is in scattered form and does not provide a collective conclusion. To fulfill that need, the present study has the following objectives: (1) to know the impact of climate change on glacial lakes, (2) to know the impact of anthropogenic activities on non-glacial lakes, and (3) to give recommendation measures for management and restoration of the Himalayan lakes. Thus, the present study brings in limelight the existing scenario of Himalayan lakes and besides, it will be helpful to the researchers in finding a gap of earlier studies for the future as well as preservation and development of the lakes.

HIMALAYAN GLACIERS AND GLACIAL LAKES

The increasing concentrations of greenhouse gases have increased the global temperature around 0.30°C–0.60°C during the 20th century and is expected to increase further 1.4°C–5.8°C up to end of the 21st century (Bajracharya et al., 2006). The temperature in the Himalayan regions in the last three decades has increased around 0.15°C–0.60°C (Shrestha et al., 2010; Jha and Khare, 2016) and it is expected that the temperature of high altitudinal regions will be increased more in future as compared to lower regions (Shrestha and Devkota, 2010; Hoy et al., 2016). The glaciers are considered as the dynamic landscape, show variability in length and mass due to changing climate, and thus are the best climate change indicator (Marzeion et al., 2014). The Himalaya is considered as one of the most vulnerable hot spots at the global level owing to its retreating glaciers because of increasing temperature and precipitation variability with changing climate (Bajracharya et al., 2008; Immerzeel et al., 2010; Hoy et al., 2016).

GLACIER RETREATING

Several studies have confirmed that the Himalayan glaciers are retreating since the end of the little ice age (Dobhal et al., 2004), but the retreating rate during the last few decades has increased several times due to climate change (Bolch et al., 2008; Pratap et al., 2016). The retreating rate of the Gangotri glacier was confirmed by Naithani et al. (2001), and they found that it receded around 850 m in the past 20 years in comparison to 2 km during the last 200 years. The frontal recession of the Gangotri glacier, Garhwal Himalaya was also monitored by Bhambri et al. (2012) through high resolution data from Corona, Hexagon, Indian satellite CartoSat-1. In that study, the glacier was found to be retreated around 819 m and lost around 0.41 km^2 from its front during 1965–2006. The Milam glacier of Kumaun Himalaya was assessed by Raj (2011) by using temporal satellite data and he found that the glacier has retreated 90 m vertically and 1,328 m laterally during 1954–2006. Raj et al. (2012) again monitored the Milam glacier by using Resourcesat-2 LISS IV data and found a 120-m shift in the snout position during 2006–2011. The field evidence and satellite data strengthened the fact that this glacier has been continuously retreating since 1906, but the average rate was increased during the last decade.

The fluctuation in 224 glaciers of Beas Basin was observed by Dutta et al. (2012) using IRS and Landsat imageries from 1972 to 2006 and it was found that the entire glaciers decreased from 419 to 371 km^2. Schmidt and Nusser (2012) monitored 121 small glaciers and 60 cirque and valley glaciers of the Kang Yatze Massif region of Trans-Himalaya Ladakh, to know the change in size as well as the retreating rate during 1969–2010. In that study, they used satellite imageries like Corona, Landsat, and SPOT and found that the glaciated area reduced from 96.4 to 82.6 km^2, i.e. 14% with a front retreat of 125 m. Pandey and Venkataraman (2012) in an earlier study observed the Chhota Shigri glacier of Chandra Basin, by using toposheet and remote sensing data and found that the glacier has retreated 950 m during 1962–2008. Glacier change in the Kashmir Alpine Himalayan region was observed by Murtaza and Romshoo (2015) and they noticed that the glacier reduced up to 17% during 1983–2013 and shifted upward from 80 to 300 m due to increasing temperature and decreasing precipitation. That study further indicated that the glacier dynamics may impact on water supply and streamflow as well as other water-dependent sectors in the concerned region. Gaddam et al. (2016) monitored 19 glaciers of Baspa Basin, western Himalaya, by using satellite data from 1962 to 2014 and found 24% deglaciation i.e. 41.1 km^2. It was also inferred that the glacier retreating was due to decreasing precipitation and increasing temperature.

The Shah glacier in the Ravi Basin of Himachal Pradesh was monitored by Chand and Sharma (2016) using satellite data from 1965 to 2013 and they noticed that the total area of the glacier has decreased around 0.21 km^2. The Dokriani glacier of the

Garhwal Himalaya was monitored by Dobhal et al. (2004) to know about the recession and morpho-geometrical changes during 1962–1995. The glacier in that study was found to have receded 550 m (20%) with an average rate of 16.6 m/yr, but from the snout position, the recession was 3,957 m^2 (10%) with an average rate of 17.4 m/yr. Mehta et al. (2011) in an earlier study at Rataban (7.4 km^2) and Tipra glaciers (7.5 km^2) of the Alaknanda River Basin of Garhwal Himalaya observed glacier snout and surface change during 1962–2008. The average retreating of these two glaciers during 1962–2002 was found to be 13.4 m/a, the snout receded 535 m, and the surface area 18%, but that rate increased thereafter during 2002–2008 and became 21.2 m/a at the Rataban and 21.3 m/a at the Tipra glacier. Besides, during 1962–2008, the total area was found to be reduced 0.028 km^2 at the Rataban glacier and 0.084 km^2 at the Tipra glacier, due to increase in temperature. Bajracharya and Mool (2009) in their earlier study at the Dudh Koshi Basin monitored 278 glaciers and among those glaciers, the Sagarmatha glacier (Mount Everest region) was found to be retreated with the average rate of 10–59 m/a during 1976–2007, but some glaciers had a retreated rate of 74 m/a during the last half decade. Besides, the retreating rates at Imja and Lumding glaciers were 42–34 m/a during 1976–2000 and 74 m/a during 2000–2007, respectively.

The change in glaciers of Pumqu River Basin, Xizang, Tibet was monitored by Che et al. (2014) using satellite data and they noticed that 160 glaciers having a total area of 276.57 km^2 reduced during 1970–2013. The increasing temperature and precipitation variability were found to be the main casual factors behind the retreating of the glaciers. Racoviteanu et al. (2015) monitored glaciers change in eastern Himalayan regions namely Arun and Tamur Basin of Nepal, Teesta Basin of Sikkim, India, and some parts of Bhutan and China. In that study, various satellite imageries like Landsat, Corona, ASTER, QuickBird, and WorldView-2 from 1962 to 2000/2006 were used. Based on elevation, altitudinal range, slope, and so on, the glaciers during 1962–2000 were found to be retreated 0.44%/yr in Sikkim, 0.23%/yr in Kanchenjunga-Sikkim, and 0.50%/yr in Nepal. The loss of these glaciers during 1962–2006 was compared with that of glaciers of other Himalayan regions and their retreating rates were found to be similar. Jiawen et al. (2006) in their study at Xiaxiabangma and Qomolangma regions of the Tibetan Plateau found that the glaciers are retreating at the rate of 4.0–5.2 and 5.5–9.5 m/a, respectively. It was inferred in that study that the glaciers are continuously retreating for the last few decades due to warmer temperature and reduced precipitation, which is similar to the observation made by several other researchers in the past.

The glaciers of the Qomolangma region in the northern slope of middle Himalaya, Tibetan Plateau were monitored by Qinghua et al. (2009) using remote sensing and GIS techniques and they found that the glaciers were retreated 3.23 km^2 during 1974 and 1976, 8.68 km^2 during 1976–1992, 1.44 km^2 during 1992–2000, 1.14 km^2 during 2000–2003, and 0.52 km^2 during 2003–2008. It was also observed that the middle Himalayan glaciers are having equal retreating rates as that of western Himalayan glaciers. Li et al. (2016) carried out a study at the glaciers of the Ayilariju region of western Himalaya, China, by using Landsat Thematic Mapper/Enhanced Thematic Mapper plus, topographic maps, GIS techniques, and so on. They found that the total glaciated area was reduced from 190.37 to 162.52 km^2, i.e. 14%, during 1980–2011. The reasons behind were inferred to be the rise in temperature and the decreasing wind speed and precipitation accelerated by weakening the south Asian summer monsoon.

The essence of earlier studies related to glaciers is that the glaciers are retreating at a fast speed for the last few decades due to increasing temperature and precipitation variability as a result of climate change.

Increasing Numbers and Size of Glacial Lakes

The size and numbers of those lakes that receive water from the melting of glaciers are increasing with time. The increasing temperature may further enhance the retreating of glaciers in the future and thus, more potentially harmful glacial lakes will be formed (Bajracharya et al., 2008). Several studies have shown the formation of glacial lakes with an equal pace of glacier retreating. In an earlier study, the water level of 13 glacial lakes in the Karakoram region of north-west Himalaya, India was observed by Srivastava et al. (2013), and they found that the size and water level of ten lakes increased at the rate of 0.173 m/yr, whereas that of the rest three lakes decreased at the rate of 0.056 m/yr. The increasing water level was due to an increase in seasonal snow and glacier melting as a result of rising temperature and on the other hand, the reduced water level was owing to precipitation variability in the north-western Himalayan region. Raj and Kumar (2016) in an earlier study in Uttarakhand identified 362 glacial lakes by using remote sensing data. In that study, all the lakes were found to be increased in size and it was confirmed that the glacial lakes are dynamic in nature and are expanding with time due to climate change.

Che et al. (2014) in an earlier study monitored glacial lakes from Pumqu River Basin, Xizang, Tibet, using satellite imagery and found no change in numbers of glacial lakes during 1970–2001, whereas the numbers of lakes increased by 55 and the size increased by 11.14 km^2 (i.e. 26.76%) during 2001–2013. The Imja Tsho Lake, located in the Mount Everest region of Nepal, was monitored by Valenzuela et al. (2014) and they found that the depth of the lake increased by ~26 m and the volume increased by ~26 million m^3 during 2002–2012, owing to retreat in the glacier terminus. The glacial lake of Tsho Chubda Chamkhar Chu Basin, Hindukush Himalaya, Bhutan was monitored by Jain et al. (2015) using satellite imagery (ASTER) and they found that the size of the lake increased at the rate of 0.03 km^2/yr during 1978–2000, 0.04 km^2/yr during 2004–2005, and again 0.03 km^2/yr during 2005–2010. During the study, it was also noticed that the increased size is not related to entry of water from outside, but due to melting of its lower portion, which indicates the impact of increasing temperature on lake body also. Wang et al. (2016a) monitored the glacial lakes of Nam Co Basin, Tibet Plateau by using satellite data and found that the total area of glacial lake expanded by 0.97 km^2 (36.7%) and besides, the number of newly formed glacial lakes increased by 36% during 1991–2011. The surface area of the Nam Co Lake increased by 72.64 km^2 (3.71%) with an average rate of 3.63 km^2/yr. It was inferred that the main factors responsible for increasing numbers and sizes of glacial lakes is the variability in temperature and precipitation owing to climate change.

The expansion, disappearance, and formation of glacial lakes in Bhutan during the 1990s, 2000s, and 2010 were noticed by Veettil et al. (2016) using satellite imagery and they found that the moraine-dammed glacial lakes are being formed and expanded due to melting of glaciers. Nie et al. (2017) monitored 4,950 glacial lakes of entire Himalayan regions by using 348 Landsat satellite imagery of 30-m resolution from 1990 to 2015. It was found that the lakes situated at 4,000–5,700 m above sea level expanded 14.1% during the said period, due to glacier melting as a result of warming the Himalayan environment. Shrestha and Balla (2011) in an earlier study monitored Lumding Tsho Lake of the Dudh Koshi Basin by using topographical maps and imageries from 1963 to 2009. The length of the lake was found to be increased from 620 to 2,140 m and its area increased by 10.36–97.61 ha, which is 5% of the total area of the lake. The size of the lake situated at the supra glacier in the Mt. Qomolangma region, Tibetan Plateau, China, was monitored by Qinghua et al. (2009) and they found that the size of the lake increased from 0.05 to 0.71 km² during 1974–2008, which is ~13 times more in the duration of the last 35 years.

The variations in the size of glacial lakes of the Boshula mountain range were observed by Wang et al. (2012) on the basis of field investigation as well as hydrodynamic modeling; they found that the glacial lakes increased in dimension during the 1970s–2009. The people downstream were also warned to face potential danger and carry out mitigation measures. Wang and Zhang (2013) monitored the changing scenario of 604 glacial lakes situated at central Chinese Himalayan, by using Landsat satellite imagery of 1990–2010 and found that the moraine-dammed 199 lakes increased at the rate of 0.45 km²/yr, whereas other lakes increased at the rate less than 0.05 km²/yr. The change in size of glacial lakes of Tarim Basin in Central Asia was observed using Landsat imagery of 1990, 2000, and 2013 by Wang et al. (2016b). In that study, the average increase in size of the lakes was found to be 0.67%/a during the last 23 years with the highest rate in the Tian Shan and Altun Shan region and the slowest in the Pamir and Karakoram mountain region. The increasing rate in the total area of glacial lakes during 1990–2000 was 0.23%/a, which increased around four times and became 0.98%/a, during 2000–2013. The impact of climate change on the size of ice-contact glacial lakes was found to be 3.35%/a, and at non-glacial lakes, it was 0.57%/a.

GLACIAL LAKE OUTBURST FLOODS IN THE HIMALAYAN REGION

The GLOFs and their associated risks are the most common in Himalayan regions, which have caused a lot of fatalities and catastrophic damages during the last decades (Nie et al., 2017). The main cause of GLOFs is the increasing size of glacial lakes due to the melting and retreating of Himalayan glaciers as a result of changing climate and besides this, the high-intensity rainfall (Das et al., 2015). It has been inferred that due to changing climate, the glacier retreating rate may be increased further in the future (Maanya et al., 2016) and generate more GLOFs. Several past studies have inferred that most of the Himalayan glacial lakes are increasing in size and having the futuristic risk of outburst potential. Worni et al. (2012b) in an earlier study observed the damage and outburst potential of 251 glacial lakes of greater Himalayan regions. During the study, 12 lakes (8 in Sikkim, 2 in Himachal Pradesh, and 2 in J&K) were found to have high risks of GLOFs, while 93 lakes had low risks. Besides, 101 lakes were found safe having no risk of outburst potential. Raj et al. (2013) monitored a moraine-dammed glacial lake of South Lhonak glacier (Sikkim Himalaya) by using Corona and LISS III satellite data from 1962 to 2008 and found the value of outburst potential to be 42% and the peak discharged at 586 m³/s. Mal and Singh (2014) in an earlier study noticed that the Hemkund, Geldhura, and Sitkeng glacial lakes of Uttrakhand, India have increased in size, showing risk of outburst. The Chorabari Lake of Kedarnath, India also burst out due to flash flood and caused death of thousands of people downstream (Rao et al., 2014). In an earlier study, Raj and Kumar (2016) had identified 362 glacial lakes in Uttarakhand (India) and among those, eight lakes were declared to have outburst potential.

In an earlier study, Bajracharya and Mool (2009) at the Mount Everest region of Nepal observed an increased size of 34 glacial lakes and among those, 12 glacial lakes had high risk of GLOFs. Besides, 24 new glacial lakes were also identified in the concerned region. The risks of outburst potential of glacial lakes in Nepal and Bhutan Himalaya were observed by Fujita et al. (2013), using visible band image as well as on the basis of steep lakefront area and potential flood volume (PFV). They found 794 glacial lakes with zero PFV having no risk of outburst, whereas 49 lakes were noticed to have 10 million m^3 PFA, which is equal to the total volume of water generated during earlier major recorded GLOFs. The glacial lakes of entire Himalayan regions are found to be increased in size and have risk of damage with time. In a previous study at the Pumqu River Basin of Xizang by Che et al. (2014), the total area of glacial lakes was found to be increased by 26.76% and besides this, 19 glacial lakes were noticed to have outburst potential. It was also confirmed that the variability in precipitation and temperature have increased the risk of GLOFs.

A previous study carried out by Nie et al. (2013) in central Himalaya, using satellite imagery from 1990 to 2010, has revealed 17.11% enhancement in the size of glacial lakes. In that study, 67 lakes were found to be increased very fast, which requires continuous monitoring to avoid the risk associated with them. Wang and Zhang (2013) in their study at the central Chinese Himalaya identified 23 potential dangerous glacial lakes, which were increased by 77.46% during 1990–2010. Shijin et al. (2015) in an earlier study had identified moraine-dammed 329 lakes having a total area of 125.43 km^2 in the Chinese Himalaya and among them, 116 lakes having 49.49 km^2 area were declared to have outburst potential. Furthermore, the lakes of the mid-eastern Himalayan region were noticed to have high risk of GLOFs, whereas the lakes of eastern Himalayan regions had very low risk. Nie et al. (2017) in a recent study identified 4,950 glacial lakes over the entire Himalayan higher altitudinal regions and among those, 118 lakes were found to be expanded at a very fast rate during 1990–2015 and have a higher risk of GLOFs.

HIMALAYAN NON-GLACIAL LAKES

The non-glacial lakes include freshwater, brackish, and salt water lakes, which are under stress of pollution, eutrophication, and bacterial contamination, due to the increasing pressure of anthropogenic activities. Several studies have been carried out on these lakes to monitor water quality, nutrient enrichment, sedimentation rate, ecological status, and so on.

WATER QUALITY

The water quality of several Himalayan lakes has been monitored on the basis of physico-chemical parameters and their comparison with standard permissible limits prescribed by the World Health Organization (WHO) and the Bureau of Indian Standard (BIS). The water quality of these lakes is determined time to time because the local people use the water from these lakes for various domestic purposes (Kumar et al. 2019c). Besides, the lakes have also been assessed to find out the water quality index, types of water, mechanism controlling water chemistry, source of major ions, phosphorus fraction, eutrophication, and so on. It has been confirmed by Khadka and Ramanathan (2012) that the major ions such as Ca^{2+}, Mg^{2+}, Na^+, K^+, Cl^-, NO_3^- and HCO_3^-, CO_3^{2-}, and so on are abundantly present in various Himalayan lakes and the studies related to them give information about the source of their origin and presence of inorganic nutrients inside the lakes. An earlier study carried out by Sarah et al. (2011) at the Manasbal Lake of Kashmir valley found that the concentrations of major ions inside the lake are within the permissible limits and besides, Mg^{2+}-HCO_3^- (02%), Ca^{2+}-HCO_3^- (13%), and hybrid (85%) types of water are present.

Singh et al. (2008) and Singh and Jain (2013) in their studies observed the water quality and eutrophic status of various western Himalayan lakes (Surinsar, Mansar, Dal, Tso Kar, Tso Moriri, and so on) (Figure 2.2) and found most of the lakes to

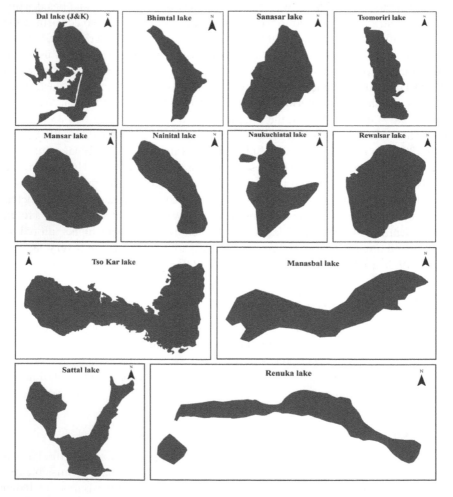

FIGURE 2.2 Aerial views of some of the famous lakes in the Indian Himalayan region.

be eutrophic and even hyper-eutrophic. Besides, various other chemical parameters were found more than the standard permissible limit prescribed by WHO. The water quality of Wular Lake was monitored by Sheikh et al. (2014) and it was found to be pollution free, but affected by natural as well as anthropogenic activities. The Wular Lake was again monitored by Bhat and Pandit (2014) and the water quality was found to be suitable for drinking, but the increased concentrations of various chemical parameters inferred the movements of the lake toward eutrophication. Najar and Khan (2012) assessed the water quality of three lakes (J&K) and found that the pollution stress is highest in Khushal Sar, moderate in Anchar, and lowest in the Dal Lake. The surrounding geology, agricultural activities, and domestic waste were noticed to be the main causal factors of eutrophication in these lakes. Kanakiya et al. (2014) in an earlier study monitored seasonal changes in water quality of Dal Lake and found the water quality index in the range of 101.35–120.33 at four different sites and thus, the water of the lake is considered unsuitable for drinking. The water quality of Dal Lake (J&K) in an earlier study has been found to be suitable for domestic use (Gaury et al., Communicated paper, a).

The Mansar Lake in the past was assessed by Kumar et al. (2006) for its quantitative and qualitative characteristics and they found the chemical compositions of various elements changing throughout the year due to variation in temperature, sunshine, wind action, and so on. Besides, the availability of phosphate (>0.03 mg/L) indicated the eutrophic nature of the lake. Anshumali and Ramanathan (2007) carried out a study at the Pandoh Lake to know seasonal variation in major ion chemistry and found that the carbonate weathering is the dominant source of major ions in the lake and the water quality was found to be free from pollution stress. Attri and Santvan (2012) assessed the Prashar Lake of Mandi (HP) to know the importance of the lake for the people, flora, and fauna and physico-chemical parameters. That study revealed that the water of the lake is suitable for domestic purposes. The water quality of Rewalsar Lake has been monitored by Gaury et al. (2018) (communicated, b) and they found that the water was polluted and unsuitable for domestic use. Kumar et al. (2020a) in their study monitored the Rewalsar Lake as hyper-eutrophic as a result of the mismanaged anthropogenic activities. Singh and Sharma (2012) assessed the trophic status of Renuka Lake and found it eutrophic. The Renuka Lake was again monitored by Kumar et al. (2019a) and they found that the water was polluted and unsuitable for drinking but suitable for irrigation (Kumar et al. 2019b). Nautiyal et al. (2012) carried out a study at Dodital Lake to know seasonal variation and found many of chemical parameters below the standard permissible limit, during all seasons.

The Mirik Lake of Darjeeling hill was evaluated by Mondal et al. (2012) and they noticed various chemical parameters and transparency above the permissible limit. A study at the Khecheopalri Lake of Sikkim Himalaya by Jain et al. (1999) revealed that its nitrogen content is lower in the rainy season and its chloride content is the highest in winter. The phosphorus concentration inside that lake was observed to be highest in the rainy season due to forest and agricultural runoff as a result of monsoon rain. Besides, the water of the lake at one site was found to be eutrophic. Khadka and Ramanathan (2012) monitored the Begnas Lake of Pokhara valley (Nepal) by using water samples collected in three different seasons and found that the water quality was within the permissible limit. The Mansarovar Basin Lake (Tibet) was assessed by Yao et al. (2015) and they found that the lake was slightly alkaline and besides, the carbonate weathering was a dominant source of major ions. Xiao et al. (2015) in an earlier study noticed the water quality, catchment weathering, and major ion chemistry of Bosten Lake China and found carbonate weathering as a dominant source of major ions and Ca^{2+}-HCO_3^- type of water in the lake. The water of Bosten Lake was found to be unsuitable for drinking, but fit for irrigation.

A pH value >7 indicates that the water of most of the Himalayan lakes is slightly alkaline (Table 2.1). The higher concentration of $Ca^{2+} + Mg^{2+}$ indicates that carbonate weathering is the dominant source of major ions in the lakes, whereas silicate weathering ($Na^+ + K^+$) is also contributing to the lake, but to a lesser extent. The presence of NO_3^- and SO_4^{2-} in most Himalayan lakes reveals the impact of anthropogenic activities like waste disposal and use of chemical fertilizer in the catchment regions (Gaury et al. 2018).

SEDIMENTATION AND ASSOCIATED CHEMISTRY

Water quality in lakes is controlled by its interaction with the sediments, which is the main process inside lakes. The input of contaminated sediments in the lakes is also one of the major factors causing water pollution in the Himalayan lakes. The lakes with static water get a long time of interaction with the sediments and as a result, the water quality gets highly affected. Besides the natural factors, the anthropogenic activities cause speedy erosion in the surrounding and stand responsible for enhancing sedimentation, which in turn degrade the natural scenario of the lakes. The sediments primarily act as a sink of chemical pollutants and thereafter continue to supply it to the above water columns. The sedimentation is also responsible for reducing the size and numbers of Himalayan lakes with time. A previous study by Handa et al. (1991) has indicated that owing to sedimentation, the total area of Dal Lake from the beginning of 19th century has reduced near $20\,km^2$. Another study at the Mansar Lake by Rai et al. (2007) has shown the sedimentation rate in the range of 0.25 cm/yr. The human activities in the catchment area of these lakes were found to be responsible for accelerating the sedimentation rate. The lakes in J&K show more sedimentation as compared to lesser Himalayan lakes like Bhimtal, Sattal, Naukuchiatal, Nainital, and so on due to higher input from the Siwalik terrain that contains silt, sandstone, and mudstone dominant lithology, which is more susceptive to erosion and siltation than the lesser Himalayan terrain. Due to this reason, several lakes like Sanasar, Khajjiar, Khurpatal and Sariatal have been reached at the verge of extinction.

Sarkar et al. (2016) in an earlier study observed sedimentation in Rewalsar Lake and found that on comparison with other north-western Himalayan lakes, the sedimentation rate in the lake was 3.35 cm/yr during the last ~50 years. This rate from 1963 to 1995 AD was 3.92 cm/yr and thereafter reduced to 2.78 cm/yr, which is more than other lakes of north-western Himalayan regions. The human activities in the surrounding of the lake were found to be responsible for the fast sedimentation in the lake. The sediments characteristic and bathymetric study at Bhimtal Lake by Malik and Panwar (2013) has shown the sedimentation at the rate of 0.27–1.05 mm/yr, which is more at Mallital than Central and Tallital zone due to land use pattern, deforestation, modern agricultural practices, and construction of roads and houses in the

TABLE 2.1

The Concentration of Different Chemical Parameters in Several Himalayan Lakes

Lake	pH	EC	Ca^{2+}	Mg^{2+}	Na^+	K^+	SO_4^{2-}	HCO_3^-	Cl^-	NO_3^-	Reference
Dal	8.6	181	24	07	04	0.9	–	88	10	0.7	Singh et al. (2008)
Surinsar	8.4	545	38	13	10	04	11	127	10	05	Singh et al. (2008)
Tso Kar	8.8	63530	1300	4010	1061	1715	26	04	9028	–	Singh et al. (2008)
Tso Moriri	8.9	3550	35	747	791	209	88	01	24	–	Singh et al. (2008)
Mansar	8.4	416.5	21.6	6.45	11.02	2.68	6.88	96.83	7.75	0.63	Kumar et al. (2006)
Wular	9.8	643.2	39.27	17.84	5.6	0.35	29.7	170	5.6	3.1	Sheikh et al. (2014)
Tso Kyo	6.5	8.2	1.3	0.3	0.9	0.5	4.2	9.5	0.6	0.8	Deka et al. (2015)
Sella	6.1	5.2	0.9	0.3	2.2	0.3	3.7	11.3	0.5	0.7	Deka et al. (2015)
Pandoh	7.1	80.8	18.0	3.3	3.8	2.1	2.7	49.2	2.4	10.3	Anshumali and Ramanathan (2007)
Renuka	8.4	540.3	56.45	37.19	7.95	2.02	145.2	43.37	0.34	–	Das and Kaur (2001)
Nainital	8.3	312.6	492.6	2235	548.8	9.9	1,024	1,803	276.7	71.0	Chakrapani (2002)
Sattal	7.4	59.7	574.7	259.1	109.2	2.8	43.9	770.3	109.8	30.8	Chakrapani (2002)
Bhimtal	7.9	101	190.3	490.2	4.6	4.6	62.6	481.9	148.0	43.1	Chakrapani (2002)
Naukuchiatal	7.4	70.0	515.5	255.8	177.6	3.6	33.9	623.8	9.8	13.4	Chakrapani (2002)
Chandra Tal	8.1	212	22.4	13.2	0.9	1.4	12.5	118	0.5	0.5	Singh et al. (2016)
Begnas	7.3	90.5	7.0	2.0	3.9	1.4	7.3	25.3	2.6	5.3	Khadka and Ramanathan (2012)
Mansrovar	–	–	27.91	30.98	51.89	5.50	32.14	323.4	14.30	–	Yao et al. (2015)
Deepak Tal	8.8	130	30.5	8.8	2.8	0.3	20.0	120	4.3	0.1	Singh et al. (2014)
Suraj Tal	9.1	140	32.1	10.5	2.6	0.3	22.0	130	5.0	0.1	Singh et al. (2014)
Sissu	8.5	200	53.0	30.0	10.5	2.7	86.0	286	56.0	4.1	Singh et al. (2014)

catchment area. An earlier study at the Nainital Lake has revealed sedimentation in the range of 0.60–1.35 cm/yr, which is very high, probably due to its large catchment area encircled by carbonate, dolomite, limestone, shales with marlite rocks, which are more susceptible to weathering as compared to other Kumaun Himalayan lakes and besides, the steep slopes and higher pressure of anthropogenic activities in the surrounding of the lake (Kumar et al., 2007).

The sedimentation rate in Indian Himalayan lakes is found to be more as compared to the Chinese Himalayan lakes (Table 2.2). The highest sedimentation rate is shown by Rewalsar Lake, which is several times more than other lakes of northwest Himalaya, probably due to steep slopes of surrounding hills and late-quaternary deposits, which are highly susceptible to erosion and weathering. The Bhimtal Lake is having more sedimentation than other lakes, but somehow equal to Dal lake of J&K. The main cause of the higher sedimentation rate at Dal Lake is the large catchment area and the impact of the Jhelum River, which brings a lot of sediment and deposits in the catchment of Dal Lake (Sarkar et al., 2016). However, the Bhimtal Lake is having small catchment area than Dal Lake, but rather than shows more sedimentation probably due to higher impact of anthropogenic activities (Malik and Panwar, 2013). The Nainital and Sattal Lake are having a minute difference in the sedimentation rate and catchment area. Although the catchment area of Nainital Lake is less than that of Sattal Lake, the sedimentation rate is more, which may be due to more pressure from anthropogenic activities. However, it can be inferred that the Naukuchiatal Lake is having least sedimentation than other Kumaun Himalayan lakes, due to its less catchment area and less human activities. Diwate et al. (2020) in their study at the Renuka Lake observed an average sedimentation at the rate of ~0.64 cm/yr during a time period of 61 years. The Chinese Himalayan lakes are having larger surface and catchment area than those of Indian Himalayan lakes, due to which the sedimentation rate is less. However, Dongtingu Lake shows the highest sedimentation rate, whereas the Longganhu Lake reveals the lowest sedimentation rate amongst all Chinese Himalayan Lakes. It is to be inferred that the sedimentation rate in lakes increases due to more anthropogenic activities, rock types and besides the large catchment area, but large surface areas of lakes also impact on reducing sedimentation rate.

Several past studies have also shown that the lake sediments are enriched with heavy metals (Cr, Co, Ni, Cu, Zn, Cd, Pb, and Hg) (Table 2.3), generated from the surrounding rocks, catchment area, or domestic and industrial waste, and thereafter disturb water quality as well as the aquatic life cycle. In a previous study, Singh and Sharma (2013) monitored the physico-chemical characteristics of Mansar Lake and found the sediment to be alkaline in nature, texture sandy loam, total phosphorus content fairly high, and organic matter varied from 1.91% to 3.45%. Das and Haake (2003) monitored Rewalsar Lake to know the effect of provenance, tectonic setting, and source area weathering on geochemistry and noticed the sediments enriched with major ions, trace, and rare earth elements. The sediments in that lake were found to be derived from metamorphic source terrain. The Rewalsar Lake has also been again monitored by Meena et al. (2017) to know heavy metal pollution record from the last five decades and it was found that the lake was enriched with heavy metals, due to natural and anthropogenic activities. The highest concentration of metals was noticed during 1990–2004, which thereafter decreased until 2012 AD.

TABLE 2.2

Sedimentation Rate of Different Himalayan Lakes

Sl No.	Name of Lake	Coordinates	Surface Area (km²)	Catchment Area (km²)	Depth (m)	Sedimentation Rate (cm/yr)	References
1	Dal	34°07′N, 74°52′E	13.39	337.17	4.07	0.93	Kumar et al. (2007)
2	Mansar	32°69′N, 75°14′E	0.59	1.67	38.25	0.25	Rai et al. (2007)
3	Manasbal	34°15′N, 74°40′E	2.81	33	13	0.44	Kusumgar et al. (1992)
4	Rewalsar	31°63′N, 76°83′E	0.026	1.73	06	3.35	Sarkar et al. (2016)
5	Nainital	29°38′N, 79°45′E	0.46	4.7	27.3	0.78	Kumar et al. (2007)
6	Sattal	29°34′N, 79°53′E	0.249	5.69	20	0.73	
7	Bhimtal	29°35′N, 79°56′E	0.478	10.77	25.8	0.95	
8	Naukuchiatal	29°32′N, 79°21′E	0.306	3.25	42.25	0.64	
9	Xingyun	24°20′N, 102°47′E	34.7	325	07	0.017–0.125	Zhang et al. (2014)
10	Erhai	25°45′N, 100°11′E	250	-	20.7	51 cm/ka	
11	Guchenghu	31°27′N, 118°93′E	24.5	248	6.5	0.13–0.23 mm	Xiang et al. (2002)
12	Poyanghu	29°07′N, 119°27′E	2.933	16.2×10^4	5.10	0.01–0.062	Xiang et al. (2002)
13	Dongtinghu	28°87′N, 112.59′E	2,432.5	25.7×10^4	6.39	0.078–0.193	
14	Chaohu	31°51′N, 117°54′E	769.55	9258	2.69	0.04–0.02	
15	Longganhu	29°93′N, 116°14′E	316.2	5511	3.78	0.005	
16	Changdanghu	31°61′N, 119°54′E	89	-	1.10	0.023–0.025	
17	Paihu	30°28′N, 113°23′E	12.4	260.3	1.5	0.055	
18	Yangchenhu	31°25′N, 120°48′E	119.0	-	1.4	0.027–0.028	

TABLE 2.3

Heavy Metals Concentration in Some of the Famous Lakes

Sr. No.	Name of Lake	Coordinates	Mn	Cr	Cu	Zn	Ni	Pb	Co	Reference
1	Nainital	29°38′N, 79°45′E	-	11–59	18–40	31–105	30–64	6–21	7–33	Purushothaman and Chakrapani (2012)
2	Bhimtal	29°35′N, 79°56′E	-	47–71	36–60	42–73	26–42	45–178	13–20	
3	Naukuchiatal	29°32′N, 79°21′E	-	12–47	80–120	66–159	26–51	48–73	41–87	
4	Rewalsar	31°63′N, 76°83′E	425.95–1,417.26	64.70–165	24.60–97	84–116	23.30–50	19.70–29	10.50–16.33	Meena et al. (2017)
5	Chilka	19°43′N, 85°19′E	-	4–74	10–101	21–63	52–143	28–59	-	Panda et al. (1995)
6	Kolleru	16°39′N, 81°13′E	-	40–66	205–572	356–671	0.18–2.21	2.54–5.6	2–4.4	Sekhar et al. (2003)
7	Varthur	12°94′N, 77°74′E	112–167	0–21.37	130.52–134	25.71–220.25	16.2–68	4.43–88.5	11.68–69.37	Jumbe and Nandini (2009)
8	Akkulam-Veli	8°52′N, 76°89′E	28.4–173.9	43.5–64.5	11.7–47.8	63.9–161.9	30–43.8	6–57.1	-	Sheela et al. (2012)
9	Jannapura	13°42′N, 75°38′E	-	-	89.75	0.034	40.05	0.0189	8.0	Puttaiah and Kiran (2007)
10	Taihu	31°20′N, 120°12′E	-	77.37	31.12	87.32	29.81	33.05	22.49	Wei and Wen (2012)
11	Dianch	24°48′N, 102°40′E	-	115.18	90.05	153.92	45.97	65.76	33.36	
12	Nansi	34°36′N, 117°12′E	-	76.44	-	91.64	-	26.93	-	Yang et al. (2015)

Besides, the lithogenic inputs were observed to dominate the metal concentration prior to 1990 AD, while anthropogenic activities after 1990 AD.

A study by Das et al. (2008) at Renuka Lake pertaining to sediment geochemistry has shown the lake sediments enriched with Al_2O_3, TiO_2, Fe_2O_3 and K_2O, Ni, Zr, Th, U, and Nb and besides, the moderate chemical weathering in the catchment area of the lake. Kumar et al. (2019d) monitored the Renuka Lake to understand the heavy metals contamination history and found that the lake has been low to considerable contaminated during ca 1839–2002 AD. The metal fractionation study at the Nainital Lake carried by Jain et al. (2007) to know the eco-toxic potential of metal ions has indicated that the lake sediments are enriched with lead (13%–26%), nickel (17%–24%), manganese (04%–13%), zinc (02%–03%), and cadmium (14%–23%) in a medium risk group.

The heavy metals concentration in Himalayan and other lakes of Indian lower regions is found to be varied with one another (Table 2.3). The Mn concentration is found to be highest in Rewalsar Lake than in others. The content of Cr is noticed to be highest in Rewalsar Lake followed by Dianch Lake, whereas the Cu and Zn concentrations are noticed to be highest in the Kolleru Lake. The Chilka Lake shows the highest concentration of Ni, while the highest concentration of Pb is found in the Bhimtal Lake. However, Co is found to be highest in the Naukuchiatal Lake and thereafter in the Varthur Lake. On the other hand, the concentrations of Mn, Cr, Cu, Zn, Ni, Pb, and Co are found to be lowest in Varthur, Varthur, Taihu, Jannapura, Kolleru, Jannapura, and Kolleru Lake, respectively. The difference in concentrations of heavy metals in all the lakes is due to higher or lower impacts of anthropogenic inputs.

ECOLOGICAL STATUS OF HIMALAYAN LAKES

The ecological studies on Himalayan lakes have shown the presence of different types of aquatic organisms harmful to the lakes and the lives dependent on them. The earlier studies also have strengthened that lakes attract other creatures and support biodiversity. Sharma et al. (2010) in an earlier study monitored Surinsar Lake and found species that could survive in polluted water such as *Anabaena* sp., *Ankistrodesmus, Achnanthes, Chroococcus, Navicula, Nitszshia, Oscillatoria, Spirulina,* and *Synedra.* The presence of these species in the lake indicates pollution stress and eutrophic status as a result of anthropogenic activities. The Surinsar Lake was again assessed by Slathia and Dutta (2013) to know the quantitative and qualitative distribution of zooplankton population and seasonal variation in number, diversity, and abundance of crustacean, rotifer, and protozoa. The dominancy of protozoan and other species capable to survive in the polluted environment was found, which thereby indicated that the lake is polluted and eutrophic as a result of excessive biological interventions. The density and diversity of zooplankton in that lake were found to be controlled by the food supply, dissolved oxygen, and temperature, which are the essential elements for growth and survival.

A taxonomic study at Wular Lake by Shah and Pandit (2014) has revealed the presence of 42 taxa of 3 Ostracoda, 16 Copepoda, and 23 Cladocera. The species like *Alona affinis, Bryocamptus minutus, Chydorus sphaericus*, and *Cyclops bicolor* were found to be most frequently distributed. The presence of the crustacean community confirmed that the Wular Lake is having initial signals of eutrophication. Ganie et al. (2015) monitored Anchar and Manasbal Lake and found 15 taxa of

Cladocera with an average density of 267 individuals per liter at Anchar Lake and 22 taxa at Manasbal Lake having an average density of 134 individuals per liter. The average density of Cladocera at both the lake was noticed to be very high and indicated a high-level contamination. Singh and Banyal (2013a) in an ecological study at the Khajjiar Lake fo und 121 insect species of 8 orders, 28 families, and 108 genera. Orthoptera was found to be the dominant order, followed by Coleoptera, Odonata, Hymenoptera, Hemiptera, Diptera, and Homoptera. Another study by Singh and Banyal (2013b) at Khajjiar Lake has shown the occurrence of two fish species as *Cyprinus carpio communis* and *Cyprinus carpio specularis*. Jindal et al. (2014) in their study at Rewalsar Lake identified 47 phytoplankton species belonging to 7 groups. Among those, the two species *Microcystis aeruginosa* and *Oscillatoria limosa* were found abundantly in the lake and besides other species like *Chlorella vulgaris, Ankistrodesmus falcatus, Euglena oxyuris, Oscillatoria princeps,* and *Scenedesmus bijugatus* were scarcely present. That study revealed that the Rewalsar Lake was highly polluted. Singh et al. (2008) in their study at the Renuka Lake found the total coliform bacterial contamination in the hypolimnion zone in the range of 0–150/MPN/100 mL, which is much higher than the standard permissible limits. The Renuka Lake was again monitored by Singh and Sharma (2012) and variations were found in total coliform at a mean rate of 61/MPN/100 mL and fecal coliform at a mean rate of 50/MPN/100 mL, which indicated that the water was polluted and unfit for human consumption. Gupta and Shukla (1996) at Nainital Lake observed the diversity, density, and species composition of benthic protozoa and its seasonal and spatial variation and found the ciliates as the dominant protozoa in terms of abundance. Besides, other species like *Penard, Branchioecetes, Bursaridium, Paramecium multimicronucleatum, Lauterborn, Spirostomum, P. aurelia,* and so on were abundantly present in the lake. The reduced level of the dissolved oxygen (DO) and the nutrients enrichment are observed as the major causal factors for the fish kills from the Rewalsar Lake during summer season, which indicated the polluted water quality of the lake (Kumar et al. 2020b).

CONCLUSIONS

- The present study concludes that the natural scenario of the greater, lesser, and outer Himalayan lakes is changing with time due to climate change and anthropogenic activities.
- Glaciers like Gangotri, Milam, Shah, Chhota Sigri, and those situated in Kashmir, eastern, western, and middle Himalaya, and Dudh Koshi Basin are retreating at a fast rate with increasing temperature and precipitation variability as a result of changing climate.
- All the glaciers of the Himalayan regions are having equal and similar trends of retreating and melting with time and are at the verge of extinction.
- The rate of glaciers melting has increased very fast from the last few decades.
- The numbers and size of glacial lakes are increasing with an equal pace of glacier melting and causing a risk of outburst and GLOFs.
- The people living in downstream regions of glacial lakes are vulnerable to GLOFs, which have caused death of thousands of people all over the world.

- Non-glacial lakes in the Himalayan regions are facing the stress of pollution, eutrophication, sedimentation, bacterial contamination, and so on.
- The parameters like pH, EC, TDS, Ca^{2+}, Mg^{2+}, Na^+, K^+, NO_3^-, and HCO_3^- in most of the Himalayan lakes have reached beyond the standard permissible limit prescribed by WHO and BIS.
- In Himalayan lakes, the anthropogenic activities include waste disposal, use of chemical fertilizers in the catchment region, fish feeding, and adding other nutritious substances in the lakes.
- Most of the Himalayan lakes are alkaline in nature and carbonate weathering of the surrounding rocks is the dominant source of major ions.
- The eutrophic and hyper-eutrophic status of non-glacial Himalayan lakes viz. Dal, Tso Moriri, Surinsar, Mansar, Renuka lake, and Tso Kar are due to the use of phosphorus and nitrogen as chemical fertilizers in the catchment regions and other nutritious substances as a result of anthropogenic activities.
- The lakes show monthly and seasonal variations in the concentration of different parameters like pH, EC, major ions, and so on due to the changing water level, difference in mixing, and degradation rate of various components with variability in precipitation and temperature.
- The acidic nature of certain lakes is due to the decomposition of carbonaceous minerals, which have been attributed by $pH < 7$, while alkaline with $pH > 7$. The alkaline nature of certain Himalayan lakes is due to less concentration of total dissolved salts, which have been observed in several lakes like Dal, Tso Moriri, Tso Kar, Khecheopalri, and Sattal, Bhimtal, Naukuchiatal, and Nainital.
- The heavy metals (Cr, Co, Ni, Cu, Zn, Cd, Pb, Hg, and so on), major ions (Ca^{2+}, Mg^{2+}, Na^+, K^+, NO_3^-, Cl^-, CO_3^{2-}, and HCO_3^-), and other toxic elements are present in higher concentrations in some of the freshwater Himalayan lakes.
- The types and concentration of several parameters are different in Himalayan lakes due to surrounding rock types, vegetation, erosion rate, environmental condition, and human interferences.
- The Himalayan lakes are also reducing in numbers, depths, and size, owing to sedimentation, siltation, weeds, and land acquisition. Some of the most alarming lakes are Dal, Mansar, Nainital, Rewalsar, Bhimtal, Naukuchiatal, and Sattal.
- Anthropogenic activities have been noticed to be responsible for increasing sedimentation inside lakes.
- The sediments in lakes absorb chemical substances coming from outside and thereafter continue to supply them for a long time to the above water.
- The siltation and sedimentation in lakes situated at the Siwalik terrain are more than the lesser Himalayan lakes due to rocks type, which are more prone to weathering. Besides, the sedimentation rate is also dependent on the catchment area and anthropogenic activities.
- The bacterial contamination, phytoplankton, zooplankton, and higher population of fishes are some of the most worrisome problems of Himalayan lakes.

MITIGATION MEASURES

- To reduce the risk of hazard associated with greater Himalayan glacial lakes, it is very necessary to monitor the melting process as well as retreating of the glaciers.
- The outburst potential of glacial lakes needs to be assessed regularly to avoid the risk of damage associated with them.
- The satellite imageries, as well as GIS and Remote Sensing technique, can assist in monitoring and mapping of glaciers and glacial lakes.
- The warning system in GLOFs prone areas should be located to reduce the loss of lives and properties. It is very important to manage the water resources and reduce futuristic damage.
- The regular analysis of water quality can provide plentiful assistance in making a proper strategy to control pollution stress and eutrophication in Himalayan lakes.
- The water quality index should be determined to know the water quality of lakes, which requires values of parameters like pH, EC, TH, and Ca^{2+} Mg^{2+}, Na^+, K^+, and NO_3^- and their comparison with standard permissible limits prescribed by BIS as well as WHO.
- The studies related to the trophic state index can provide information about the oligotrophic, mesotrophic, eutrophic, or even hyper-eutrophic status of≈the lakes. This index can be used for trophic measurement and is beneficial for the preservation and development of lakes.
- The oligotrophic stage of eutrophic lakes can be well returned by reducing the supply of phosphorus and other nutritious substances inside the lakes. Fish feeding at lakes must be stopped.
- Several Himalayan lakes like Dal, Khajjiar, Rewalsar, Renuka, and Nainital are facing the stress of pollution owing to tourism and related activities, which should be allowed in a controlled manner.

Thus, the Himalayan glaciers, glacial lakes, and non-glacial lakes are a tremendous source of freshwater upon which billions of people from different Asian countries depend for drinking, irrigation, and other domestic and industrial purposes. It is very clear that the present scenario of Himalayan glacial and non-glacial lakes is under the stress of loss due to changing climate and mismanaged anthropogenic activities and may create water scarcity in the near times to come. The glaciers are melting at a fast speed as a result of which the numbers and size of glacial lakes are increasing at a fast rate and showing the risk of outburst in the form of GLOFs downstream. The non-glacial lakes are facing problems of pollution, eutrophication, sedimentation, bacterial contamination, and so on and are reached at the verge of extinction. The anthropogenic activities are considered to be the primary factors responsible for changing the natural scenario of these lakes. However, this study can also provide a new and further platform for searching the gap of the previous for new studies for preservation and further development of Himalayan lakes.

REFERENCES

Anderson DM, Glibert PM, Burkholder JM (2002). Harmful algal blooms and eutrophication: Nutrients sources, composition and consequences, *Estuarine Journal*, 25(4), 704–726.

Anshumali, Ramanathan AL (2007). Seasonal variation in the major ion chemistry of Pandoh Lake, Mandi district, Himachal Pradesh, India, *Applied Geochemistry*, 22, 1736–1747.

Attri PK, Santvan VK (2012). Assessment of socio- cultural and ecological consideration in conserving wetland - A case study of Parashar Lake in Mandi District, Himachal Pradesh, *International Journal of Plant, Animal and Environmental Sciences*, 2(1), 131–137.

Bajracharya SR, Mool P (2009). Glaciers, glacial lakes and glacial lake outburst floods in the Mount Everest region, Nepal, *Annals of Glaciology*, 50(53), 81–86.

Bajracharya SR, Mool PK, Shrestha BR (2006). The impact of global warming on the glaciers of the Himalaya, *International Symposium on Geo-Disaster, Intra Structural Management and Protection of World Heritage Sites*, 25–26 November, NEC, EC& NSET Nepal, Proceeding, pp. 231–242.

Bajracharya SR, Mool PK, Shrestha BR (2008). Global climate change and melting of Himalayan glacier. In Ranade PS (ed.), *Melting glacier and rising sea level: Impact and implications*, ICFAI University Press, Hyderabad, pp. 28–46.

Banshtu RS, Prakash C (2013). Application of remote sensing and GIS in hazard assessment of glacial lakes outburst floods in Himalayan region of Himachal Pradesh, India, *Journal of Earth Science and Climate Change*, 4, 4. doi: 10.4172/2157-7617.S1.011.

Benn DI, Evans DJA (2010). *Glaciers and glaciation*, 2nd edn. Hodder Arnold Publications, London.

Bhambri R, Bolch T, Chaujar RK (2012). Frontal recession of Gangotri glacier, Garhwal Himalaya, from 1965 to 2006, measured through high resolution remote sensing data, *Current Science*, 102(3), 489–494.

Bhat SA, Pandit AK (2014). Surface water quality assessment of Wular lake, A Ramsar Site in Kashmir, Himalaya, using discriminate analysis and WQI, *Journal of Ecosystem*, 1–18.

Bolch T, Buchroithner M, Pieczonka T, Kunert A (2008). Planimetric and volumetric glacier changes in the Khumbu Himal, Nepal, since 1962 using Corona, Landsat TM and ASTER data, *Journal of Glaciology*, 54(187), 592–600.

Carey M, Huggel C, Bury J, Portocarrero C, Haeberli W (2012). An integrated socio-environmental framework for climate change adaptation and glacier hazard management: Lessons from lakes 513, Cordillera Blanca, Peru, *Climate Change*, 112, 733–767.

Chakrapani GJ (2002). Water and sediment geochemistry of major Kumaun Himalayan lakes, India. *Journal Environmental Geology*, 43, 99–107.

Chand P, Sharma MC (2016). Monitoring frontal changes of Shah glacier in the Ravi Basin, Himachal Himalaya (India) from 1965 to 2013, *National Academy Science Letters*, 39, 109–114. doi: 10.1007/s40009-016-0420-x.

Che T, Xiao L, Liou YA (2014). Changes in glaciers and glacial lakes and the identification of dangerous glacial lakes in the Pumqu River Basin, Xizang (Tibet), *Advances in Meteorology*, 2014, 1–8. doi: 10.1155/2014/903709.

Chen W, Doko T, Fukui H, Yan W (2013). Change in Imja Lake and Karda Lake in the Everest region of Himalaya, *Natural Resources*, 4, 449–455.

Clague JJ, Evan SG (2000). A review of catastrophic drainage of moraine- dammed lake in British Columbia, *Quaternary Science Reviews*, 19, 1763–83.

Dahal KH (2008). *Hazard and risk: Perception of glacial lake outburst flooding from Tsho Rolpa Lake*, www.geo.mtu.edu.

Das BK, Gaye B, Kaur P (2008). Geochemistry of Renuka Lake and wetland sediments, lesser Himalaya (India): Implication for source-area weathering, provenance and tectonic setting, *Environmental Geology*, 54, 147–163.

Das BK, Haake BG (2003). Geochemistry of Rewalsar Lake sediment, Lesser Himalaya, India: Implications for source-area weathering, provenance and tectonic setting, *Geosciences Journal*, 7(4), 299–312.

Das BK, Kaur P (2001). Major ion chemistry of Renuka Lake and weathering processes, Simaur District, Himachal Pradesh, India. *Environmental Geology*, 40, 908–917.

Das S, Kar NS, Bandyopadhyay S (2015). Glacial lake outburst flood at Kedarnath, Indian Himalaya: A study using digital elevation models and satellite images, *Natural Hazards*, doi: 10.1007/s11069-015-1629-6.

Deka JP, Tayengi G, Singh S, Hoque RR, Prakash A, Kumar M (2015). Source and seasonal variation in the major ion chemistry of two eastern Himalayan high altitude lakes, India, *Arabian Journal of Geosciences*, 8, 10597–10610.

Diwate P, Meena NK, Bhushan R, Pandita S, Chandana KR, Kumar P (2020). Sedimentation rate (210Pb and 137Cs), grain size, organic matter and bathymetric studies in Renuka Lake, Himachal Pradesh, India, *Himalayan Geology*, 41(1), 51–62.

Dixit S, Tiwari S (2005). Nutrient overloading of freshwater lake in Bhopal, India, *Earth Day* (21), ISSN-1076-7975.

Dobhal DP, Gergan JT, Thayyen RJ (2004). Recession and morphogeometrical changes of Dokriani glacier (1962–1995) Garhwal Himalaya, India, *Current Science*, 86(5), 692–693.

Dutta S, Ramanathan AL, Linda A (2012). Glacier fluctuation using Satellite Data in Beas Basin, 1972–2006, Himachal Pradesh, India, *Journal of Earth System Science*, 121(5), 1105–1112.

Fujita K, Sakai A, Nuimura T, Yamaguchi S, Sharma RR (2009). Recent change in Imja glacial lake and its damming moraine in Nepal Himalaya revealed by in situ surveys and multi-temporal ASTER imagery, *Environmental Research Letters*, 4, 1–7.

Fujita K, Sakai A, Takenaka S, Nuimura T, Surazakov B, Sawagaki T, Yamanokuchi T (2013). Potential flood volume of Himalayan glacial lakes, *Natural Hazards and Earth System Sciences*, 13, 1827–1839.

Gaddam VK, Kulkarni AV, Gupta AK (2016). Estimation of glacial retreat and mass loss in Baspa Basin, Western Himalaya, *Spatial Information Research*, 24, 257–266. doi: 10.1007/s41324-016-0026-x.

Ganie MA, Parveen M, Balkhi MH, Khan MI (2015). Structure and diversity of cladoceran communities in two lakes with varying nutrient compositions in the Jhelum River Basin, Kashmir, *International Journal of Fisheries and Aquatic Studies*, 3(2), 456–462.

Gaury PK, Meena NK, Mahajan AK (2018). Hydrochemistry and water quality of Rewalsar Lake of Lesser Himalaya, Himachal Pradesh, India, *Environmental Monitoring and Assessment*, 190(2), 84.

Ghimire M (2005). A review of studies on glacier lake outburst flood and associated vulnerability in the Himalaya, *The Himalaya Review*, 35–36, 49–64.

Gupta PK, Shukla U (1996). Seasonal and spatial distribution of some benthic Protozoa in an eutrophic freshwater lake of Central Himalaya, *Oecologia Montana*, 5, 100–105.

Gupta S, Nayek S, Saha RN (2012). Major ion chemistry and metal distribution in coal mine Pit Lake contaminated with industrial effluents: Constraints of weathering and anthropogenic inputs, *Environmental Earth Sciences*, 67, 2053–2061.

Guyer GT, Ilhan F (2011). Assessment of pollution profile in Buyukcekmece watershed, Turkey, *Environmental Monitoring and Assessment*, 173, 211–220.

Guyuan L, Faping B, Xiaoyi X, Jia C, weiqun S (2011). Seasonal variation of dissolved inorganic nutrients transported to the linjiang bay of the three gorges reservoir China, *Environmental Monitoring and Assessment*, 73, 55–64.

Hairston NGJ, Fussmann GF (2002). Lake ecosystem, In *Encyclopaedia of life sciences*, Macmillan Publishers Ltd., Nature publishing group, New York, www.els.net.

Hambrey M, Quincey DJ, Glasser NF, Reynolds JM, Richardson SJ, Clemmens S (2008). Sedimentological, geomorphological and dynamic context of debris-mantled glaciers, Mount Everest (Sagarmatha) region, Nepal, *Quaternary Science Reviews*, 27(25–26), 2361–2389.

Handa BK, Kumar A, Bhardwaj AK (1991). Study on Dal lake, Srinagar, J&K, eutrophication studies, *Bhujal News, Central Ground Water Board, New Delhi*, 6, 15–20.

Hasler AD (1947). Eutrophication of lake by domestic drainage, *Ecology*, 28, 383–395.

Hewitt K, Liu J (2010). Ice-dammed lakes and outburst floods, Karakoram Himalaya: Historical perspectives on emerging threats, *Journal of Physical Geography*, 31(6), 528–551.

Hoy A, Katel O, Thapa P, Dendup N, Matschullat J (2016). Climatic changes and their impact on socio-economic sectors in the Bhutan Himalayas: An implementation strategy, 16, 1401–1415, doi: 10.1007/s10113-015-0868-0.

Immerzeel WW, van Beek LPH, Bierkens MFP (2010). Climate change will affect the Asian water towers, *Science*, 328(5984), 1382–1385.

Jain A, Rai SC, Pal J, Sharma E (1999). Hydrology and nutrient dynamics of a sacred lake in Sikkim Himalaya, *Hydrobiology*, 416, 13–22.

Jain CK, Malik DS, Yadav R (2007). Metal fractionation study on bed sediments of lake Nainital Uttaranchal, India, *Environmental Monitoring and Assessment*, 130, 129–139.

Jain SK, Rishitosh KS, Chaudhary A, Shukla S (2015). Expansion of glacial lake, Tsho Chubda, Chamkhar Chu Basin, Hindukush Himalaya, Bhutan, *Natural Hazards*, 75, 1451–1464.

Jha LK, Khare D (2016). Detection and delineation of glacial lakes and identification of potentially dangerous lakes of Dhauliganga Basin in the Himalaya by remote sensing techniques, *Natural Hazards*, 85, 301–327, doi: 10.1007/s11069-016-2565-9.

Jiawen REN, Zhefan JING, Jianchen PU, Xiang QIN (2006). Glacier variations and climate change in the central Himalaya over the past few decades, *Annals of Glaciology*, 43, 218–222.

Jindal R, Thakur RK, Singh UB, Ahluwalia AS (2014). Phytoplankton dynamics and species diversity in a shallow eutrophic, natural mid-altitude lake in Himachal Pradesh (India): Role of physicochemical factors, *Chemistry and Ecology*, 30, (4), 328–338.

Jumbe AS, Nandini N (2009). Impact assessment of heavy metals pollution of Vartur Lake, Bangalore, *Journal of Applied and Natural Science*, 1(1), 53–61.

Kanakiya RS, Singh SK, Sharma JN (2014). Determining the water quality index of an urban water body Dal Lake, Kashmir, India, *IOSR Journal of Environmental Science, Toxicology and Food Technology*, 8(12), 64–71.

Kang SC, Xu YW, You QL, Flugel WA, Pepin N, Yao TD (2010). Review of climate and cryospheric change in the Tibetan Plateau. *Environmental Research Letters*, 5, 015101. doi: 10.1088/1748-9326/5/1/015101.

Khadka UR, Ramanathan AL (2012). Major ion composition and seasonal variation in the lesser Himalayan Lake: Case of Begnas Lake of the Pokhara valley, Nepal, *Arabian Journal of Geosciences*, 6, 4191–4206.

Kumar B, Rai SP, Nachiappan RP, Kumar US, Singh S, Diwedi VK (2007). Sedimentation rate in North Indian lakes estimated using 137 Cs and 210Pb dating techniques, *Current Sciences Journal*, 92(10), 1416–1420.

Kumar P, Mahajan AK (2020a). Trophic status and its regulating factors determination at the Rewalsar Lake, northwest Himalaya (HP), India, based on selected parameters and multivariate statistical analysis, *SN Applied Sciences*, 2, 1266.

Kumar P, Mahajan AK, Kumar A (2019c). Groundwater geochemical facie: implications of rock-water interaction at the Chamba city (HP), northwest Himalaya, India, *Environmental Science and Pollution Research*, 1–15.

Kumar P, Mahajan AK, Kumar P (2020b). Determining limiting factors influencing fish kills at Rewalsar Lake: a case study with reference to Dal Lake (Mcleodganj), western Himalaya, India, Arabian Journal of Geosciences, 13(17), 1–21.

Kumar P, Mahajan AK, Meena NK (2019a). Evaluation of trophic status and its limiting factors in the Renuka Lake of Lesser Himalaya, India, *Environmental Monitoring and Assessment*, 191(2), 105.

Kumar P, Meena NK, Diwate P, Mahajan AK, Bhushan R (2019d). The heavy metal contamination history during ca 1839–2003 AD from Renuka Lake of Lesser Himalaya, Himachal Pradesh, India, *Environmental Earth Sciences*, 78(17), 549.

Kumar P, Meena NK, Mahajan AK (2019b). Major ion chemistry, catchment weathering and water quality of Renuka Lake, north-west Himalaya, India, *Environmental Earth Sciences*, 78(10), 319.

Kumar V, Rai SP, Singh O (2006). Water quantity and quality of Mansar Lake located in Himalayan foothills, India, *Journal of Lake and Reservoir Management*, 22(3), 191–198.

Kusumgar S, Agrawal DP, Bhandari N, Deshpande RD, Raina A, Sharma C, Yadava MG (1992). Lake sediments from the Kashmir Himalaya: Inverted ^{14}C chronology and its implications, *Radiocarbon Journal*, 34(3), 561–565.

Lama S, Devkota B (2009). Vulnerability of mountain communities to climate change and adaptation strategies, *The Journal of Agriculture and Environment*, 10, 65–71.

Leng R (2009). *The impact of cultural eutrophication on lakes: A review of damages and nutrient control measures, writing 20, freshwater system and society*, pp. 33–39.

Li Z, Tian L, Fang H, Zhang S, Zhang j, Li X (2016). Glacial evolution in the Ayilariju region, Western Himalaya, China: 1980–2011, *Environmental Earth Sciences*, 75, 460.

Litke DW (1999). Review of phosphorus control measures in the US and their effect on water quality. *National Water Quality Assessment Program: Water- Resources Investigation Report*, Report nr. 99-4007.

Liu J, Wang S, Yu S, Zhang L (2009). Climate warming and growth of high elevation inland lakes on the Tibetan Plateau, *Global Planetary Change*, 67, 209–217.

Lohani AK, Jain SK, Singh RD (2015). Assessment and simulation of glacial lake outburst floods for Dhauliganga Basin in Northwestern Himalayan Region, In *Dynamics of Climate Change and Water Resources of Northwestern Himalaya*, Society of Earth Scientists Series, Joshi R. et al. (eds.), Springer International Publishing, Switzerland, pp. 45–55.

Maanya US, Kulkarni AV, Tiwari A, Bhar ED, Srinivasan J (2016). Identification of potential glacial lake sites and mapping maximum extent of existing glacier lakes in Drang Drung and Samudra Tapu glaciers, Indian Himalaya, *Current Science*, 111(3), 553–560.

Mal S, Singh RB (2014). Changing glacial lakes and associated outburst floods risks in Nanda Devi Biosphere Reserve, Indian Himalayan, *Proceeding of ICWRS2014*, IAHS Publication, Bologna, Italy, pp. 255–260.

Malik DS, Panwar S (2013). Study on bathymetric and sediment characteristics of Bhimtal Lake in Kumaun region, *International Journal for Environmental Rehabilitation and Conservation*, 4(2), 50–57.

Marzeion B, Cogley JG, Richter K, Parkes D (2014). Attribution of global glacier mass loss to anthropogenic and natural causes. *Science* 345(6199), 919–921.

Massey C, Manville V, Hancox GT, Keys HJR, Lawrence C, McSaveney MJ (2010). Out-burst flood (Lahar) triggered by retrogressive landsliding, 18 March 2007 at Mt. Ruapehu, New Zealand- a successful early warning, *Landslides*, 07, 303–315.

Meena NK, Prakasam M, Bhushan R, Sarkar S, Diwate P, Banerji U (2017). Last-five-decade heavy metal pollution records from the Rewalsar Lake, Himachal Pradesh, India, *Environmental Earth Sciences*, 76, 39.

Mehta M, Dobhal DP, Bisht MPS (2011). Change of Tipra Glacier in the Garhwal Himalaya, India, between 1962 and 2008, *Progress in Physical Geography*, 35(6) 721–738.

Mondal D, Pal J, Ghosh TK, Biswas AK (2012). Abiotic characteristics of Mirik Lake water in the hills of Darjeeling, West Bengal, India, *Advances in Applied Science Research Journal*, 3(3), 1335–1345.

Mool PK (2005). *Monitoring of glaciers and glacial lakes from 1970s to 2000 in Poiqu Basin, Tibet autonomous region, PR China*, Unpublished report of International Centre for Integrated Mountain Development (ICIMOD), Kathmandu and Cold and Arid Regions Environmental and Engineering Research Institute, Gansu, China.

Murtaza KO, Romshoo SA (2015). Recent glacier changes in the Kashmir Alpine Himalayas, India, *Geocarto International*, 32, 188–205. doi: 10.1080/10106049.2015.1132482.

Naithani AK, Nainwal HV, Sati KK, Parsad C (2001). Geomorphological evidences of retreat of the Gangotri glacier and its characteristics, *Current Science*, 80(1), 87–94.

Najar IA, Khan AB (2012). Assessment of water quality and identification of pollution sources of three lakes in Kashmir, India, using multivariate analysis, *Environmental Earth Sciences*, 66, 2367–2378.

Nautiyal H, Bhandari SP, Sharma RC (2012). Physico-chemical study of Dodital Lake in Uttarkashi district of Garhwal Himalaya, *International Journal of Scientific & Technology Research*, 1(5), 58–60.

Nie Y, Liu Q, Liu S (2013). Glacial lake expansion in the Central Himalayas by Landsat images, 1990–2010, *PLoS One*, 8(12), e83973. doi: 10.1371/journal.pone.0083973.

Nie Y, Sheng Y, Liu Q, Liu L, Liu S, Zhang Y, Song C (2017). A regional-scale assessment of Himalayan glacial lake changes using satellite observations from 1990 to 2015, *Remote Sensing of Environment*, 189, 1–13.

Panda D, Subramanian V, Panigraphy RC (1995). Geochemical fraction of heavy metals in Chilka Lake (east cost of India) - A tropical coastal lagoon. *Environmental Geology*, 26, 199–210.

Pandey P, Venkataraman G (2012). Climate change effect on glacier behaviour: A case study from the Himalaya, *Earthzine Monthly Newsletter*, pp. 21–29.

Pathak H, Pathak D (2012). Eutrophication: Impact of excess nutrient status in lake water ecosystem, *Journal of Environmental and Analytical Toxicology*, 2, 148. doi: 10.4172/2161-0525.1000148.

Post A, Mayo LR (1971). *Glacier dammed lakes and outburst floods in Alaska*, Hydrologic Investigations Atlas HA-455, U.S. Geological Survey, pp. 1–10.

Pratap B, Dobhal DP, Bhambri R, Mehta M (2016). Four decades of glacier mass balance observations in the Indian Himalaya, *Regional Environmental Change*, 16, 643–658, doi: 10.1007/s10113-015-0791-4.

Purushothaman P, Chakrapani GJ (2012). Trace metals biogeochemistry of Kumaun Himalayan Lakes, Uttarakhand, India, *Environmental Monitoring and Assessment*, 184(5), 2947–2965.

Puttaiah ET, Kiran BR (2007). Heavy metal transport in a sewage fed lake of Karnataka, India, *Proceeding of Taal: 12th world lake conference*, pp. 347–354.

Qinghua YE, Zhenwei Z, Shichang K, Stein A, Qiufang WEI, Jingshi LIU (2009). Monitoring glacier and supra-glacier lakes from space in Mt. Qomolangma region of the Himalayas on the Tibetan Plateau in China, *Journal of Mountain Science*, 6, 211–220.

Racoviteanu AE, Arnaud Y, Williams MW, Manley WF (2015). Spatial patterns in glacier characteristics and area changes from 1962 to 2006 in the Kanchenjunga–Sikkim area, eastern Himalaya, *The Cryosphere*, 9, 505–523.

Rai SP, Kumar V, Kumar B (2007). Sedimentation rate and pattern of a Himalayan foothill lake using ^{137}Cs and ^{210}Pb, *Hydrological Sciences Journal*, 52(1), 181–191.

Raj KBG (2011). Recession and reconstruction of Milam Glacier, Kumaon Himalaya, observed with satellite imagery, *Current Science*, 100(9), 120–1425.

Raj KBG, Kumar KV, Mishra R (2012). Recession of Milam Glacier, Kumaon Himalaya, *Current Science*, 102(10), 1351–1352.

Raj KBG, Kumar V (2016). Inventory of glacial lakes and its evolution in Uttarakhand Himalayan using time series satellite data, *Journal of the Indian Society of Remote Sensing*, 44, 959–976. doi: 10.1007/s12524-016-0560-y.

Raj KBG, Remya SN, Kumar KV (2013). Remote sensing based hazard assessment of glacial lakes in Sikkim Himalaya, *Current Science*, 104(3), 359–364.

Rana B, Shrestha RB, Reynolds JM, Aryal R, Pokhrel A, Budhathoki KP (2000). Hazard assessment of the Tsho Rolpa glacier lake and ongoing remediation measures, *Journal of Nepal Geological Society*, 22, 563–570.

Rao KHVD, Rao VV, Dadhwal VK, Diwakar PG (2014). Kedarnath flash floods: A hydrological and hydraulic simulation study, *Current Science*, 106(4), 598–603.

Rasul G, Chaudhry QZ, Mahmood A, Hyder KW, Dahe Q (2011). Glaciers and glacial lakes under changing climate in Pakistan, *Pakistan Journal of Meteorology*, 8(15), 1–8.

Riaz S, Ali A, Baig MN (2014). Increasing risk of glacial lake outburst floods as a consequence of climate change in the Himalayan region, *Jamba: Journal of Disaster Risk Studies*, 6(1), 1–7.

Saleem M, Jeelani G, Shah RF (2015). Hydrogeochemistry of Dal Lake and the potential for present, future management by using facies, ionic ratios, and statistical analysis, *Environmental Earth Sciences*, 74, 3301–3313, doi: 10.1007/s12665-015-4361-3.

Sarah S, Jeelani GH, Ahmed S (2011). Assessment variability of water quality in a ground water-fed perennial Lake of Kashmir Himalaya using linear geo-statistics, *Journal of Earth System Science*, 120(3), 399–411.

Sarkar S, Prakasam M, Banerji US, Bhushan R, Gaury PK, Meena NK (2016). Rapid sedimentation history of Rewalsar Lake, Lesser Himalaya, India during the last fifty years - Estimated using Cs and Pb dating techniques: A comparative study with other North-Western Himalayan Lakes, *Himalayan Geology*, 37(1), 1–7.

Schindler DW (2012). The dilemma of controlling cultural eutrophication of lake, *Proceedings of the Royal Society B*, 279(1746), 4322–4333. doi: 10.10.1097/rspb.1032.

Schmidt S, Nusser M (2012). Changes of high altitude glaciers from 1969 to 2010 in the Trans-Himalayan Kang Yatze Massif, Ladakh, Northwest India, *Arctic, Antarctic, and Alpine Research*, 44(1), 107–121.

Sekhar KC, Chary NS, Kamala CT, Raj DSS, Rao AS (2003). Fractionation studies and bioaccumulation of sediment-bound heavy metals in Kolleru Lake by edible fish, *Environment International*, 29, 1001–1008.

Shah JA, Pandit AK (2014). Taxonomic survey of crustacean zooplankton in Wular Lake of Kashmir Himalaya, *Journal of Evolutionary Biological Research*, 6(1), 1–4.

Sharma KK, Verma P, Shvetambri, S (2010). Phytoplankton diversity and seasonal fluctuation in Surinsar wetland, Jammu (J&K) - A Ramsar site, *International Journal of Applied Environmental Sciences*, 5(1), 39.

Sheela AM, Letha J, Joseph S, Thomas J (2012). Assessment of heavy metal contamination in coastal lake sediments associated with urbanization: Southern Kerala, India, lakes & reservoirs, *Research and Management*, 17, 97–112.

Sheikh JA, Jeelani G, Gavali RS, Shah RA (2014). Weathering and anthropogenic influences on the water and sediment chemistry of Wular Lake, Kashmir Himalaya, *Environmental Earth Sciences*, 71, 2837–2846, doi: 10.1007/s12665-013-2661-z.

Shijin W, Dahe Q, Cunde X (2015). Moraine-dammed lake distribution and outburst flood risk in the Chinese Himalaya, *Journal of Glaciology*, 61(225), 115–126, doi: 10.3189/2015JoG14J097.

Shrestha AB, Devkota LP (2010). *Climate change in the eastern Himalayas: Observed trends and model projections; Technical Report 1*, ICIMOD, Kathmandu, p. 20. ISBN: 978 92 9115 153 0.

Shrestha BB, Nakagawa H, Kawaike K, Baba Y, Zhang H (2010). Glacial lake outburst moraine dam failure by seepage and overtopping with impact of climate change, *Annuals of Disaster Prevention Research Institute, Kyoto University*, 53B, 569–582.

Shrestha NM, Balla MK (2011). Temporal change detection of Lumding Tsho glacial lake in Dudh-Koshi Basin, Nepal, *Journal of Environmental Research and Development*, 5(3), 795–800.

Singh C, Sharma KK (2013). Assessment of physico-chemical characteristics of sediments of a lower Himalayan lake, Mansar, India. *International Research Journal of Environmental Sciences*, 2(9), 16–22.

Singh O, Jain CK (2013). Assessment of water quality and eutrophication of lakes, *Journal of Environmental Nanotechnology*, 2, 46–52.

Singh O, Rai SP, Kumar V, Sharma MK, Choubey VK (2008). Water quality and Eutrophication status of some lakes of the western Himalayan region (India), Sengupta M and Dalwani R (eds.), *Proceeding of Taal-2007: The 12th world lake conference*, pp. 286–291.

Singh O, Sharma MK (2012). Water quality and eutrophication status of the Renuka lake, District Sirmaur (H.P.), *Journal of Indian Water Resources Society*, 32 (3–4), 1–7.

Singh V, Banyal HS (2013a). Insect fauna of Khajjiar Lake of Chamba district, Himachal Pradesh, India, *Pakistan Journal of Zoology*, 45(4), 1053–1061.

Singh V, Banyal HS (2013b). Study on fish species recorded from Khajjiar lake of Chamba district, Himachal Pradesh, India, *International Journal of Science and Nature*, 4(1), 96–99.

Singh VB, Ramanathan AL, Mandal A (2016). Hydro-geochemistry of high-altitude lake: A case study of the Chandra Tal, Western Himalaya, India, *Arabian Journal of Geosciences*, 9, 308.

Singh Y, Khattar JIS, Singh DP, Rahi P, Gulati A (2014). Limnology and cyanobacterial diversity of high altitude lakes of Lahaul-Spiti in Himachal Pradesh, India. *Journal of Biosciences*, 39, 643–657.

Slathia D, Dutta SPS (2013). Hydro-biological study of a subtropical Shiwalik lake, Jammu, (J&K) (India), *International Journal of Chemical, Environmental and Biological Sciences*, 1(1), 143–148.

Srivastava P, Bhambri R, Kawishwar P, Dobhal DP (2013). Water level change of high altitude lakes in Himalaya-Karakoram from ICESat altimetry, *Journal of Earth System Sciences*, 122(6), 1533–1543.

Valenzuela MAS, Mckinney DC, Rounce DR, Byers AC (2014). Changes in Imja Tsho in the Mount Everest region of Nepal, *The Cryosphere*, 8, 1661–1671.

Veettil BK, Bianchini N, Andrade AMD, Bremer UF, Simoes JC, Junior EDS (2016). Glacier changes and related glacial lake expansion in the Bhutan Himalaya, 1990–2010, *Regional Environmental Change*, 16, 1267–1278.

Wang SJ, Zhang T (2013). Glacial lakes change and current status in the central Chinese Himalaya from 1990-2010, *Journal of Applied Remote Sensing*, 7(1), 073459. doi: 10.1117/1.JRS.7.073459.

Wang W, Yao T, Yang W, Joswiak D, Zhu M (2012). Methods for assessing regional glacial lake variation and hazard in the southeastern Tibetan Plateau: A case study from the Boshula mountain range, China, *Environmental Earth Sciences*, 67, 1441–1450, doi: 10.1007/s12665-012-1589-z.

Wang X, Liu Q, Liu S, Wei J, Jiang Z (2016b). Heterogeneity of glacial lake expansion and its contrasting signals with climate change in Tarim Basin, Central Asia, *Environmental Earth Sciences*, 75, 695.

Wang X, Zhou A, Sun Z (2016a). Spatial and temporal dynamics of lakes in Nam Co Basin, 1991–2011, *Journal of Earth Science*, 27(1), 130–138.

Wei C, Wen H (2012). Geochemical baselines of heavy metals in the sediments of two large freshwater lakes in China: Implications for contamination character and history, *Environmental Geochemistry and Health*, 34, 737–748.

Worni R, Huggel C, Stoffel M (2012b). Glacial lake in the Indian Himalaya - From an area wide glacial lake inventory to on site and modelling based risk assessment of critical glacial lakes, *Science of the Total Environment*, 468–469(1), S71–S84.

Worni R, Stoffel M, Huggel C, Volz C, Casteller A, Luckman BH (2012a). Analysis and dynamic modelling of a moraine failure and glacier lake outburst flood at ventisquero, Negro, Patagonian Andes (Argentina), *Journal of Hydrology*, 444, 134–145.

Xiang L, Lu XX, Higgitt DL, Wang SM (2002). Recent lake sedimentation in the middle and lower Yangtze Basin inferred from ^{137}Cs and ^{210}Pb measurements, *Journal of Asian Earth Sciences*, 21, 77–86.

Xiao J, Jin Z, Wang J, Zhang F (2015). Major ion chemistry, weathering process and water quality of natural waters in the Bosten Lake catchment in an extreme arid region, NW China, *Environment Earth Sciences*, 73, 3697–3708.

Xu J, Grumbine E, Shrestha A, Eriksson M, Yang X, Wang Y, Wilkes A (2009). The melting Himalaya: Cascading effects of climate change on water, biodiversity and livelihoods, *Conservation Biology*, 235, 20–30.

Yang L, Wang L, Wang Y, Zhang W (2015). Geochemical speciation and pollution assessment of heavy metals in surface sediments from Nansi Lake, China, *Environmental Monitoring and Assessment*, 187, 261.

Yannawar VB, Bhosle AB (2013). Cultural eutrophication of Lonar lake, Maharashtra, India, *International Journal of Innovative and Applied Sciences*, 3(2), 504–510.

Yao Z, Wang R, Liu Z, Wu S, Jiang L (2015). Spatial-temporal patterns of major ion chemistry and its controlling factors in the Manasarovar Basin, Tibet, *Journal of Geographical Sciences*, 25(6), 687–700. doi: 10.1007/s11442-015-1196-5.

Zhang W, Ming Q, Shi Z, Chen G, Niu J, Lei G, Chang F, Zhang F (2014). Lake sediment records on climate change and human activities in the Xingyun Lake Catchment, SW China. *PLoS One*, 9(7), e102167.

Shen J, Yang L, Yang X, Matsumoto R, Tong G, Zhu Y, Zhang Z, Wang S. (2005 Mar). Lake sediment records on climate change and human activities since the Holocene in Erhai catchment, Yunnan Province, China. *Science in China Series D: Earth Sciences*, 48(3), 353–363.

3 Holocene to Present Monsoon Variability in Peninsular India as Reflected in Lake and Marine Sediment Records
A Review

Hema Achyuthan
Anna University

CONTENTS

INTRODUCTION

SIGNIFICANCE OF RECONSTRUCTING PAST CLIMATE

To assess recurrences of droughts and floods in Peninsular India, we need sources that include paleorecords of floods, famine, drying up of lakes, ponds, and rivers, the spatial extent of vegetational cover, and the dates of flowering of plants. Combined with the present data, historical records can provide an insight into the past climate conditions (Achyuthan et al., 2017). However, to date, documentary evidence is generally limited to regions with long literary traditions, such as Britain and parts of Europe, China, and to a lesser extent North America. In addition, ship logs from Spanish, Dutch, and English ships crossing the World's oceans from the 16th through 20th centuries have provided new insights into weather patterns and how these have changed over time (Garcia-Herrara et al., 2005). Additionally, accounts, artistic depictions, paintings, and photographs of advancing and retreating mountain glaciers during recent centuries have also provided evidence of climate change on more recent timescales (for example, the retreat of the Rhône and Grindelwald glaciers in the Swiss Alps). Furthermore, eyewitness accounts of extreme weather events such as storms or droughts combined with information about famine have also been used to assess the impact of climate fluctuations on societies. Natural disasters such as volcanic eruptions can provide valuable information to assess the nature of climate change. A famous example of such an event is the eruption of the Laki volcano in Iceland in 1783 (Brayshay and Grattan 1999), which caused a dry, sulfurous fog in Europe, blocking the sun with low and variable temperatures. The eruption of the Vesuvius in 79 AD was one of the deadliest (Rolandi et al., 2008), completely burying the cities of Pompeii and Herculaneum and causing climate change and destruction of the society. So, how do we reconstruct past climates when there are no written records available at all? Therefore, to work out how the climate has changed over time, climate researchers need long-term records. Historical records provide short time scale data and that is why other past climate indicators such as growth bands in trees, corals, deposits in lakebeds, icecores, and dating peat layers provide such valuable evidence. These sources are called "proxy" or indirect data. Tools and various methodologies applied to reconstruct past climate are lake and marine sedimentology, palynology, and dendrochronology, reconstructing long-term changes in vegetation and climate in the hinterlands. Moreover, it is noted that the state of paleolimnology in the tropics was analyzed using articles published between 1997 and 2015 in the Journal of Paleolimnology, and in other international and tropical-country-based journals. Results showed that most paleolimnological studies have been carried out in high-latitude regions. About 40% of the lakes on Earth, representing almost one-third of the global lake surface area, lie within tropical latitudes. Yet in comparison to the number of paleolimnological investigations in higher-latitude lakes, there have been relatively few studies in the tropics. The goal of Escobar et al. (2020) was to evaluate whether there has been a shift in the relative amount of effort directed toward paleolimnological work in tropical regions over the last quarter century, and if not, to call for more paleolimnological studies in the tropics and suggest ways to remedy the geographic disparity. The study of Escobar et al. (2020) showed that paleolimnological studies in the tropics still lag far behind at higher latitudes, prompting us to encourage more work focusing on the tropical regions. To do so, we will require more funding from local research agencies to support paleolimnological work and train local scientists. We recommend that funded investigators from extra-tropical, developed countries should work in close collaboration with local scientists in these nations where studies are carried out also involving local students. Steps should also be taken to encourage students from tropical countries to attend international scientific meetings that focus on paleolimnology. Lastly, more such symposia should be held in tropical countries. Paleolimnological research at low latitudes can address pressing environmental issues in tropical environments, such as the effects of rapid land-use change, the eutrophication and pollution of local water bodies, and recent climate change (Escobar et al., 2020).

LAKE AND MARINE SEDIMENTS AS PROXIES FOR RECONSTRUCTING PALEOMONSOON RECORDS

Every year, billion tons of sediments get deposited and stack up continuously on the ocean floor and in lake basins. Ocean and lake sediments consist of biological and other materials that are produced in the lake/ocean or that get washed in from the nearby land. These materials are deprived of oxygen and are thus preserved as tiny fossils with accumulation or formation of carbon and other geochemicals in the sediments. Freshwater bodies such as lakes and ponds are carbon sinks. Sediment cores retrieved from ponds or the ocean floor can be used to reconstruct past climate. Especially, sediment cores collected from the lake or coastal margins help in understanding the lake or sea-level rise and fall owing to climate shifts and reveal continuous sediment. However, only on the application of radiometric dating methods, the chronology, rate, and duration of sediment deposition can be inferred. For example, a study based on the sediment grain size analysis was carried out on the Parsons Valley Lake deposit in Nilgiris, India to determine the depositional environments and paleoflood events since the late Pleistocene period (~29,838 years BP). A 72-cm lacustrine core was collected from the lake and eight organic carbon-rich sediment samples were AMS radiocarbon dated. The study reveals variations in the grain size distribution chiefly influenced by regional climatic conditions. Paleoflood events have been identified by sedimentary flood signatures of varying changes in the magnitude of sediment supply from the background silty sediment matrix around ~29,838 and ~8,405 years BP (Raja et al., 2018).

THE INDIAN MONSOON REGIME

The Indian monsoon can be defined as a major wind system and is a current and dynamic feature of the tropical atmospheric circulation systems. The seasonal drive of the intertropical convergence zone (ITCZ) (Figures 3.1a, b, and 3.2) over the equatorial region (Schneider et al., 2014) causes the Indian monsoon. In the northern hemisphere, the ITCZ shifts northward during summer and causes southwest (summer) monsoons, while it shifts toward the south where it warms the southern hemisphere resulting in the northeast (winter) monsoons. The ITCZ migrates between latitudes of 20°N and 8°S depending on the boreal summer and winter, sponsoring the rainfall over the south Asian region (Schneider et al., 2014). A southward shift of the ITCZ also suggests the weakening of the summer monsoon (Haug et al., 2001). The Indian summer monsoon (ISM) or the southwest monsoon (SWM) is a major component of the Asian monsoon system, which has a significant impact on the hydrological cycle during summer months (Cane, 2010; Gadgil, 2003; Ota et al., 2017; Rashid et al., 2011; Wang, 2002). During summer, pressure differences over the warm Tibetan Plateau and the cooler Indian Ocean form the southwesterly winds, which bring rainfall that spreads over the southern and southwest Asia (Rashid et al., 2007; Wang et al., 2001; Webster et al., 1998). During September, the air pressure over northern India increases because of rapid cooling following the retreating sun into the southern hemisphere. However, air over the adjacent Indian Ocean holds the heat as it does not cool fast enough resulting in a low-pressure zone. Thus, owing to

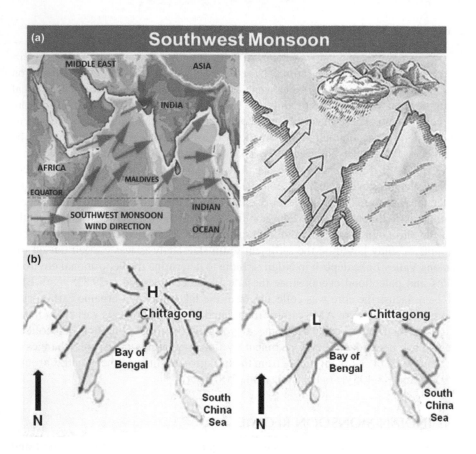

FIGURE 3.1 (a) Map showing the spread of the Indian southwest monsoon during June–September (map not to the scale). (b) Map showing the path of the northeast rains during October–December (map not to the scale).

the pressure gradient, cold air masses move from the Himalayan and Indo-Gangetic Plain regulating toward the Indian Ocean through the south, the Coromandel coast, Tamil Nadu, and this is referred to as the northeast monsoon (NEM) or the retreating monsoon (Clift and Plumb, 2008; Gupta et al., 2006; Rajmanickam et al., 2017). The development of the NEM is not only affected by the strengthening of high-pressure cells over Tibetan and Siberian Plateaus in winter, but also by the westward migration and subsequent weakening of the high-pressure cell in the southern Indian Ocean while shifting the ITCZ to the south of India. However, it is also noted that in the tropical Indian Ocean, in shorter time scales, the intensity of SWM varies owing to the sea surface temperatures and extension of Eurasian snow cover (Philander et al., 1996; Rashid et al., 2011; Robock et al., 2003; Webster et al., 1998).

Tropical monsoon or the SWM is one of the most important climate systems that affect the livelihood and socioeconomy of over 60% of the Asian population (Jianping and Qingcun, 2003). The fluctuations in its duration and intensity of the monsoonal

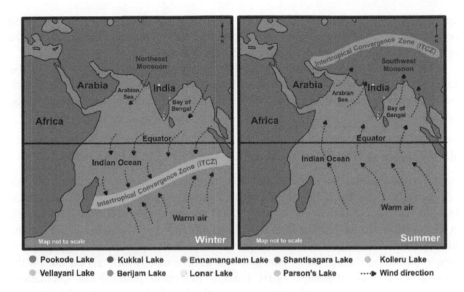

FIGURE 3.2 ITCZ and the location of the lakes in Peninsular India.

system over a period of time have a profound socio-economic effect on the population, agriculture, industry, and in general human civilization. Even an early onset or late withdrawal of monsoons often causes extreme effects such as floods leading to droughts and landslides.

The period from ~3,850 to 3,300 years BP is often inferred to be a transition toward "deurbanization", coinciding with a decline in the Indus Valley civilization and mass migration toward the eastern region due to decreased monsoon precipitation (Possehl, 2002). During the same period, i.e., the middle to late Holocene period, several river valley civilizations such as the Sumerian (~6,000–4,000 years BP) and Mesopotamian (~5,100 years BP) civilizations developed. Early Bronze Age: *Early Dynastic period* (~4,900–4,350 years BP), *Akkadian Empire* (~4,350–4,100 years BP) (Wilkinson, 1997; Matthew, 2003), and Babylon civilization (~3,900 years BP), also collapsed owing to the decline in the strength and intensity of the rains (this being one of the important reasons) and throughout the region, rain-fed agriculture was supplemented by nomadic pastoralism. Periodic breakdowns in the cultural system also occurred for several reasons. Periods of climatic instability led to periods of trade collapse and neglect of irrigation systems.

Thus, interest in the reconstruction of past climate and paleomonsoon research to predict future climate change using climate models has ignited several minds to explore and understand this unique 'monsoon' or 'mausam' atmospheric circulation phenomenon. For this purpose, several proxies such as the marine and lake sediment cores, fluvial archives, geoarchaeology tools (Fontugne and Duplessy, 1986; Chauhan et al., 1993, Achyuthan et al., 2014) are often applied.

Peninsular India experiences dominantly the SWM, lying within the core of the Indian tropical monsoonal zone (Figures 3.1a, b, and 3.2). The landscape is largely an agricultural landscape growing varieties of crops and supporting various agro-industries and practices. Therefore, reconstruction of the past paleomonsoon record and its variability is imperative to predict future setups and these data can be applied to reduce the impacts of extreme events if any.

This review aims to synthesize the Holocene paleomonsoon record of peninsular India in particular with a comparison with regional records and also outline priorities for future research to improve the understanding of the Holocene climate of the island.

RELATIONS WITH REGIONAL LAKE AND MARINE RECORDS

Based on the regional coherence of the Indian monsoon domain, paleoclimate records of Peninsular India, the Indian Ocean, the Arabian Sea, and the Bay of Bengal provide insights into the highly dynamic nature of the Holocene monsoon climate fluctuations. A short core retrieved from the Parsons Valley Lake in Nilgiris revealed a long climate history since ~30,000 years BP. A multiproxy study involving sedimentology, palynology, radiocarbon dating, stable isotopes, and geochemistry was carried out on the Parsons Valley Lake deposit, Nilgiris, India, to determine paleoclimate fluctuations and their possible impact on vegetation since the late Pleistocene. The 72-cm-deep sediment core that was retrieved reveals five distinct paleoclimatic phases: (1) warm and humid conditions with a high lake stand before the last glacial maximum (LGM; ~29,800 years BP), subsequently changing to a relatively cool and dry phase during the LGM. (2) Considerable dry conditions and lower precipitation occurred between ~16,300 and 9,500 years BP (Figure 3.3) (Raja et al., 2019). During this period, the

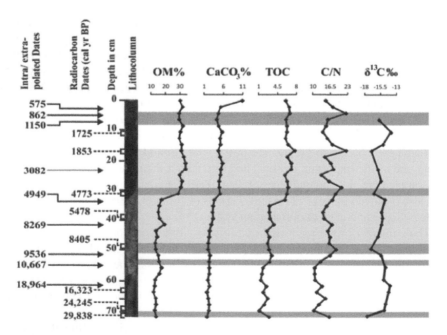

FIGURE 3.3 Down sediment core variation of OM, CaCO$_3$, TOC, C/N, and δ^{13}C analyzed from Parsons Valley Lake (Raja, 2018).

vegetation shrank and perhaps was confined to moister pockets or was a riparian forest cover. (3) An outbreak in the shift of monsoonal precipitation was witnessed in the beginning of the mid-Holocene, around 8,400 years BP, implying alteration in the shift toward warm and humid conditions, resulting in relatively high pollen abundance for evergreen taxa. (4) This phase exhibits a shift to heavier $\delta^{13}C$ values around ~1,850 years BP, with an emergence of moist deciduous plants pointing to drier conditions. (5) Human activities contributed to the exceedingly high percentage of Acacia and Pinus pollen during the Little Ice Age.

Similarly, analyses of a 17.37-m long sediment core obtained from the Kolleru Lake, east coast of India, were studied to reconstruct the climate, paleoenvironmental, and sea-level history of the region. Sedimentology and pollen data, supported by nine ^{14}C AMS dates, revealed Holocene-relative sea-level changes and provided paleoclimate data of the region from the LGM to the present. An anhydrous calcium sulfate layer collected at the core end was dated to 18,400 years BP, devoid of pollen in mottled yellowish clay, indicating a desiccated lake surface, reflecting a dry LGM climate condition. Palynomorphs in the overlying calcareous-concretion-bearing light brown silty clay showed a change from arid terrestrial herbaceous plants to freshwater taxa, indicating a change from dry to wet climate after the LGM and before 8,000 years BP. Further, upward in the core, black, sticky silty clay with abundant mangrove pollen and mollusk shells indicated a marine environment in the Kolleru Lake and aggradation of sediment stacking related to the middle Holocene sea-level rise from 8,000 to 4,900 years BP. The uppermost sandy/silty clay, with terrestrial/aquatic pollen and a ^{14}C age of 3,700 years BP indicated a freshwater environment during the late Holocene. The results of the study of Nageswara Rao et al. (2020) indicate that Kolleru Lake shifted from a dry lakebed during the LGM to a brackish lagoon during the middle Holocene, and then subsequently forming a freshwater lake by the late Holocene period. These changes occurred owing to the influence of climate and relative sea-level change along the east coast of India (Nageswara Rao et al., 2020).

Early Holocene (~12,000–8,000 Years BP)

The Guliya icecore (Tibetan Plateau) record indicates an enriched $\delta^{18}O$ ratio varying from −20‰ to −16‰, indicating an increase in temperature and higher rainfall during ~13,100–12,800 years BP (Yao et al., 1996). Based on the chronological data, alluvial ridges, occurrence of ponds, and pedogenic carbonates formed within the Holocene soils in the Ganga Basin reveal the development of a dense network of channels due to high precipitation depositing coarser fragments and drying of this system intermittently (Srivastava, 2001; Srivastava et al., 2003) during this period. These observations can be correlated with the periods of intensified monsoon as exhibited in the Ocean 189 Drilling Program Site 723, off the Oman margin and from the central Himalaya (Juyal et al., 2010) and the Talchappar Salt Lake, Rajasthan (Achyuthan et al., 2007). Similar observations were made by Dixit et al. (2014) on lake sediment records from Kotla Dahar, indicating progressive strengthening of the Indian summer monsoon from ~11,000 to 9,400 years BP identified in the northwest of India. Enhanced monsoon activity phases were also identified from 10,000 to

7,000 years BP (Nair et al., 2010) and from 10,700 to 8,600 years BP (Sandeep et al., 2017) in the lake records of Shantisagara Lake in southern India. Comparable paleo-climate records were also noted in the Eastern Arabian Sea (Sarkar et al., 2000), Oman and Yemen (Fleitmann et al., 2003, 2007), and Andaman Sea (Rashid et al., 2011), also supporting the monsoon intensification observed in Sri Lanka during the early Holocene period.

A rapid northward displacement of the mean position of the summer ITCZ and ISM rain belt during the early Holocene period was identified by the rapid decrease of $\delta^{18}O$ values in a cave stalagmite from Oman and Yemen (Fleitmann et al., 2007), while a wetter and humid climate was observed in a marine record from Andaman Sea (Rashid et al., 2011). Strengthening of monsoon from 11,000 to 7,000 years BP was also observed in sediments from Ganges–Brahmaputra river delta in the east coast of India (Goodbred and Kuehl, 2000). These records are consistent with the observed enhanced monsoon activity recorded in Sri Lanka during the early Holocene. Paleoclimate archives from south India and surrounding regions show single step development of the Indian monsoon during the early Holocene (Figure 3.4). However, a two-step progression of Indian summer monsoon during the early Holocene was also observed (He et al., 2018; Overpeck et al., 1996; Thamban et al., 2007). One from 13,000 to 12,500 years BP and the other from 10,000 to 9,500 years BP were recorded, implying that the monsoon progression did not follow astronomi-cal forcing during the early Holocene period (Overpeck et al., 1996).

Based on marine records from the eastern Arabian Sea, Thamban et al. (2007) also identified monsoon strengthening and corroborated this view with two abrupt events at 9,500 and 9,100 years BP. Similar events were also recorded in the Central Tibetan Plateau around 11,700 and 10,000 years BP, also marked by decreased $\delta^{18}O$ values in Ostracods and a significant increase of sedimentary Ti concentration (He et al., 2018). During the second event, which extended from 10,000 to 7,000 years BP, δD wax values decreased progressively and reached a minimum indicating enhanced precipitation. Sandeep et al. (2017) observed a weakening of the summer monsoon from 11,100 to 10,700 years BP in the Shantisagara lake record. Weak monsoon

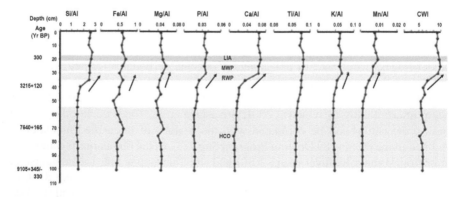

FIGURE 3.4   Down sediment core variation of the Kukkal Lake exhibiting variations in chemical ratios and CWI since ~8,500 years BP (Rajmanickam et al., 2017).

phases that occurred from 10,800 to 10,200, 9,800 to 8,800, and 8,400 to 8,000 years BP were observed in marine sediments from the Arabian Sea and these phases coincided with 10,300, 9,400, and 8,200 Bond events, respectively (Gupta et al., 2005). The proxy record of oxygen isotope ratio from the Mawmluh cave, northeastern India suggests a decrease in ISM strength during 9,800–9,200 and 8,700–8,052 years BP also indicated by enriched $\delta^{18}O$ values (Dutt et al., 2015). A 1.5°C rise in temperature with higher precipitation conditions was observed from ~7,000 to 6,000 years BP (Yao et al., 1996).

However, based on the fluvial archive from the Palar river basin, South India, a dry phase owing to a reduced trend in winter monsoon precipitation (NEM) was noticed during 10,000–4,830 years BP (Resmi and Achyuthan, 2018).

The sediment cores retrieved from the Kukkal and Berijam lakes were radiocarbon dated and the ages reveal non-linear phases of deposition. The Berijam lake core dates from 2,400 years BP to the present while older ages were obtained from the Kukkal lake sediment core that gave ages of 9,000 years BP to the present. At the Kukkal lake core, the "Holocene Climatic Optimum" that lasted till the mid-Holocene (9,000–5,000 years BP) was noted during which period the lake continued to receive sediments. The lake was deeper and the lake margins expanded forming deeper lacustrine facies.

The C/N ratios of Berijam and Kukkal lake sediment cores range from 7.87 to 2.47 and 14.02 to 8.31, indicating that the organic carbon originates mainly from lacustrine algae and aquatic weeds and that the TOC content reflects the primary productivity within the lake. Detailed geochemical analyses and chemical weathering indices applied using the Al_2O_3 - $(CaO + Na_2O)$ - K_2O diagram, scatter plots, and vertical profiles suggest that the sediments have been derived from the hinterland due to the extreme silicate weathering since the early Holocene wet period when the area around the lakes received intense ISM and the NEM rains. Pollen data exhibit savanna vegetation around the Kukkal Lake during the early Holocene, but shifting to more open vegetation cover due to the dry and arid environment since the middle Holocene to the present. The lakes supported deeper water columns and the lake extended covering large areas. Subsequently, since the last 5,000 years BP, the climate has shifted to warm and dry conditions due to the decrease in the SW and NE rains.

The results of the biogeochemical and mineralogical analyses of Lonar lake sediment core collected from the core monsoon zone in central India spanned the Holocene period. A long-term climate transition from wetter conditions during the early Holocene to drier environments during the late Holocene, delineating the insolation curve, was presented by Menzel et al. (2014).

Mid Holocene

The two natural freshwater basins situated in two different geomorphic and rainfall regimes in the Kerala state, tropical Southern India are the Pookode Lake (elliptical) and the Vellayani Lake (linear in shape). The Pookode Lake is located at an altitude of ~770 m.a.s.l. and the Vellayani Lake lies in the coastal plain and both these lakes were studied in detail for paleomonsoon reconstruction. Radiocarbon dates were obtained on the organic carbon-rich sediments for the Pookode lake and the dates range in age from the mid-Holocene to recent (6,240–565 years BP), while the radiocarbon dates obtained from the Vellayani Lake core range in age from 3,025 to 1,210 years BP revealing two post bomb inversion of ages (Veena et al., 2014a, 2014b). Phytolith and pollen records supported by continuous sedimentation indicate that during the overall dry phase of the mid-Holocene period (6,240 years BP) to present, very wet intervals of short duration were caused by intense SWM. This caused the water levels to rise and the lake margin expanded over the surrounding low-lying region characterized by lateritic soils and contracted during the long dry phases (Veena et al., 2014a, b).

The southern peninsula of the Indian subcontinent is characterized by moisture sources from both the SWM and NEM. However, the long-term climate variability associated with these two moisture sources and their relative contribution in the region is not well studied (Mishra et al., 2019). Hence, to delineate the moisture sources, a study was carried out by Mishra et al. (2019) using a multiproxy approach (geochemistry, clay mineralogy, and end-member mixing analyses of the grain size parameters) on the radiocarbon-dated sediment profile from Ennamangalam Lake, southern India. They identified three hydrological stages in the lake region: stage 1 (4,800–3,150 years BP), relative drier condition, marked by low detritus content, and higher contribution of relatively fine-grains; stage 2 (3,150–1,640 years BP), a transition phase, high sedimentation rate as compared to the preceding stage; and stage 3 (1,640 years BP to present), represented by higher detritus contribution into the lake system, higher CIA values followed by a continuous declining ratio of Mg/Al and deposition of higher coarse grain sediments indicating high energy condition probably due to higher precipitation (Mishra et al., 2019). However, a regional comparison of paleoclimate records demonstrates that the increase in precipitation observed in the Ennamangalam region during the late Holocene is in contrast to the records from the core monsoon zone. The overview of regional records indicates an inverse relationship between the SWM and the NEM strength during the late Holocene probably affected by the increasing ENSO variability (Mishra et al., 2019).

Based on the multidisciplinary data studied from the Lonar Lake sediment core, Menzel et al. (2014) identified periods of extended drought during 4,600–3,900 and 2,000–600 years BP that have been attributed to temperature changes in the Indo-Pacific Warm Pool and these correlate well with cold phases in the North Atlantic region. The most pronounced climate deteriorations occurred during 6,200–5,200, 4600–3,900, and 2,000–600 years BP. The strong dry phase between 4,600 and 3,900 years BP at the Lonar Lake corroborates the severe climate deterioration (Menzel et al., 2014).

Nearer home, a reconstruction of the Late Holocene climate and environmental history from the North Bolgoda Lake, Sri Lanka, using lipid biomarkers and pollen records revealed shifts in precipitation, salinity, and vegetational cover between 3,000 years and the present, with arid conditions ca. 2,334 and 2,067 years BP. This extreme dry period was preceded and followed by wetter conditions (Gayantha et al., 2019).

A palynological study coupled with magnetic susceptibility and stratigraphy of subsurface sediment from the South Kerala sedimentary basin (Kumaran et al., 2008) revealed that SWM became gradually reduced after ~5,000 or 4,000 years BP, thereby experiencing a late Holocene dry climate. A noted significant weakening of

ISM after ~5,000 kyr BP was also recorded in the Andaman Sea and southern India (Rashid et al., 2011; Sandeep et al., 2017). A declining trend of ISM precipitation was identified from cave stalagmite obtained from northeast India during the same period (Berkelhammer et al., 2012). In addition, abrupt weakening of SWM was observed in the lake sediment record from Haryana, India around 4,100 years BP following 4,200 years BP Bond event (Dixit et al., 2014).

During ca. 3,740–3,077 years BP, the period observed in the Deoria Tal reveals a decrease in oak forest cover and a dry period in the central Ganga Basin (Sharma and Gupta, 1997). This dry phase in the Ganga Basin is marked by an increase in agricultural practices (mainly winter crops), domestication of animals, and is also known as the early phase of Red Ware-dominated culture in the western Ganga plain (Sharma et al., 2004). During this period, the $\delta^{13}C$ ratio shows enrichment as compared to the preceding stage. The increased $\delta^{13}C$ values lie in the range of CAM plants (Meyers and Ishiwatari, 1993).This observation is consistent with the records from Kukkal (Figure 3.4) (Rajmanickam et al., 2017) and Shantisagara lakes (Sandeep et al., 2017).

Fleitmann et al. (2003) noted a gradual decrease in the weakening of monsoon precipitation that probably began around ~8,000 years BP as apparent from the cave stalagmite records from southern Oman. A cave stalagmite record from Oman and Yemen showed a further decreasing monsoon precipitation in the mid-Holocene due to the southward migration of the ITCZ (Fleitmann et al., 2007). This was followed by weak to moderate precipitation conditions mainly from 5,000 to 4,200 years BP (Fleitmann et al., 2007). Marine sediment core retrieved from the Arabian Sea and Riwasa lake sediments from northwest India also recorded similar events (Dixit et al., 2014; Thamban et al., 2007). However, east Arabian records indicate a more arid episode from 6,000 to 3,500 years BP and a significantly weaker monsoon period from 6,000 to 5,500 years BP (Sarkar et al., 2000; Thamban et al., 2007). A short-wet period occurred during 6,500–5,400 years BP owing to the higher winter monsoon precipitation, which is consistent with the warmer and wetter conditions (Achyuthan et al., 2014; Dixit et al., 2014; Sarkar et al., 2000; Thamban et al., 2007; Fleitmann et al., 2007).

Using a sediment core raised from the offshore, from the mouth of the Godavari River, Ponton et al. (2012) reconstructed the Holocene paleoclimate of the Indian peninsula and argued for the decline of the Indus valley civilization. Carbon isotopes of sedimentary leaf waxes indicated a gradual increase in aridity-adapted vegetation from ~4,000 until 1,700 years ago and were subsequently trailed by the aridity-adapted plants strongly affecting the river valley civilization causing cultural changes across the Indian subcontinent as the climate became more arid after ~4,000 years. Sedentary agriculture took hold in the drying central and south India, while the urban Harappan civilization collapsed in the already arid Indus basin. The establishment of a more variable hydroclimate over the last ca. 1,700 years BP may have led to the rapid proliferation of water-conservation technology in south India.

Late Holocene

Understanding paleoclimate shifts is important during the late Holocene period since the expansion of human settlements occurred during this period. An extremely strong SWM phase was recorded in-between 1,780 and 1,300 years BP (Gayantha et al., 2017) that is consistent with the enhanced hydrological conditions observed during 1,800–1,000 years BP in higher central Himalayas (Bhushan et al., 2018) and increased winter monsoon precipitation recorded from 1,880 to 1,440 years BP in South India (Resmi and Achyuthan, 2018). Based on the Pookode lake record, Veena et al. (2014a) observed a short phase of enhanced SWM activity during the 1,400–760 and 420–140 years BP.

There have been several rapid phases of significant variability in the intensity of SWM precipitation as observed in the Kukkal and Berijam lake records. These were characterized by several wetter events of shorter duration of smaller magnitudes resulting in the expansion of lake margins for a short duration. The strong phases were followed by weak phases causing drier conditions mostly around the RWP and MWP. The LIA event is observed in the Kukkal lake core (Figure 3.4). The Berijam lake core holds the record of the sedimentation events of the MWP and the LIA. The lag effect of these events has been noted. CaO/MgO ratios and chemical weathering intensity (CWI) values of the Vellayani Lake sediment core indicate overall dry conditions for 3,000 years BP (late Holocene).

The period 2,200–1,800 years BP coincides with the RWP, previously thought to be restricted to higher latitude regions (Vollweiler et al., 2006; Desperat et al., 2003; Martínez-Cortizas et al., 1999; Laskar et al., 2011, 2013a, b). The Bay of Bengal sediment core data indicate that the effect of RWP is evident in the tropics also. Weakening of the monsoon has also been observed around 1500 and 400–800-years BP, the latter period is the transition from MWP to LIA. A significant reduction in the SWM during this period was also reported by several others (Fleitmann et al., 2004; Sinha et al., 2007).

Gupta et al. (2003) observed a weak monsoon phase recorded in the continental margin sediments off Oman, during 1,900–1,400 years BP. An extended but a steady intensification of the Indian summer monsoon was observed from 400 years BP to the present in the northern Andaman Sea record (Ota et al., 2017). A colder phase with suppressed monsoon activities from 600 to 250 years BP (Srivastava et al., 2017), decreased winter monsoon precipitation from about 1,440 years BP to the present (Resmi and Achyuthan, 2018), and wetter to drier transition around ~680 years BP were observed. This event was continued till 340 years BP as identified in Kedarnath peat records, from Central Himalayas, river sediments of Palar River basin, South India, and in a Cave Stalagmite record from Southern Oman (Fleitmann et al., 2004). During the late Holocene period, intense monsoon activities were recorded in many parts of Sri Lanka as revealed by different paleoclimate proxies. A strengthening of monsoon activity extending from ~3,600 to 1,500 years BP also recorded in Sri Lanka is comparable to the records from both eastern Arabian Sea and northern Andaman Sea (Achyuthan et al., 2014; Sarkar et al., 2000). Xu et al. (2002) studied plant cellulose δ^8O variations in sediment cores retrieved from peat deposits at the northeastern edge of the Qinghai-Tibetan Plateau in China. Following the decline of the RWP, their data revealed the existence of three particularly cold intervals centered at approximately 500, 700, and 900 AD during the Dark Ages (cold period).

It is also noted in all these studies (Figure 3.5) that dissimilarities in the timing of the onset and retreating monsoons are obvious in both summer and winter monsoons in different geographical localities in Peninsular India and other continental records also implying the local response of the climate system to the tropical atmospheric circulation phenomenon.

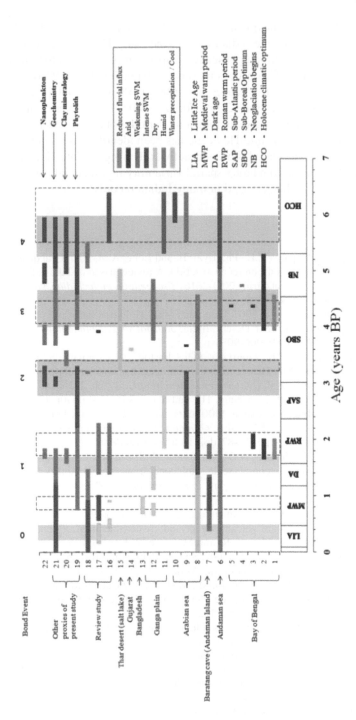

FIGURE 3.5 A comparison of monsoonal shifts with global climate events, Bond events (shaded), and the Arabian Sea, Bay of Bengal, Andaman Sea, and Indian sub-continent. [Reference.(1) Chauhan et al. (2004), (2) Chauhan and Vogelsang (2006), (3) Chauhan and Suneethi (2001), (4) Mathien and Bassinot (2008), (5) Chauhan et al. (2000), (6) Rashid et al. (2007), (7) Laskar et al. (2013a, 2013b), (8) LuÈckge et al. (2001), (9) Sarkar et al. (2000), (10) Thamban et al. (2002), (11) Sharma et al. (2006), (12) Saxena et al. (2013), (13) Masud Alam et al. (2009), (14) Singh et al. (2007), (15) Enzel et al. (1999), (16) Patnaik et al. (2012), (17) Kuppusamy and Ghosh (2012), (18) Thamban et al. (2007), and (19–21) are from the Landfall island, Bay of Bengal (Nagasundaram et al., 2020)].

CONCLUSIONS

Based on the palaeoclimate information available in Peninsular India as well as the Arabian Sea, Bay of Bengal, and the Indian Ocean, the Early Holocene was characterized by a strong, wetter monsoon phase. A comparison of stable isotope data reveals that the beginning of the Holocene is marked with much lighter $\delta^{18}O$ values, which subsequently persisted with two pauses at ~5,000–4,300 and ~2,000 years BP when compared to the LGM (magnitude 2.44% for *G. ruber*). This enhanced monsoon activity phase was followed by a weak monsoon period around 8,000 years BP. In most palaeoclimatic records, the mid-Holocene has been identified as a semi-arid to arid phase or a period of decreased monsoon activity. During the late Holocene period, highly variable climates with larger spatial variability in millennial and centennial time scales have been characterized in many Indian records.

REFERENCES

Achyuthan, H., Farooqui, A., Gopal, V.E.E., Phartiyal, B. and Lone, A.M. 2017. Late quaternary to Holocene southwest monsoon reconstruction: A review based on lake and wetland systems (studies carried out during 2011-2016). *Proceedings of the Indian National Science Academy*, 82(3), pp. 847–868.

Achyuthan, H., Kar, A. and Eastoe, C.J. 2007. Late quaternary-Holocene lake-level changes in the eastern margin of the Thar Desert, India. *Journal of Paleolimnology*, 38(4), pp. 493–507. doi: 10.1007/s10933-006-9086-6.

Achyuthan, H., Nagasundaram, M., Gourlan, A.T., Eastoe, C., Ahmad, S.M. and Padmakumari, V.M. 2014. Mid-Holocene Indian summer monsoon variability off the Andaman Islands, Bay of Bengal. *Quaternary International*, 349, pp. 232–244.

Berkelhammer, M., Sinha, A., Stott, L., Cheng, H., Pausata, F.S.R. Yoshimara, K. 2012. An abrupt shift in the Indian monsoon 4000 years ago. In: Giosan, L., Fuller, D.Q., Nicoll, K., Flad, R.K. and Clift, P.D. (eds.), *Climates, Landscapes, and Civilizations*, Geophysical Monograph Series. pp. 75–87. doi: 10.1029/2012GM001207.

Bhushan, R., Sati, S.P., Rana, N., Shukla, A.D., Mazumdar, A.S. and Juyal, N. 2018. High-resolution millennial and centennial scale Holocene monsoon variability in the Higher Central Himalayas. *Palaeogeography, Palaeoclimatology, Palaeoecology*, 489, 95–104.

Brayshay, M. and Grattan, J. (1999). Environmental and social responses in Europe to the 1783 eruption of the Laki fissure volcano in Iceland: A consideration of contemporary documentary evidence. In: Firth, C.R. and McGuire, W.J. (eds.), *Volcanoes in the Quaternary*, Geological Society, London, UK, Special Publication 161, pp. 173–187.

Cane, M.A. 2010. Climate: A moist model monsoon. *Nature*, 463(7278), pp. 163.

Chauhan, O.S., Borole, D.V., Gujar, A.R., Antonio, M., Mislanker, P.G. and Rao, Ch.M. 1993. Evidences of climatic variations during late Pleistocene - Holocene in the eastern Bay of Bengal. *Current Science*, 65(7), pp. 558–562.

Chauhan, O.S., Patil, S.K. and Suneethi, J. 2004. Fluvial influx and weathering history of the Himalayas since Last Glacial Maxima; isotopic, sedimentological and magnetic records from the Bay of Bengal. *Current Science*, 87(4), pp. 509–515.

Chauhan, O.S., Sukhija, B.S., Gujar, A.R., Nagabhushanam, N. and Paropkari, A.L. 2000. Late Quaternary variations in clay mineral along the SW continental margin of India: evidence of climatic variations, *Geo Marine Letters*, 20, pp. 118–122,

Chauhan, O.S. and Suneethi, J. 2001. 18 ka BP records of climatic changes, Bay of Bengal: Isotopic and sedimentological evidences. *Current Science*, 81, pp. 1231–1234.

Chauhan, O.S. and Vogelsang, E. 2006. Climate induced changes in the circulation and dispersal patterns of the fluvial sources during late Quaternary in the middle Bengal Fan. *Journal of Earth System Science*, 115(3), pp. 379–386.

Clift, P.D. and Plumb, R.A. 2008. *The Asian Monsoon: Causes, History and Effects*, Cambridge University Press, Cambridge, UK.

Desperat, S., Sánchez-Goñi, M.F. and Loutre, M.F. 2003. Revealing climatic variability of the last three millennia in northwestern Iberia using pollen influx data. *Earth and Planet Science Letters*, 213, pp. 63–78.

Dixit, Y., Hodell, D.A., Sinha, R. and Petrie, C.A., 2014. Abrupt weakening of the Indian summer monsoon at 8.2 kyr BP. *Earth and Planetary Science Letters*, 391, pp. 16–23.

Dutt, S., Gupta, A.K., Clemens, S.C., Cheng, H., Singh, R.K., Kathayat, G. and Edwards, R.L. 2015. Abrupt changes in Indian summer monsoon strength during 33,800 to 5500 years B.P. *Geophysical Research Letters*, 42(13), 5526–5532. Doi: 10.1002/2015GL064015.

Enzel, Y., Ely, L.L., Mishra, S., Ramesh, R., Amit, R., Lazar, B., Rajaguru, S. N., Baker, V.R. and Sandler, A. 1999. High-resolution Holocene environmental changes in the Thar Desert. Northwestern India. *Science*, 284(5411), pp. 125–128.

Escobar, J., Serna, Y., Hoyos, N., Velez, M.I. and Correa-Metrio, A. 2020. Why we need more paleolimnology studies in the tropics. *Journal of Paleolimnology*, 64, pp. 47–53. doi: 10.1007/s10933-020-00120-6.

Fleitmann, D. Burns S.J., Mangini A., Mudelsee M., Kramers J., Villa I., Neff U., Al-Subbary A.A., Buettner A., Hippler D. and Matter A. 2007. Holocene ITCZ and Indian monsoon dynamics recorded in stalagmites from Oman and Yemen (Socotra). *Quaternary Science Reviews*, 26, pp. 170–188.

Fleitmann, D., Burns, S.J., Mudelsee, M., Neff, N., Kramers, J., Mangini, A. and Matter, A. 2003. Holocene forcing of the Indian monsoon recorded in a stalagmite from Southern Oman. *Science*, 300, pp. 1737–1739.

Fleitmann, D., Burns, S.J., Neff, U., Mudelsee, M., Mangini, A. and Matter, A. 2004. Palaeoclimatic interpretation of high-resolution oxygen isotope profiles derived from annually laminated speleothems from Southern Oman. *Quaternary Science Reviews*, 23, pp. 935–945.

Fontugne, M.R. and Duplessy, J.C. 1986. Variations of the monsoon regime during the upper Quaternary: evidence from carbon isotopic record of organic matter in North Indian Ocean sediment cores. *Palaeogeography, Palaeoclimatology, Palaeoecology*, 56(1–2), 69–88.

Gadgil, S. 2003. The Indian monsoon and its variability. *Annual Review of Earth and Planetary Sciences*, 31(1), pp. 429–467.

Garcia-Herrera, R., Können, G.P., Wheeler, D.A., Prieto, M.R., Jones, P.D. and Koek, F. 2005. CLIWOC a climatological database for the World's Oceans 1750–1854. *Climatic Change*, 73(1), 1–12. doi: 10.1007/s10584-005-6952-6.

Gayantha, K., Routh, J. and Chandrajith, R. 2017. A multi-proxy reconstruction of the late Holocene climate evolution in Lake Bolgoda, Sri Lanka. *Palaeogeography, Palaeoclimatology, Palaeoecology*, 473, pp. 16–25.

Gayantha, K., Routh, J., Krishnamurthy, A., Lazar, J., Prasad, S., Chandrajith, R., Roberts, P. and Gleixner, G. 2019. Biomarker and pollen approach to reconstruct late Holocene climate and environmental history in Western Sri Lanka, *29th International Meeting on Organic Geochemistry*, Durbin, Ireland.

Goodbred, S.L., Jr. and Kuehl, S.A. 2000. Enormous Ganges-Brahmaputra sediment discharge during strengthened early Holocene monsoon. *Geology*, 28(12), pp. 1083–1086.

Gupta, A.K., Anderson, D.M. and Overpeck, J.T. 2003. Abrupt changes in Asian southwest monsoon during the Holocene and their links to the North Atlantic Ocean. *Nature*, 421, pp. 354–356.

Gupta, A.K., Anderson, D.M. and Pandey, D.N. 2006. Adaptation and human migration, and evidence of agriculture coincident with changes in the Indian summer monsoon during the Holocene. *Current Science*, 90(8), pp. 1082–1090.

Gupta, A.K., Das, M. and Anderson, D.M. 2005. Solar influence on the Indian summer monsoon during the Holocene. *Geophysical Research Letters*, 32, pp. L17703.

Haug, G.H., Hughen, K.A., Sigman, D.M., Peterson, L.C. and Rohl, U. 2001. Southward migration of the intertropical convergence zone through the holocene. *Science*, 293(5533), pp. 1304–1308. doi: 10.1126/science.1059725.

He, Y., Hou, J., Brown, E.T., Xie, S. and Bao, Z. 2018. Timing of the Indian summer monsoon onset during the early Holocene: Evidence from a sediment core at Linggo Co, central Tibetan Plateau. *The Holocene*, 28(5), pp. 755–766.

Jianping, L. and Qingcun, Z. 2003. A new monsoon index and the geographical distribution of the global monsoons. *Advances in Atmospheric Sciences*, 20(2), pp. 299–302.

Juyal, N., Sundriyal, Y., Rana, N., Chaudhary, S. and Singhvi, A.K. 2010. Late quaternary fluvial aggradation and incision in the monsoon-dominated Alaknanda valley, Central Himalaya, Uttrakhand, India. *Journal of Quaternary Science*, 25(8), pp. 1293–1304. doi: 10.1002/jqs.1413.

Kumaran, K.P.N., Limaye, R.B., Nair, K.M. and Padmalal, D. 2008. Palaeoecological and palaeoclimate potential of subsurface palynological data from the Late Quaternary sediments of South Kerala Sedimentary Basin, southwest India. *Current Science*, 95(4), pp. 515–526.

Kuppusamy, M. and Ghosh, P. 2012. Cenozoic climatic record for monsoonal rainfall over the Indian Region. In: Wang, S.-Y. (ed.), *Modern Climatology*, In Tech, London, UK, pp. 257–288. doi: 10.5772/36206.

Laskar, A.H., Raghav, S., Yadava, M.G., Jani, R.A., Narayana, A.C. and Ramesh, R. 2011. Potential of stable carbon and oxygen isotope variations of speleothems from Andaman Islands, India, for palaeomonsoon reconstruction. *Journal of Geological Resources*, 2011, 7. doi: 10.1155/2011/272971.

Laskar, A.H., Yadava, M.G., Ramesh, R., Polyak, V.J. and Asmerom, Y. 2013a. A 4 kyr stalagmite oxygen isotopic record of the past Indian Summer Monsoon in the Andaman Islands. *Geochemistry, Geophysics Geosystems*, 14, pp. 3555–3566.

Laskar, A.H., Yadava, M.G., Sharma, N. and Ramesh, R. 2013b. Late Holocene climate in the Lower Narmada valley, Gujarat, Western India, inferred using sedimentary carbon and oxygen isotope ratios. *The Holocene*, 23(8), pp. 1115–1122.

Lückge, A., Doose-Rolinski, H., Khan, A.A., Schulz, H. and Von Rad, U. 2001. Monsoonal variability in the northeastern Arabian Sea during the past 5000 years: Geochemical evidence from laminated sediments. *Palaeogeography, Palaeoclimatolgy, Palaeoecology*, 167, pp. 273–286.

Martínez-Cortizas, A., Pontevedra-Pombal, X., García-Rodeja, E., Nóvoa-Muñoz, J.C. and Shotyk, W. 1999. Mercury in a Spanish peat bog: Archive of climate change and atmospheric metal deposition. *Science*, 284(5416), pp. 939–942. doi: 10.1126/science.284.5416.939.

Masud Alam, A.K.M., Xie, S. and Wallis, L.A. 2009. Reconstructing late Holocene palaeoenvironments in Bangladesh: Phytolith analysis of archaeological soils from Somapura Mahavihara site in the Paharpur area, Badalgacchi Upazila, Naogaon district, Bangladesh. *Journal of Archaeological Science*, 36, pp. 504–512.

Matthews, R. 2003. *The Archaeology of Mesopotamia: Theories and Approaches.* eBook Pub. London, Routledge, pp. 256, doi: 10.4324/9780203390399.

Mathien, E. and Bassinot, F. 2008. *Abrupt hydrographic changes in the Bay of Bengal during the Holocene. Geophysical Research Abstracts*, 10, EGU2008-A-09423.

Menzel, P., Gaye, B., Mishra, P.K., Ambili, A., Basavaiah, B., Marwan, N. Plessen, B., Prasad, S., Riedel, N., Stebich, M. and Wiesner, M. 2014. Linking Holocene drying trends from Lonar Lake in monsoonal central India to North Atlantic cooling events. *Palaeogeography Palaeoclimatology Palaeoecology*, 410, pp. 164–178.

Meyers, P.A. and Ishiwatari, R. (1993/09). Lacustrine organic geochemistry—An overview of indicators of organic matter sources and diagenesis in lake sediments. *Organic Geochemistry*, 20(7), 867–900.

Mishra, P.K., Yadav, A., Gautamc, P.K., Lakshmidevi, C.G., Singh, P. and Ambili, A. 2019. Inverse relationship between south-west and north-east monsoon during the late Holocene: Geochemical and sedimentological record from Ennamangalam Lake, southern India. *Catena*, 182, pp. 104–117.

Nagasundaram, M., Achyuthan, H. and Rai, J. 2020. Mid to late Holocene reconstruction of the southwest monsoonal shifts based on a marine sediment core, off the Landfall Island, Bay of Bengal. In: *The Andaman Islands and Adjoining Offshore: Geology, Tectonics and Palaeoclimate*. Society of Earth Scientists Series, Springer, Cham, Switzerland, pp. 315–400.

Nageswara Rao, K., Pandey, S., Kubo, S., Saito, Y., Naga Kumar, K. Ch. V. Demudu, G., Bandaru, H., Nagumo, N., Nakashima, R. and Sadakata, N. 2020. Paleoclimate and Holocene relative sea-level history of the east coast of India. *Journal of Paleolimnology*, 64, pp. 71–89. doi: 10.1007/s10933-020-00124-2.

Nair, K.M., Padmalal, D., Kumaran, K.P.N., Sreeja, R., Limaye, R.B. and Srinivas, R. 2010. Late Quaternary evolution of Ashtamudi–Sasthamkotta lake systems of Kerala, south west India. *Journal of Asian Earth Sciences*, 37(4), 361–372.

Ota, Y., Kawahata, H., Murayama, M., Inoue, M., Yokoyama, Y., Miyairi, Y., Aung, T., Hossain, H.M.Z., Suzuki, A. and Kitamura, A. 2017. Effects of intensification of the Indian Summer Monsoon on northern Andaman Sea sediments during the past 700 years. *Journal of Quaternary Science*, 32(4), pp. 528–539.

Overpeck, J., Anderson, D., Trumbore, S. and Prell, W. 1996. The southwest Indian Monsoon over the last 18 000 years. *Climate Dynamics*, 12(3), pp. 213–225.

Patnaik, R., Gupta, A.K., Naidu, P.D., Yadav, R.R., Bhattacharyya, A. and Kumar, M. 2012. Indian monsoon variability at different time scales: Marine and terrestrial proxy records. *Proceedings of the Indian National Science Academy*, 78(3), pp. 535–547.

Philander, S.G.H., Gu, D., Lambert, G., Li, T., Halpern, D., Lau, N.C. and Pacanowski, R.C. 1996. Why the ITCZ is mostly north of the equator. *Journal of Climate*, 9(12), pp. 2958–2972.

Ponton, C., Giosan, L., Eglinton, T.I., Fuller, D.Q., Johnson, J.E., Kumar, P. and Collett, T.S. 2012. Holocene aridification of India. *Geophysical Research Letter*, 39, L03704. doi: 10.1029/2011GL050722.

Possehl, G.L. 2002. *The Indus Civilization: A Contemporary Perspective*. Altamira Press, Walnut Creek, CA.

Raja, P. 2018. Tropical rainforest dynamics and climate implications Nilgiris India: A multi proxy record since the Late Pleistocene. Unpublished Ph.D. thesis, Anna University, Chennai, India.

Raja, P., Achyuthan, H., Farooqui, A., Ramesh, R., Pankaj, K. and Chopra, S. 2019. Tropical rainforest dynamics and palaeoclimate implications since the late Pleistocene, Unpublished Ph.D. thesis, Nilgiris, India. *Quaternary Research*, 91(1), pp. 367–382.

Raja, P., Achyuthan, H., Geethanjali, K., Baghel, P.K. and Chopra, S. 2018. Late Pleistocene paleoflood deposits identified by grain size signatures, Parsons valley Lake, Nilgiris, Tamil Nadu. *Journal of the Geological Society of India*, 91, pp. 547–553.

Rajmanickam, V., Achyuthan, H., Eastoe, C. and Farooqui, A. 2017. Early-Holocene to present palaeoenvironmental shifts and short climate events from the tropical wetland and lake sediments, Kukkal Lake, Southern India: Geochemistry and palynology. *The Holocene*, 27(3), pp. 404–417.

Rashid, H., England, E., Thompson, L. and Polyak, L. 2011. Late Glacial to Holocene Indian Summer Monsoon variability based upon sediment records taken from the Bay of Bengal. *Terrestrial. Atmospheric and Ocean Science*, 22(2), pp. 215–228.

Rashid, H., Flower, B.P., Poore, R.Z. and Quinn, T.M. 2007. A 25 Ka Indian Ocean monsoon variability record from the Andaman Sea. *Quaternary Science Reviews*, 26, pp. 2586–2597.

Resmi, M.R. and Achyuthan, H. 2018. Northeast monsoon variations during the Holocene inferred from palaeochannels and active channels of the Palar River basin, Southern Peninsular India. *The Holocene*, 28(6), pp. 895–913.

Robock, A., Mu, M., Vinnikov, K. and Robinson, D. 2003. Land surface conditions over Eurasia and Indian summer monsoon rainfall. *Journal of Geophysical Research: Atmospheres*, 108(D4), 4131.

Rolandi, G., Paone, A., De Lascio, M., and Stefani, G. 2008. The 79 AD eruption of Somma: the relationship between the date of the eruption and the southeast tephra dispersion. *Journal of Volcanology and Geothermal Research*, 169(1), pp. 87–98. doi: 10.1016/j.jvolgeores.2007.08.020.

Sandeep, K., Shankar, R., Warrier, A.K., Yadava, M.G., Ramesh, R., Jani, R.A., Weijian, Z. and Xuefeng, L. 2017. A multi-proxy lake sediment record of Indian summer monsoon variability during the Holocene in southern India. *Palaeogeography, Palaeoclimatology, Palaeoecology*, 476, pp. 1–14.

Sarkar, A., Ramesh, R., Somayajulu, B.L.K., Agnihotri, R., Jull, A.J.T. and Burr, G.S. 2000. High resolution Holocene monsoon record from the eastern Arabian Sea. *Earth and Planetary Science Letters*, 177(3–4), pp. 209–218.

Saxena, A., Prasad, V. and Singh, I.B. 2013. Holocene palaeoclimate reconstruction from the phytoliths of the lake-fill sequence of Ganga Plain. *Current Science*, 104(8), pp. 1054–1062.

Schneider, T., Bischoff, T. and Haug, G.H. 2014. Migrations and dynamics of the intertropical convergence zone. *Nature*, 513(7516), pp. 45–53.

Sharma, C. and Gupta, A. 1997. Vegetation and climate in Garhwal Himalaya during Early Holocene: Deoria tal. *Paleobotanist*, 46, pp. 11–116.

Sharma, S., Joachimski, M.M., Tobschal, H., Singh, I., Tewari, D.P. and Tewari, R. 2004. Oxygen isotopes of bovid teeth as archives of paleoclimatic variations in archaeological deposits of the Ganga plain, India. Quaternary Research, 62, pp. 19–28.

Sharma, S., Joachimski, M.M., Tobschall, H.J., Singh, I.B., Sharma, C. and Chauhan, M.S. 2006. Correlative evidence of monsoon variability, vegetation change and human habitation in Senai lake deposit, Ganga plain. *Current Science*, 90, pp. 973–978.

Singh, V., Prasad, V. and Chakraborty. S. 2007. Phytoliths as indicators of monsoonal variability during mid-late Holocene in main land Gujarat, western India. *Current science*, 92(12), pp. 1754–1759.

Sinha, A., Cannariato, K.G., Stott, L.D., Cheng, H., Edwards, L.R., Yadava, M.G., Ramesh, R. and Singh, I.B. 2007. A 900-year (600 to 1500 A.D.) record of the Indian summer monsoon precipitation from the core monsoon zone of India. *Geophysical Research Letters*, 36(4), L16707. doi: 10.1029/2007GL030431.

Srivastava, P. 2001. Palaeoclimatic implications of pedogenic carbonates in Holocene soils of the Gangetic plains, India. *Palaeogeography, Palaeoclimatology, Palaeoecology*, 172, pp. 207–222.

Srivastava, P., Agnihotri, R., Sharma, D., Meena, N., Sundriyal, Y.P., Saxena, A., Bhushan R., Sawlani, R., Banerji, U.S., Sharma, C., Bisht, P., Rana, N. and Jayangondaperumal R. 2017. 8000-year monsoonal record from Himalaya revealing reinforcement of tropical and global climate systems since mid-Holocene. *Scientific Reports* 7, 14515. doi: 10.1038/s41598-017-15143-9.

Srivastava, P., Singh, I.B., Sharma, M. and Singhvi, A.K. (2003). Luminescence chronology and Late Quaternary geomorphic history of Ganga plain, India. *Palaeogeography, Palaeoclimatology, Palaeoecology*, 197, pp. 15–41.

Thamban, M., Kawahata, H. and Rao, V.P. 2007. Indian summer monsoon variability during the Holocene as recorded in sediments of the Arabian Sea: timing and implications. *Journal of Oceanography*, 63, pp. 1009–1020.

Thamban, M., Rao, V.P. and Schneider, R.R. 2002. Reconstruction of late Quaternary monsoon oscillations based on clay mineral proxies using sediment cores from the western margin of India. *Marine Geology*, 186, pp. 527–539.

Veena, M.P., Achyuthan, H., Eastoe, C., and Farooqui, A. 2014a. A multi-proxy reconstruction of monsoon variability in the late Holocene, South India. *Quaternary International*, 325, pp. 63–73.

Veena, M.P., Achyuthan, H., Eastoe, C., and Farooqui, A. 2014b. Human impact on low-land Vellayanii Lake, south India: A record since 3000yrs BP. *Anthropocene*, 8, pp. 83–91.

Vollweiler, N., Scholz, D., Muhlinghaus, C., Mangini, A. and Spotl, C. 2006. A precisely dated climate record for the last 9 kyr from three high alpine stalagmites, Spannagel Cave, Austria. *Geophysical Research Letters*, 33, L20703. doi: 10.1029/2006GL027662.

Wang, B. 2002. Rainy season of the Asian–Pacific summer monsoon. *Journal of Climate*, 15(4), pp. 386–398.

Wang, B., Wu, R. and Lau, K.M. 2001. Interannual variability of the Asian summer monsoon: Contrasts between the Indian and the Western North Pacific-East Asian Monsoons. *Journal of Climate*, 14(20), pp. 4073–4090.

Webster, P.J., Magana, V.O., Palmer, T.N., Shukla, J., Tomas, R.A., Yanai, M.U. and Yasunari, T. 1998. Monsoons: Processes, predictability, and the prospects for prediction. *Journal of Geophysical Research: Oceans*, 103(C7), pp. 14451–14510.

Wilkinson, T.J. 1997. Holocene environments of the high plateau, Yemen. Recent geological investigations. *Geoarchaeology*, 12, pp. 833–864.

Xu, H., Hong, Y., Lin, Q., Hong, B., Jiang, H., and Zhu, Y. 2002, Temperature variations in the past 6000 years inferred from $\delta^{18}O$ of peat cellulose from Hongyuan, China. *Chinese Science Bulletin*, 7, pp. 1578–1584.

Yao, T. D., Jiao, K.Q., Tian, L.D., Yang, Z.H., Shi, W.L. and Thompson, L.G. 1996. Climatic variations since the Little Ice Age recorded in the Guliya Ice Core. *Science China*, 39D, pp. 587–596.

4 Speleothems, Monsoon, and Holocene– Quaternary Climate
A Review from the Indian Subcontinent

V.C. Tewari
Sikkim Central University

CONTENTS

INTRODUCTION

Speleothems or cave deposits are natural continental archives of the paleoclimate record and are abundantly found in all the continents. In the Indian subcontinent, speleothems have been recorded from NW to NE Himalaya, Meghalaya, and Peninsular India, where calcium carbonate is precipitated as stalagmites, stalactites, flowstones, and moonmilk cave deposits, Figure 4.1. Caves in peninsular India are mainly found in the Proterozoic Mahanadi, the Kaladgi, and the Cuddapah basins. Dandak, Jhumar, Gupteshwar, Kailash, and Kotumsar caves are a part of the Mahanadi basin (Kaushal et al., 2018 and references therein). The Mawmluh cave from Meghalaya, northeast currently covers the Last Glacial Maximum (LGM) and Heinrich stadials. Records from peninsular Valmiki cave cover the later phase of Heinrich event 1 and the last deglaciation. Bittoo, Kalakot, and Timta caves from the Lesser Himalaya, north India, and Mawmluh cave from Meghalaya, northeast India (Figures 4.1 and 4.2)

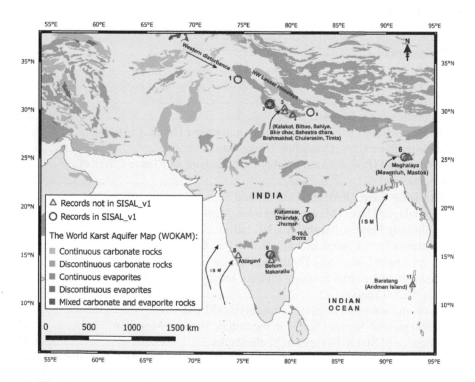

FIGURE 4.1 Map of India showing the locations of the caves (see map legend). NW Himalayan caves (1: Kalakot; 2: Bittoo, Sahiya, Bhir Dhar; 3: Sahastradhara, Brahmakhal; 4: Chulerasim; 5: Timta), NE Meghalaya (6: Mawmluh, Mastos), Peninsular India (7: Kutumsar, Dandak, Jhumar; 8: Akalagavi; 9: Belum, Nakarallu; 10: Borra), and Baratang (11: Andaman Island).(Modified after Kaushal et al., 2018 and adapted from World Karst Aquifer Map (WKAM, Chen et al., 2017).

cover the Bølling–Allerød and Younger Dryas periods. The Sahiya cave from
Garhwal Lesser Himalaya, north India, Mawmluh and Umsynrang caves from northeast India, and Kotumsar cave from peninsular India encompass significant periods
of the Holocene (Kathayat et al., 2017, 2018). Speleothems may provide an important
record of the paleoclimate and paleomonsoon since they are continental archives and
not subjected to diagenesis, erosion, and terrestrial deposits. Miocene to Holocene
paleoclimatic, paleooceanographic, and paleo-monsoonal studies from the tropical
and monsoonal regions of the Indian ocean and subcontinent and SE Asia have been
attempted and reviewed in recent years by various researchers (Kaushal et al., 2018
and references therein; Dutt et al., 2015, 2018; Tewari, 2008, 2009, 2011, 2013b,
2015; Kathayat et al.,2018; Kotlia et al., 2015, 2017; Fleitmann et al., 2003; Johnson
et al., 2006; Sinha et al., 2015; Yadav and Ramesh, 2005). Oxygen isotopic variations
in speleothems especially stalagmite growth laminae are used for interpreting the
amount of rainfall (Cheng et al., 2012, 2013; Wang et al., 2005; Sinha et al., 2015,
2016; Kaushalet al., 2018). In Chapter 10 of this book, the authors own research
on the speleothems from Himalaya and Meghalaya and a review of the work done
by other workers on Quaternary–Holocene climate and monsoon (Figure 4.1) are
incorporated. All these studied speleothems are recorded from the important caves
of India, which lie in the high monsoonal regions and the ITCZ passes over them
(Figure 4.2). Therefore, it is quite obvious and significant to study the strength of
the Indian summer monsoon (ISM) and decadal-scale seasonal variations (Kaushal
et al., 2018). As one of the most prominent seasonally recurring atmospheric circulation patterns, the Asian summer monsoon (ASM) plays a vital role in the life and
livelihood of about one-third of the global population. Changes in the strength and

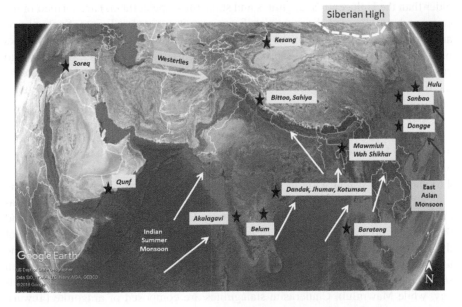

FIGURE 4.2 Map showing locations of important caves in South and Southeast Asia (source
Google Earth) with arrows indicating moisture trajectories of the Indian summer monsoon.
(Compiled from various sources.)

seasonality of the ASM significantly affect the ASM region, yet the drivers of change and the varied regional responses of the ASM are not well understood. In the last two decades, there were several studies reconstructing the ASM using stalagmite-based proxies such as oxygen isotopes (δ^{18}O). Such reconstructions allow examination of ASM drivers and responses, increasing monsoon predictability. In this chapter, stalagmite δ^{18}O records from northeast India at the center of the larger ASM region have been focused on. Northeast India has particularly well-decorated caves providing several high resolution and good quality stalagmite-based time series (Kaushal et al., 2018; Tewari et al., 2020). The δ^{18}O records from this region are also uniquely supported by an extensive field monitoring program in addition to multiple proxies including δ^{13}C and trace element ratios providing robust climate and environmental interpretations. We have examined northeast Indian stalagmite records collated in the Speleothem Isotope Synthesis and Analysis version 1 (SISAL_v1) database and support the database with a summary of record quality and climatic interpretations of the δ^{18}O record during different climate states (Kaushal et al., 2018). The stalagmites from this region show high amplitude changes in response to dominant drivers of the ASM on orbital to multi-centennial timescales and indicate the magnitude of monsoon variability in response to these drivers.

We suggest the most useful time periods and locations for further work to understand the regional climatic response to monsoon circulation changes (Tewari et al., 2020). The significance of the northeast Indian speleothem records is highlighted in this chapter and the need for conservation of these invaluable paleoarchives is emphasized. In recent years, speleothem records have received attention, as speleothems represent one of the few continental archives that can match the icecore records (as in the Arctic and Antarctic polar icecores) in resolution and age control for time periods older than the Holocene. Stalagmites and stalactites (speleothems) are formed in the caves and are regarded the most significant archives of paleoclimate. Speleothems are limestone deposits within the caves. The groundwater trickling through cracks in the roof of caves contains dissolved calcium bicarbonate, which is transferred into calcium carbonate, precipitates out of water solution, and forms a ring of calcite on the roof of the cave. Stalagmites are formed as a result of evaporation and precipitation from the solution after the trickling water falls from the stalactites (Figures 4.3 and 4.4). Speleothems are the least altered deposits and are very reliable to study the monsoon pattern of the past. The long-term rainfall and precipitation variability over the Himalayas is not well established. Negative oxygen isotope values have been recorded for stalagmites from the Lesser Himalayan caves (Figures 4.5 and 4.6) and are consistent with the data available from other caves in the Asian region indicating higher monsoon rainfall. Microscopic studies of the stalagmites and stalactites show wavy laminations of alternating dark and light bands of calcite and the possible presence of bacteria to form calcite precipitation. The microfacies of the speleothems show radiaxial fibrous calcite. Possibly the Mg has triggered the formation of radiaxial fabric (Tewari, 2009, 2011; Figure 4.7). Some northern Indian stalagmites Sahastradhara and Brahmakhal show mixed calcite–aragonite mineralogy, while Mawmluh, Chulerasim stalagmites are composed of aragonite (Tewari, 2011; Baskar et al., 2011, Duan et al., 2012). X-ray diffraction (XRD) and scanning electron microscopy (SEM) investigations were carried out for a well-laminated stalagmite from Sahastradhara cave, Dehradun, Lesser Himalaya India, to identify

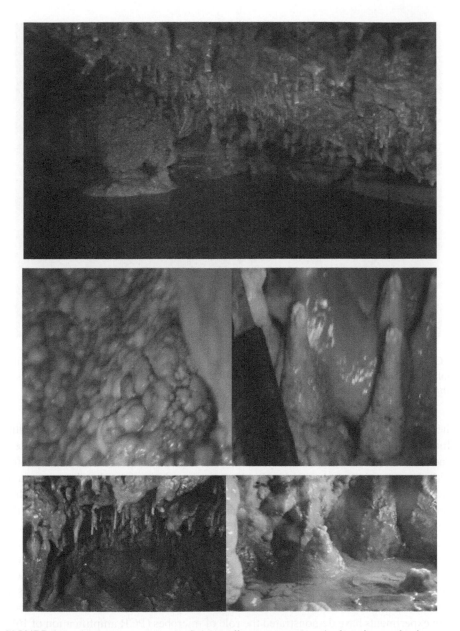

FIGURE 4.3 Speleothems from the Sahastradhara cave system (stalagmites, stalactites top; middle; flowstone (left), stalagmite growth (right); soda straw (thin stalactites and stalagmite, right).

the mineral composition and the fabric of the stalagmite laminae (Tewari and Jaiswal, 2014). Jaiswal and Tewari (2012) have described the speleothems from the Sahastradhara cave, Dehradun and interpreted the paleoclimatic conditions based on various carbonate microfacies. The laminae of this stalagmite are composed of alternating compact and porous sublayers. The XRD results confirm that the stalagmite is composed mainly of primary aragonite needles. The SEM results show that

FIGURE 4.4 (a) Dripping water from stalactites. (b) Stalactite column and moonmilk, Sahastradhara cave. (c) Large hanging stalactites, Brahmakhal cave. (d)Speleothem pillars in Mawsmai cave, Meghalaya.

the compact sublayer is composed of elongated columnar aragonites with a longitudinal orientation (Figure 4.7). Some dark organic laminae have also been recorded. Yadava et al. (2004) considered the dark layers in speleothems as $CaCO_3$ precipitated with the trapped detrital particles. Studies on annually laminated stalagmites from India suggest that the thickness of annual bands is primarily controlled by the drip rate (Yadava and Ramesh, 2005). The Sahastradhara cave system is located in the Neoproterozoic Ediacaran Krol carbonates of the Uttarakhand Lesser Himalaya. These carbonate deposits are rich in stromatolites and Ediacaran metaphytes and metazoans (Tewari, 2012). The geotechnical investigations have shown that these rocks in the Surbhi area, Mussoorie Syncline are highly sheared and fractured, and frequent landslides are common due to heavy precipitation in the monsoon season (Tewari, 2014; Venkateswarlu and Tewari, 2014; Bryanne and Tewari, 2014).The culture experiments have demonstrated the role of microbes (PCR amplification of 16Sr RNA genes (16Sr DNA) in the stalactite formation) (Baskar et al., 2011 in Tewari and Seckbach, 2011). Laboratory experiments have shown that bacterial species isolated from stalactites are able to precipitate carbonates under controlled conditions from Sahastradhara cave in Dehradun (Baskar et al., 2006). Another carbonate mineral present is aragonite formed in freshwater. Various types of light (carbonate) and dark (organic) laminae are related to the microclimatic decadal-scale seasonal variations (Tewari, 2009, 2011, 2015). The morphology, internal structure, composition, and formation of various speleothems are described below.

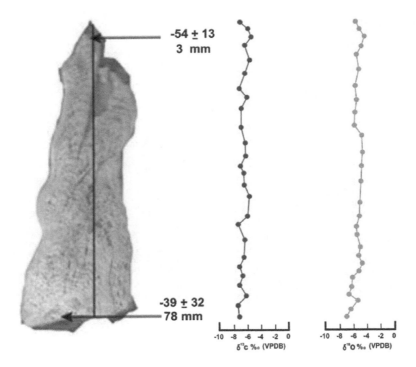

FIGURE 4.5 Oxygen and carbon isotope variation recorded from a stalagmite from the Sahastradhara cave (dated between 39 ± 32 and 54 ± 13 years), Dehradun, Uttarakhand Lesser Himalaya.

STALAGMITES

Stalagmites are formed by water dripping from a stalactite and grow from the floor toward the ceiling. Stalagmites show a variety of morphologies, from cylindrical to conical forms (Figures 4.3, 4.6, and 4.8).A great fall height may also give rise to forms resembling stacked dishes or cylindrical stalagmites with a sunken central splash cup (Fairchild et al., 2007). Kaufmann (2003) demonstrated that the growth rate and radius of stalagmites are controlled by the climate. In particular, the stalagmite diameter is largely controlled by the drip rate, which is related to rainfall, and large stalagmite diameters form during wet periods. In stalagmites, crystals grow with their growth axis perpendicular to the substrate. Repeated fabric patterns or cyclic mineralogical changes have been observed in stalagmites formed in climates characterized by seasonal contrasts and are interpreted as reflecting seasonal variations in the composition of the drip waters. Laminated stalagmites are, therefore, very accurate archives of the past climate and environmental changes. Carbonate speleothems may serve as proxy indicators of paleoclimate, and provide information on annual temperature, rainfall, atmospheric circulation and vegetation changes (McDermott, 2004). Stalagmites are commonly used for uranium series dating (Figures 4.5 and 4.6).

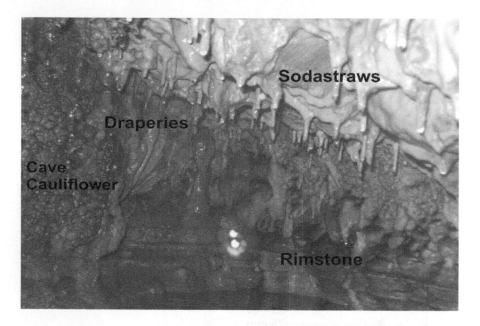

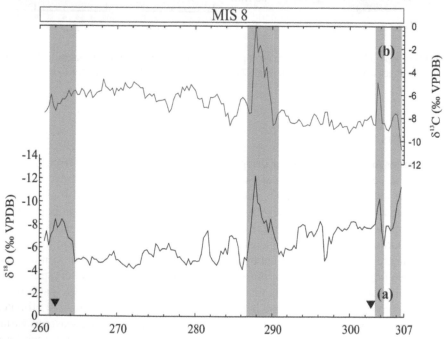

FIGURE 4.6   Speleothems in Bhir Dhar cave, Chakrata (above), and oxygen and carbon isotope variation during the Marine Isotope Stage (MIS 8).

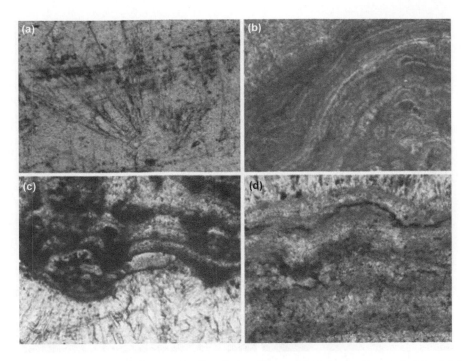

FIGURE 4.7 (a) Photomicrograph of radiating fibrous fabric (aragonite needles) and (b) banding in speleothem, Brahmakhal cave, (c) mixed mineralogy (aragonite needles below and calcite above), and (d) microbially laminated dark organic layer in Sahastradhara cave, Dehradun, Uttarakhand.

STALACTITES

Stalactites are cylinder or cone-shaped speleothems that extend downwards from the cave ceiling. They commonly form from dripping water (Figures 4.4, 4.6, and 4.8). Conical stalactites may be several tens of centimeters wide and taper downward, and they commonly consist of concentric layers of crystals elongated perpendicular to the central vertical growth axis (the axis, which extends from the ceiling toward the floor of the cave). The original hollow center of the stalactite may be clogged by sparite crystals.

SODA STRAW STALACTITES

Soda straws or cylinder-shaped stalactites commonly show a hollow center and relatively thin walls (average of about 0.2 mm). Water is drawn slowly to the tip of the soda straw and exits from these hollow stalactites in the form of single droplets (Figures 4.3, 4.6, and 4.8). The diameter of soda straws is thus determined by the surface tension of the water drop at their tip. Soda straws, therefore, accrete downwards, not outwards. Their outer surface is relatively smooth, whereas the inner wall is characterized by growth steps, which are formed by the emergence of crystal terminations. Soda straws commonly show a regularly spaced banding, which suggests an annual origin (Fairchild et al., 2007).

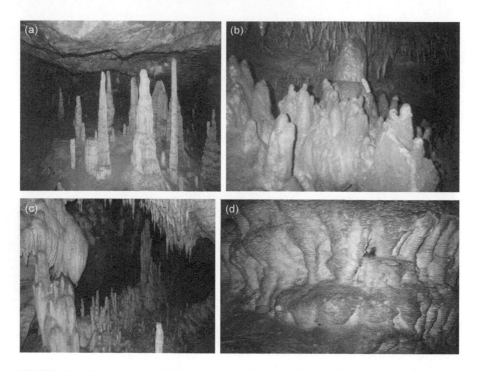

FIGURE 4.8 Stalagmites in Mawmluh (a) and Phyllut caves (b) in Meghalaya, NE India and stalagmites (growing from the floor) and stalactites (hanging from the roof). (c) Kutumsar cave, Peninsular India and speleothems from Baratang cave (d), Andaman Island, India.

FLOWSTONES

Stalagmitic flowstones or flowstones commonly consist of stacked layers of crystals elongated normal to the substrate. Flowstones may be several meters thick, tens or hundreds of meters long in the downstream direction, and are formed through degassing of a thin water current, which is slightly supersaturated with respect to calcium carbonate (Figure 4.3). The growth rate of stalagmitic flowstones varies from a few micrometers (mm) up to about 100 mm/yr. They can grow over tens of thousands of years when water flows over their surface and cease to grow during dry and (or) very cold periods (Fairchild et al., 2007). Stalagmitic flowstones are, therefore, ideal for addressing the chronology of warm events during glacials, and for the timing of glacial to interglacial transitions.

MOONMILK CAVE DEPOSIT

Moonmilk is a cave deposit and it forms stalagmites, stalactites, and flowstones, which commonly consist of carbonate minerals calcite, aragonite, and hydromagnesite. Microorganisms are also found associated with moonmilk. Calcite moonmilk is a porous, plastic deposit consisting of calcite fibre crystals and water (Borsato et al., 2000). Calcite moonmilk may form thick deposits along the roof and walls of caves

(Figures 4.3 and 4.4). Moonmilk forms under environmental conditions where other speleothems may not form. It has not been observed in the karst surface environment, although it may develop at the boundary between the light and the dark zone in a cave.

SPELEOTHEMS FROM THE NORTHERN INDIAN UTTARAKHAND HIMALAYAN CAVES

SAHASTRADHARA CAVE, DEHRADUN, UTTARAKHAND

The caves are located at Sahastradhara in the Dehradun valley, a crescent-shaped intermountain valley that is formed in the carbonates of the Krol belt (Tewari, 2011, 2012, Figure 4.3). The age of the Krol belt has been very controversial and now has been established as Neoproterozoic (Ediacaran; Tewari, 2012) and Sahastradhara lies between 300 23′07.6″N and 780 07′44.9″E with an altitude of 830.5 m above the sea level. It shows best-developed speleothems such as stalactites, stalagmites, pillars, moonmilk, and flowstones (Figure 4.3). This cave also shows maximum percolating water from the ceiling and the stalactites and stalagmites are showing carbonate precipitation. Stalagmites (speleothems growing on the cave floor; Figures 4.9 and 4.10) are powerful paleoclimatic archives, as they contain a multitude of environment-dependent proxies (e.g. $\delta^{18}O$, $\delta^{13}C$, trace elements, humic acids). They have been established as one of the best terrestrial archives of past climate because they can be accurately and precisely dated by absolute U/Th methods. Figure 4.5 shows a stalagmite from the Sahastradhara cave that has been dated and the high-resolution oxygen and carbon isotope data are generated (Tewari and Jaiswal, 2014). Speleothems (stalagmites, stalactites, flowstones, and so on) grow in caves as the result of precipitation of calcium carbonate (either as calcite or aragonite) from drip waters that enter the cave's atmosphere. Mineralogical studies (XRD and SEM-EDAX) of the stalactite and stalagmite samples revealed that calcite is the dominant mineral. The ratios of the stable isotopes oxygen and carbon of the carbonate ($CaCO_3$) are analyzed as they depend on environmental factors, such as rainfall amount, temperature, vegetation cover and type, and so on. As speleothems grow, they incorporate trace elements into their structure, and the concentration and ratios of those might be related to environmental conditions (atmosphere, soil, host rock, and cave air) at the time of deposition. Where banding in stalagmites can be demonstrated to be annual by dating methods, stalagmites have the potential to record long-term, high-resolution climatic data. The $\delta^{18}O$ isotope data of drip water from Sahastradhara cave in Dehradun during the monsoon season (August and September 2007–2010) vary from−4.58% to 5.14% (VPDB; Tewari, 2009, 2011, 2012). Tewari (2009, 2011) has interpreted that in the western part of the ISM in Oman and Yemen, the oxygen isotope ratios of stalagmite calcite primarily show variations in the amount of rainfall with more negative $\delta^{18}O$ indicating higher monsoon rainfall. Therefore, it is concluded that the speleothems are very significant for the paleoenvironmental and paleoclimatic records. The U/Th dating of stalagmite, oxygen isotope data, and the interpretation of the monsoon are discussed separately.

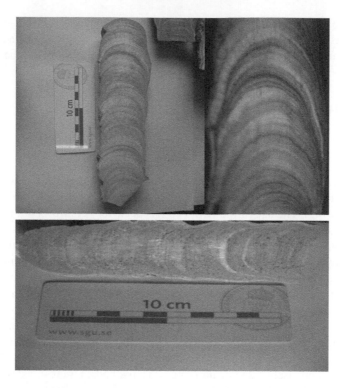

FIGURE 4.9 Polished slabs of the Stalagmites from the Mawmluh and Mastos caves of Meghalaya showing growth lamination pattern radiaxial fabric and calcite in laminations.

Brahmakhal (Prakateshwar), Sahiya, Chulerasim and Timta Caves of Uttarakhand

The Brahmakhal (Prakateshwar) caves are situated near the village Mehar Gaon in the Uttarkashi district of the Garhwal Himalaya (30°23′145″N; 78°07′743″E). Stalactites, flowstones, and stalagmites are well developed in the cave (Figure 4.4). The thin sections of the speleothems have been studied for isotopic, mineralogical, and microfacies analysis. Radiating fibrous calcite and microlamination of calcite and organic-rich microbial laminae have been recorded in the Prakateshwar and Sahastradhara speleothems and described earlier (Figure 4.4; Tewari, 2008, 2009, 2011, 2015). The other more important caves are found in the Chakrata, Chulerasim, Patalbhuvaneshwar, and Timta in the Pithoragarh district of Uttarakhand Himalaya (Sinha et al., 2005, 2007). Stalagmites are well developed in the Sahiya cave near Chakrata and well dated and interpreted for the ISM and migration of ancient civilizations in the Indian subcontinent region (Kathayat et al., 2017).

Bhir Dhar Cave, Chakrata, Uttarakhand

The Bhir Dhar cave (N 30.79, E 77.77) is located in the Meso-Neoproterozoic Deoban Limestone, Chakrata area, Garhwal Lesser Himalaya (Figure 4.6). The Deoban

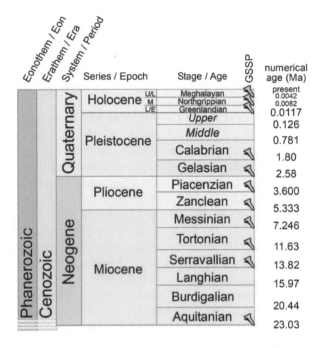

FIGURE 4.10 Meghalayan Stage (a new stage from the Mawmluh Cave), Shillong Plateau, Meghalaya, India. (Adapted from the International Commission on Stratigraphy.)

Limestone is characterized by the stromatolitic–cherty–oolitic limestone rich in cyanobacterial microfossils (Tewari, 2011, 2015). The stalagmites are well developed in the cave and constrained by two absolute U/Th series dates 302 (bottom) to 262 ka (top) (Jaiswal, 2015). The area receives 75–90% of the rainfall during the SW monsoon season from June to September (Mooley et al., 1981). Indian monsoon variation during the 302–262 ka falls under MIS 8 (Lisiecki and Raymo, 2005). The $\delta^{18}O$ and $\delta^{13}C$ values vary from−4.15 to −12.20% and 0.12 to −8.83% relative to the VPDB respectively (Figure 4.6, Jaiswal, 2015). The $\delta^{18}O$ signatures reflect the variation in the intensity of ISM as well as the position of the ITCZ (Fleitmann et al., 2007; Tewari, 2011, 2015). The ITCZ has been located over the study area during the wetter/warmer season. Figure 4.6 shows the speleothems in the Bhir Dhar cave (above) and variation in oxygen and carbon isotopes and paleoprecipitation in the NW Himalayan region during Marine Isotope Stage (MIS 8, below).

Meghalayan Caves (Mawmluh, Mastos, Mawsmai, Phyllut and Puri Caves in Shillong Plateau, Meghalaya, NE India)

Speleothems from the Mawsmai, Mawmlu, Mastos, and Krem Phyllut caves have been recorded and the data on mineralogy, geochemistry, and geomicrobiological aspects have been studied (Tewari, 2011, 2013, 2015) and Baskar et al. (2011). The richly fossiliferous Lakadong Limestone (Paleocene–Eocene) hosts these caves in the Shillong Plateau. The larger benthic foraminifera assemblage *Discocyclina, Assilina,*

Nummulites, *Alveolina*, and Coralline algae *Districhoplax* and *Lithophyllum* are the main fossils recorded from the caves (Tewari, 2019). The Mawsmai and Mustos caves are located in a thickly forested area and are quite small (160 m long, 15 m high, and 4–10 m wide). The main entrance to these caves is located close to the Mawsmai village and the entry is a fairly narrow (1.8 m) vertical opening. The cave is totally aphotic and has many stalagmites and stalactites (Figure 4.8). The average annual temperature of the inner cave is ~15°C–19°C. The stalactites range in sizes from 7–10 cm length and 8–15 cm diameter (small), to 50–150 cm length and 50–100 cm diameter (large). The Krem Phyllut and Mawmluh cave have a large passageway and very large stalagmites are found (Figure 4.8). The Krem Phyllut cave is relatively large (with a total length of 1,003 m, a width of 4.5 m, and a height of 15 m). The deep aphotic inner cave wall had an average annual temperature of ~15°–17°. The stalactites range from 6–7 cm in length and 25–30 cm in diameter (small) to 30–40 cm in length and 50 cm in diameter (large). The columns are 40–45 cm long and the diameter of the upper end is 40 cm and that of the lower end is 60 cm. The speleothems are larger at the entrance and smaller toward the interior. The Mastos cave shows spectacular development of the stalagmites and stalactites. Figure 4.9 (below) is a polished slab of the stalagmite and the variation in oxygen isotope is shown in a stalagmite from the same cave (Figure 4.11). Figure 4.14 shows the [230]Th dating results of a stalagmite from Mastos cave (MA). Three U series dates are a (base), 1,946 ± 57 (208 mm), b (middle), 1,386 ± 97 (155 mm), and c (top) 237 ± 55 (8 mm). The ages are reported in years BP (years before the present/AD 1950). The method is based on Cheng et al. (2013). The stalagmites were studied in detail for the microfacies and the mineralogy of the micro laminae. The photograph of the polished slabs clearly demonstrates

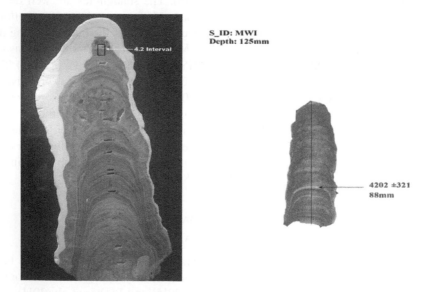

FIGURE 4.11 Meghalayan Stage stalagmite, Mawmluh cave, dated by ICS (4,200 years, left) and present work (MW1, 4,202 ± 321 years).

the thin and thick microlaminae (Figure 4.9). Calcite in the stalagmites is the dominant mineral in all the caves in Mawmluh and Mastos areas. The polished slab of the stalagmite from Mastos cave shows dark and light color laminae and radiaxial fabric due to microclimatic fluctuations (Figure 4.9). So far, only the Uttarakhand and Meghalayan caves have been examined for geomicrobiological as well as paleo-climatic studies (Tewari, 2011; Baskar et al., 2011). Monitoring of the other remote Himalayan caves is required for establishing speleothem proxy data in the future for climate modeling and reconstructions. The Umsynrang record from northeast India is the only record that covers the entire Holocene from India (but is not in SISAL_v1). The world's longest sandstone cave named Krem Puri has been recently discovered in Meghalaya. It has a staggering length of 24.5 km, almost three times the height of Mount Everest, and contains some dinosaur fossils from 66 to 76 million years ago as well. The cave system, which is 24,583 m long, was discovered near the Laitsohum village, located in the Mawsynram area in Meghalaya, East Khasi district.

SPELEOTHEMS FROM PENINSULAR INDIA (KUTUMSAR, CHHATISGARH, BORRA, BARTANG CAVE, ANDAMAN ISLAND) AND THEIR SIGNIFICANCE

Kutumsar cave, Bastar district, Chhatisgarh from peninsular India covers significant periods of the Holocene (Kausal et al., 2018 and references therein). Band et al. (2018) have shown that the Kotumsar $\delta18O$ stalagmite record from Peninsular India indicate a gradual decrease in the amount of rainfall from 8500 to 7300 years BP followed by an increasing trend from 7300 to 5600 years BP. Dandak, Jhumar, Gupteshwar, and Kailash caves are a part of the Mahanadi basin (Figures 4.1 and 4.2). The high-resolution Dandak and Jhumar cave records from peninsular India provide the most reliable record of the Little Ice Age (LIA) for this region. $\delta^{18}O$ data of all or a part of the LIA beginning around 750–450 years BP in the peninsular region are obtained. Liang et al. (2015) reviewed the existing stalagmite and other proxy data from south and east Asia which revealed a broad spatial pattern in precipitation over south and east Asia during the Little Ice Age. Sanwal et al. (2013) have suggested weaker Indian Summer Monsoon during the Little Ice Age and stronger Western disturbances. Borra caves are the second largest caves in the Indian subcontinent (Figure 4.1) situated in Araku valley, near Vishakhapatnam(85 15N: 83 3 E). Speleothems are well developed in the cave and the geomicrobiological studies of Borra cave show the presence of filamentous bacteria *Leptothrix* (Baskar et al., 2011). The Baratang cave is located in limestone formations of the Cenozoic age in the Andaman Island in the Bay of Bengal (Figures 4.1 and 4.8, below right). There are a few records from the peninsular and NW Lesser Himalayan regions of India that are not available in SISAL_v1 (Table 1; Kaushal et al., 2018). However, these caves (Sahastradhara and Bhir Dhar, Figures 4.3, 4.5, and 4.6) are now well dated and the oxygen and carbon isotopes and petrography of carbonate minerals (mixtures of calcite and aragonite) are published earlier (Tewari, 2011, 2013; Tewari and Jaiswal, 2014; Jaiswal, 2015). Besides this, the Panigarh and Sainji stalagmites show mixed calcite–aragonite mineralogy, while Akalagavi, Chulerasim, Umsynrang, and Dharamjali stalagmites are composed of aragonite. A few records provide $\delta^{18}O$ measurements for certain time intervals, but lack age control to examine events closely, such as the Tityana, Akalagavi, Belum, and Baratangcave records (Kaushal et al., 2018).

MEGHALAYAN STAGE FROM SHILLONG PLATEAU, INDIA: YOUNGEST PHASE IN EARTH'S HISTORY (4,200 YEARS)

The International Commission on Stratigraphy (ICS) is the official keeper of geological time scale and it has created a new phase in Earth's geological history and named it the Meghalayan Stage (Figure 4.10) after a stalagmite (Figure 4.12) from the Mawmluh cave in Meghalaya, near Cherrapunji, Shillong Plateau, India. The Meghalayan Stage marks the beginning of the climatic event at 4,200 years and continues till today. The Meghalayan Age began with a global drought that devastated the agricultural civilizations from Egypt to China. Dr. Stanley Finney, the Secretary General of the International Union of Geological Science (IUGS), has confirmed it and included it in the revised Geological Time Scale for the Holocene Period (Figure 4.10). Geological Periods are divided into Series (Epoch) that are again divided into Stage/Ages based on the animal and plant forms or species originating in them. The Quaternary Period is divided into two Series (Epochs) – Pleistocene and Holocene. The current age in which we live is called the Holocene Epoch, which reflects everything that has happened over the past 11,700 years. The Holocene Series that covers approximately 11,700 years in Earth's history is coincident with

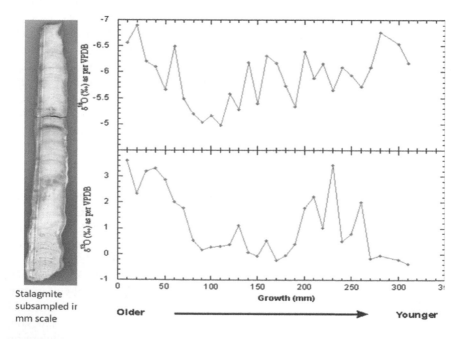

Stalagmite subsampled ir mm scale

FIGURE 4.12 Oxygen isotope variation in the stalagmite from Mastos cave, Meghalaya, NE India.

the late Stone Age. The Holocene Series is divided into three Stages of which the Meghalayan Stage is the Upper (youngest) Holocene. It starts at 4,200 years ago when agricultural societies around the world experienced an abrupt and critical mega-drought that resulted in the collapse of civilizations and prompted human migrations in Egypt, Greece, Syria, Palestine, Mesopotamia, the Indus Valley, and the Yangtze River Valley. The Lower Holocene Stage is known as Greenlandian and the Middle Holocene is designated as Northgrippian. It is shown in the revised Chronostratigraphic Geological Time Scale (Figure 4.10). The Mawmluh cave record covering the 4.2 ka event provides sufficient age control with low errors of ±40–60 years. The Mawmluh cave $\delta^{18}O$ records are supported by robust monitoring studies that significantly increase confidence in the interpretation of the stalagmite $\delta^{18}O$ records from this cave (Figure 4.13). The Meghalayan Stage is unique among the many intervals of the geological time scale. Its beginning coincides with a global cultural event produced by a global climatic event. For the Meghalayan, the spike is epitomized in specific chemical characters, the finest example of which can be seen in the layers of stalagmites on the floors of caves in the Mawmluh area, Meghalaya. A new recent study reveals that a long-lasting drought gradually caused the collapse of the Indus Valley Civilization. The dry spell, lasting 900 years, slowly destroyed its irrigation systems, devastated agriculture, and eventually forced Indus people at places such as Harappa and Mohenjo-Daro to abandon the thriving urban culture. Looking for favorable climate and water resources, the drought-hit people migrated toward the Ganga–Yamuna plains and tried to hang on by resorting to village culture. Clues to such a long drought came from the sediments of the Tso Moriri Lake, located in the Ladakh region of Jammu and Kashmir in India (Dutt et al., 2018). Analysis of the lake's sediments revealed that the ISM began to weaken around

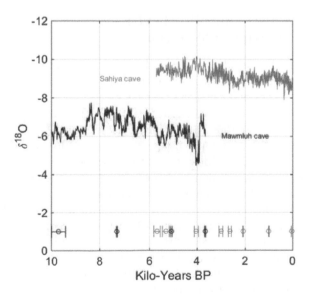

FIGURE 4.13 Stalagmite $\delta^{18}O$ records from the Mawmluh and Sahiya caves plotted against time. Meghalayan Stage corresponds to 4.2 kyr BP.

4,350 years ago. This decreased the moisture transport and the snow deposition in the northwest Himalaya, which, in turn, considerably reduced the water supply in the Indus River and its tributaries. The mysterious displacement of the Indus Civilization is generally attributed to socio-economic and political turmoil as well as climate change in South Asia. A recent study linked its fall to a 200-year-long arid phase that reduced the discharge of the Indus River around 4,200 years ago. The high-resolution stratigraphic boundaries in the Earth's history have to reflect global events of mass extinction, climate change, and extra-terrestrial signatures (Tewari, 2019). Interestingly, the Mawmluh cave in which the stalagmites are found and isotopically dated is the extension of the Lakadong Limestone. The Lakadong Limestone was just deposited after the Cretaceous–Tertiary mass extinction (65 million years ago) and characterized by the Paleocene–Eocene microfossils of foraminifera (Tewari et al., 2010; Tewari, 2019 and references therein). The Cretaceous–Tertiary golden spike is represented by traces in sediments of the element iridium, variation in carbon, oxygen, and mercury isotopes, and extinction of dinosaurs in the Meghalaya across the boundary (Tewari, 2019).

SPELEOTHEM ISOTOPES SYNTHESIS AND ANALYSIS (SISAL)

The SISAL database was created by the SISAL Working Group supported by Past Global Changes (PAGES; http://pastglobalchanges.org/ini/wg/sisal). SISAL aims to compile and synthesize stalagmite $\delta^{18}O$ and $\delta^{13}C$ records to develop a global database that can be used to explore past climate changes and to enable climate model evaluation. The first version of the database, SISAL_v1, is available at Atsawawaranunt et al. (2018), Kaushal et al. (2018), Figure 4.1, modified after Kaushal et al. (2018).

SPELEOTHEMS AND ABRUPT CHANGE IN
THE INDIAN SUMMER MONSOON

The rapid rise in the global temperature has had profound effects on the global climate including the ISM, leading to an increase in the frequency of extreme events globally (Intergovernmental Panel on Climate Change, 2013; Kitoh et al., 2013; Dutt et al., 2015; Tewari, 2014; Kaushal et al., 2018). A rapid rise in global temperature has been recorded after the mid-19th century as a result of natural climate cycle, industrial, other developmental activities, etc. (Mann and Jones, 2003; Mann et al., 2009; Thompson et al. 2020). Some studies from the Indian monsoon region predict strong summer monsoon conditions and more frequent occurrence of extreme events with a rise in the global temperature (Anderson et al., 2002; Goswami et al., 2006; Kitoh et al., 2013; Dutt et al., 2015). Others suggest a weakening of the ISM due to a decrease in land–sea thermal contrast between the Indian Ocean and Indian landmass (Roxy et al., 2015). ISM is a very important phenomenon in the Indian subcontinent and the entire Asian region. Speleothem records suggest a general strong ISM condition during the Medieval Climate Anomaly (MCA) in the Indian sub-continent (Singh et al. 2015). The lifeline of billions of people living in this region like agriculture mainly depends on the ISM. The ISM also controls the climate of the South Asian region. The scientific study of the speleothem stalagmites (cave deposits formed by the calcium carbonate precipitation due to dripping water) is quite significant in dating and establishing the intensity and abrupt change in the monsoonal conditions by measuring the oxygen isotope ratios of the speleothems. In India, there are thousands of caves very well preserved and found very useful in the study of paleoclimate and paleomonsoon. In northeast India, the largest caves are developed in Meghalaya around the town of Cherrapunji (Mawsmai, Mawsynram, Mawmluh, Mastos caves, and so on, Figures 4.1, 4.8, and 4.9) where the heaviest monsoon rain in the world is recorded. The Meghalayan region presently receives most of its annual precipitation (~80%) during the summer monsoon season (June–September) (Parthasarathy, 1960; Murata et al., 2007; Dutt et al., 2015; Figure 4.15). The cave carbonates from Meghalaya, NE India are one of the best and widely used climate proxies to track changes in the ISM strength on decadal to subdecadal time scales (Kaushal et al., 2018; Berkelhammer et al., 2012; Dutt et al., 2015; Myers et al., 2015). Earlier studies from this region reflected wet ISM during the Bølling–Allerød, early Holocene warming, and Marine Isotope Stage (MIS-3), whereas a dry ISM was observed during the Henrich events, LGM, and Younger Dryas cold intervals (Berkelhammer et al., 2012; Dutt et al., 2015).The stalagmites from these caves have given important clues about the intensity of the ISM in the past, an abrupt change in monsoon conditions since prehistoric time and now well dated by the uranium-series technique and oxygen isotope variation (Figure 4.11, 4.12, and 4.14).

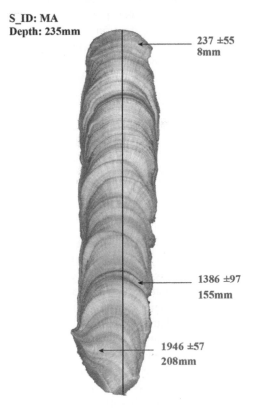

FIGURE 4.14 Stalagmite from Mastos cave (MA), Meghalaya showing three U series dates: $1,946 \pm 57$ (208 mm), $1,386 \pm 97$ (155 mm), and 237 ± 55 (8 mm).

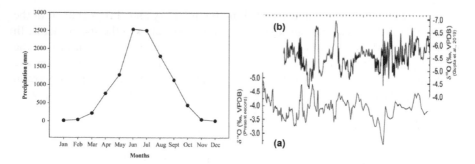

FIGURE 4.15 (a) Precipitation in Meghalaya, Mawmluh area, NE India and (b) $\delta^{18}O$ record from stalagmite dating 1,000 to 2,000 years AD (Dutt et al., 2018).

SIGNIFICANCE OF STABLE ISOTOPES IN SPELEOTHEMS

Changes in climate and environment at the surface are transmitted to the cave through the physical and chemical properties of infiltrated water. The ratio of stable C and O isotopes incorporated in speleothem carbonates mostly reflects soil dynamics (carbon) and atmospheric (oxygen) phenomena. Trace elements are additional environmentally significant proxy data incorporated in speleothems and their variations in concentration are commonly a function of soil pH changes. Aerosols formed by natural or anthropogenic emissions, such as sulfates, are also transported from the surface and incorporated in speleothems. Growth rate, fabrics, and fluid inclusions complement the pool of environmental and climate proxy data from cave deposits. In particular, speleothem fabrics provide information on the deviation from the equilibrium of the depositional system, which plays a role in equilibrium isotope incorporation. Several complex processes influence the incorporation of C and O isotopes in speleothems and, consequently, the variability of $\delta^{13}C$ and $\delta^{18}O$ values of speleothem layers deposited at different times. These processes have been thoroughly discussed by Fairchild et al. (2007).

Oxygen Isotopes, Monsoon, and Paleoclimate
during the Quaternary Period

The variation of $\delta^{18}O$ in stalagmites is generally related to the precipitation amount during the monsoon season. The oxygen isotopic compositions of speleothem calcite from tropical and monsoon locations are primarily controlled by the $\delta^{18}O$ value of precipitation. $\delta^{18}O$ values of regional precipitation and the changes in calcite $\delta^{18}O$ over time primarily reflect changes in the amount of monsoonal precipitation. Cave calcite also contains information about the isotopic composition of meteoric precipitation, is widespread, and can be dated with ^{230}Th. Fleitmann et al. (2003) and Tewari (2009, 2011) have interpreted that in the western part of the ISM in Oman and Yemen, the oxygen isotope ratios of stalagmite calcite primarily show variations in the amount of rainfall with more negative $\delta^{18}O$ indicating higher monsoon rainfall.

Oxygen Isotope Variation in Stalagmite from Sahastradhara Cave, Dehradun, Uttarakhand, NW Himalaya

The high-resolution carbon and oxygen isotopic composition of speleothem (stalagmite) from the Sahastradhara cave, Dehradun, Uttarakhand, Lesser Himalaya, India has been recorded by Tewari and Jaiswal (2014, Figure 4.5). Isotopic measurements from a 9.2 cm high stalagmite from the Sahastradhara cave allowed us the reconstruction of the monsoon precipitation, paleoclimate, and past vegetation. Stalagmites are widely formed in the global continental environment (caves) and are very useful in accurate dating by micro lamination counting and uranium-series dating for exact paleoclimatic interpretations (Figure 4.5). The oxygen and carbon isotope ratio variation in stalagmite can be interpreted to know the intensity of monsoon precipitation, paleoclimate, and past vegetation. The increase in rainfall is indicated by the lighter values of the oxygen isotope ratio and decrease in the carbon isotope ratio in the studied stalagmite as shown in Figure 4.5. A highly depleted $\delta^{18}O$ (−7.83‰ VPDB) in the present study from a 9.2 cm high stalagmite from the Sahastradhara cave has been reported. Less negative $\delta^{18}O$ values (−4. 58‰ V-PDB to −5.14‰ V-PDB) have been recorded in the Himalayan speleothems from the Uttarakhand and the Meghalaya, NE India (Tewari, 2011, 2013, 2015). The carbon isotope $\delta^{13}C$ ratio of the present stalagmite from the Sahastradhara cave shows a much lighter signal (−8.02‰ VPDB). The lighter $\delta^{13}C$ values in the stalagmites generally suggest a wet and cool climate (Figure 4.5).

INDIAN SUMMER MONSOON, SOUTH ASIAN MONSOON, INTERTROPICAL CONVERGENCE ZONE, AND TIBETAN PLATEAU

The $\delta^{18}O$ record from the Mawmluh cave indicates a significant ISM variability in the Meghalaya, NE India during 112 BC–1752 AD (Dutt et al., 2018, 2020; Figure 4.15). Strong ISM conditions were observed between 112 BC and ~440 AD. Speleothem record from the Dongge cave, China also suggests increased monsoonal precipitation during 212 BCE to 400 CE (Wang et al., 2001). This strong monsoon interval has also been reflected as an interval of high precipitation in other cave and lake records from China and India (Wang et al., 2005; Sinha et al., 2015; Dutt et al., 2018), and increased *Globigerina bulloides* abundance in the NW Arabian Sea indicates an enhanced intensity of summer monsoon winds (Gupta et al., 2003). From ~440 to 575 AD, a sudden rise in $\delta^{18}O$ values from −4.66‰ to −3.33‰ was observed indicating an abrupt weakening of the ISM strength, which persisted for more than a century (lasting ~135 years) until ~AD 575. The other cave records of ISM variability also suggest a gradual weakening of the ISM during this time (Wang et al., 2005; Sinha et al., 2015). Therefore, to conclude, the ISM is one of the Earth's most dynamic features interacting with the atmosphere, lithosphere, and hydrosphere. The intensity of ISM has varied greatly during the Quaternary. ISM strength is highly sensitive to the mean latitudinal position of the ITCZ and controls the $\delta^{18}O$ of stalagmite significantly. The climate in eastern and southern Asia is dominated by the marked seasonal cycle with relatively dry conditions in winter and heavy rain in summer (Figure 4.16, below). The South Asian monsoon includes the regional monsoons of India, the Indochina, and the South China Sea, which serve as classical

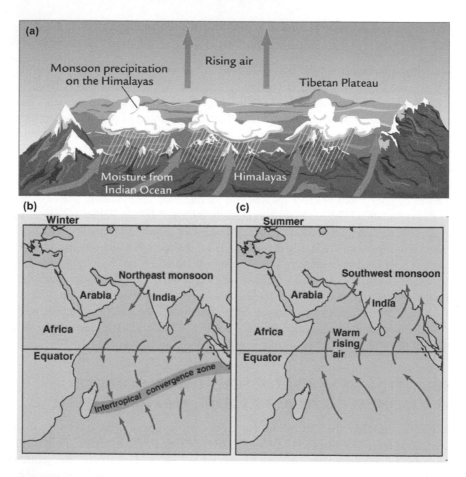

FIGURE 4.16 Relationship between Monsoon precipitation on the Himalaya, moisture from the Indian Ocean and Tibetan Plateau (a) and intertropical convergence zone (b), northeast monsoon during winter and southwest monsoon in summer, (c).

tropical monsoon circulations in which rain falls almost entirely in the ITCZ (Figure 4.16) displaced from the equator (Gadgil, 2003). Tibet may affect the South Asian monsoon more by serving as a barrier to cool, dry air from the north than by serving as a source of heat to the atmosphere above it. The heating of the atmosphere over the Tibetan Plateau has long been held to drive the ISM and strongly influence the broader-scale South Asian summer monsoon (Yanai and Wu, 2006). The relationship among the ISM, Himalaya, and Tibetan Plateau is shown in Figure 4.16 (above) and the ITCZ during northeast monsoon in winter and southwest monsoon in summer (below). Recent work by Molnar et al. (2010) on Asian monsoon suggests that the high Tibetan Plateau may affect the South Asian monsoon less by heating the overlying atmosphere than by simply acting as an obstacle to southward flow of cool, dry air. India–Eurasia collision took place around 50 million years ago (Tewari, 2019 and references therein) and the high rise of the Tibetan Plateau must have taken place immediately after it in the north. The impact of the Tibetan Plateau on northern-hemisphere atmospheric circulation and the close association of Tibet with the South Asian monsoon must have controlled the climate in this region as shown in Figure 4.16 (above). Molnar et al. (2010) have suggested that a rise of Tibet of ~1,000–2,000 m near 8 Mya led to a strengthening of the South Asian monsoon. Abrupt changes in the ISM have also displaced the ancient civilizations historically in the South Asian region (Gadgil, 2003). The weak summer monsoon had effects on the socio-economic conditions in South Asia during the last millennium (Yadava et al., 2016). Most importantly, the newly established Meghalayan Stage begins at 4,200 years ago when agricultural societies around the world experienced an abrupt and critical mega-drought that resulted in the collapse of civilizations and prompted human migrations.

ACKNOWLEDGMENTS

I am grateful to Dr. Niloy Khare, Advisor, Scientist G, Ministry of Earth Sciences, Government of India, New Delhi for inviting me to contribute the chapter on speleothems, monsoon, and quaternary climate. I thank Dr. Hai Cheng, Department of Earth Sciences, University of Minnesota, Minneapolis, USA for kindly providing the data on uranium-series dating of the stalagmites from Sahastradhara, Mawmluh, and Mastos caves from India. Prof. A.N. Sial, NEG-LABISE, University of Pernambacu, Recife, Brazil, is thanked for analyzing the speleothems for C and O isotopes in his laboratory. This is a contribution to the International Project Speleothem Isotopes Synthesis and Analysis (SISAL).

REFERENCES

Anderson, D., Overpeck, J., and Gupta, A.K. 2002. Increase in the Asian Southwest Monsoon during the past four centuries. *Science*, 297, 596–599.

Atsawawaranunt, K., Harrison, S., and Comas Bru, L. 2018. *SISAL (Speleothem Isotopes Synthesis and AnaLysis Working Group) Database Version 1.0*. Available online: http://researchdata.reading.ac.uk/139/ (accessed 4 December 2018).

Band, S., Yadava, M.G., Lone, M.A., Shen, C.-C., Sree, K., and Ramesh, R. 2018. High-resolution mid-Holocene Indian Summer Monsoon recorded in a stalagmite from the Kotumsar Cave, Central India. *Quaternary International*, 479, 19–24. doi: 10.1016/j.quaint.2018.01.026.

Baskar, S., Baskar, R., Mauclaire, L., and McKenzie, J.A. 2006. Microbially induced calcite precipitation in culture experiments: Possible origin for stalactites in Sahastradhara caves, Dehradun, India. *Current Science*, 90, 58–64.

Baskar, S., Baskar, R., Tewari, V.C., Thorseth, I.H., Ovrias, L., Lee, N.M., and Routh, J. 2011. Cave geomicrobiology in India: Status and prospects. In: V.C. Tewari and J. Seckbach (eds.), *Stromatolites: Interaction of Microbes with Sediments, Cellular Origin, Life in Extreme Habitats and Astrobiology*, Vol. 18 Springer Science + Business B.V., Dordrecht, The Netherlands, pp. 541–570. ISBN 978-94-007-0396-4.

Berkelhammer, M., Sinha, A., Stott, L., Cheng, H., Pausata, F.S.R., and Yoshimura, K. 2012. An abrupt shift in the Indian monsoon 4000 years ago. In: L. Giosan, D.Q. Fuller, K. Nicoll, R.K. Flad, and P.D. Clift (eds.), *Climates, Landscapes, and Civilizations: American Geophysical Monograph, 198*, pp. 75–87, American Geophysical Union. doi: 10.1029/2012GM001207.

Borsato, A., Frisia, S., Jones, B., Van der Borg, K., 2000. Calcite moonmilk: Crystal morphology and environment of formation in caves in the Italian Alps. *Journal of Sedimentary Research*, 70, 1173–1182.

Bryanne, V.Z. and Tewari, V.C. 2014. Geotechnical and geological investigations of the Surbhi landslide, Mussoorie Syncline, Uttarakhand Lesser Himalaya. *Journal of the Indian Geological Congress*, 6 (2), 57–77.

Chen, Z., Auler, A.S., Bakalowicz, M., Drew, D., Griger, F., Hartmann, J., Jiang, G., Moosdorf, N., Richts, A., Stevanovic, Z., et al. 2017. The World Karst Aquifer Mapping project: Concept, mapping procedure and map of Europe. *Hydrogeology Journal*, 25, 771–785.

Cheng, H., Edwards, R.L., Shen, C.C., Polyak, V.J., Asmerom, Y., Woodhead, J., 2013. Improvements in ^{230}Th dating, ^{230}Th and ^{234}U half-life values, and U–Th isotopic measurements by multi-collector inductively coupled plasma mass spectrometry. *Earth Planetary Science Letters*, 371–372, 82–91. doi: 10.1016/j.epsl.2013.04.006.

Cheng, H., Sinha, A., Wang, X., Cruz, F.W., Edwards, R.L. 2012. The Global Paleomonsoon as seen through speleothem records from Asia and the Americas. *Climate Dynamics*, 39, 1045–1062. doi: 10.1007/s00382-012-1363-7.

Duan, W., Kotlia, B.S. and Tan, M. 2012. Mineral composition and structure of the stalagmite laminae from Chulerasin cave, Indian Himalaya, and the significance for paleoclimatic reconstruction. *Quaternary International*, 298, 93–97.

Dutt, S., Gupta, A.K., Cheng, H., Clemens, S.C., Singh, R.K., Tewari, V.C. 2020. Indian summer monsoon variability in northeastern India during the past two millennia. *Quaternary International*. doi: 10.1016/j.quaint.2020.10.021.

Dutt, S., Gupta, A.K., Clemens, S.C., Cheng, H., Singh, R.K., Kathayat, G., and Edwards, R.L. 2015. Abrupt changes in Indian summer monsoon strength during 33,800 to 5500 years B.P. *Geophysical Research Letters*, 42, 5526–5532. doi: 10.1002/2015GL064015.

Dutt, S., Gupta, A.K., Wünnemann, B., and Yan, D. 2018. A long arid interlude in the Indian summer monsoon during ~4,350 to 3,450 cal. yr BP contemporaneous to displacement of the Indus valley civilization. *Quaternary International*, doi: 10.1016/j.quaint.2018.04.005.

Fairchild, I.J., Frisia, S., Borsato, A., Tooth, A. 2007. Speleothems. In: D.J. Nash and S.J. McLaren (eds.), *Geochemical Sediments and Landscapes*. Blackwell Publishing, Oxford, 200–245.

Fleitmann, D., Burns, S.J., Mangini, A., Mudelsee, M., Kramers, J., Villa, I., and Matter, A. 2007. Holocene ITCZ and Indian monsoon dynamics recorded in stalagmites from Oman and Yemen (Socotra). *Quaternary Science Reviews*, 26(1), 170–188.

Fleitmann, D., Burns, S.J., Mudelsee, M., Neff, U., Kramers, J., Mangini, A., and Matter, A. 2003. Holocene forcing of the Indian monsoon recorded in a stalagmite from southern Oman. *Science*, 300, 1737–1739, doi: 10.1126/science.1083130.

Gadgil, S. 2003. The Indian monsoon and its variability. *Annual Review of Earth and Planetary Sciences*, 31, 429–467.

Goswami, B.N., Venugopal, V., Sengupta, D., Madhusoodanan, M.S., and Xavier, P.K. 2006. Increasing trend of extreme rain events over India in a warming environment. *Science*, 314, 1442–1445. doi: 10.1126/science.1132027.

Gupta, A.K., Anderson, D.M., and Overpeck, J.T. 2003. Abrupt changes in the Asian southwest monsoon during the Holocene and their links to the North Atlantic Ocean. *Nature*, 421, 354–357. doi: 10.1038/nature01340.

Intergovernmental Panel on Climate Change. 2013. Climate change 2013: The physical science basis. In: *Contribution of Working Group I to the Fifth Assessment Report of the Intergovernmental Panel on Climate Change*, Stocker, T.F. et al. (eds.), Cambridge University Press, Cambridge, UK, pp. 35–118.

Jaiswal, J. 2015. Indian monsoon and ITCZ dynamics recorded in stalagmites from Chakrata, NW Himalaya. In: *30th Himalaya-Karakoram-Tibet Workshop*, Wadia Institute of Himalayan Geology, Dehradun, India, October, 6–8, (poster, abstract volume), pp. 201–202.

Jaiswal, J. and Tewari, V.C. 2012. Study of the Sahastradhara Caves, Dehradun with Special Reference to Paleoclimate. *Technical Report*, WIHG, Dehradun, India, 22p.

Johnson, K.R., Hu, C., Belshaw, N.S., and Henderson, G.M. 2006. Seasonal trace-element and stable-isotope variations in a Chinese speleothem: The potential for high-resolution paleomonsoon reconstruction. *Earth and Planetary Science Letters*, 244, 394–407.

Kathayat, G., Cheng, H., Sinha, A., Berkelhammer, M., Zhang, H., Duan, P., Li, H., Li, X., Ning, Y., and Edwards, R.L. 2018. Evaluating the timing and structure of the 4.2ka event in the Indian summer monsoon domain from an annually resolved speleothem record from Northeast India. *Climate of the Past*, 14, 1869–1879.

Kathayat, G., Cheng, H., Sinha, A., Yi, L., Li, X., Zhang, H., Li, H., Ning, Y., and Edwards, R.L. 2017. The Indian monsoon variability and civilization changes in the Indian subcontinent. *Science Advances*, 3, e1701296. doi: 10.1126/sciadv.1701296.

Kaufmann, G. 2003. Stalagmite growth and palaeo-climate: A numerical perspective. *Earth and Planetary Science Letters*, 214, 251–266.

Kaushal, N., Breitenbach, S.F.M., Lechleitner, F.A., Sinha, A., Tewari, V.C., Ahmad, S.M., Berkelhammer, M., Band, S., Yadava, M., Ramesh, R., Hende, G. 2018. The Indian summer monsoon from a speleothem $\delta^{18}O$ perspective – A review. *Quaternary*, 2018, 1, 1–29. doi: 10.3390/quat 1030029.

Kitoh, A., Endo, H., Kumar, K.K., Cavalcanti, I.F.A., Goswami, P., and Zhou, T. 2013. Monsoons in a changing world: A regional perspective in a global context. *Journal of Geophysical Research: Atmospheres*, 118, 3053–3065. doi: 10.1002/jgrd.50258.

Kotlia, B.S., Singh, A.K., Joshi, L.M., and Dhaila, B.S. 2015. Precipitation variability in the Indian Central Himalaya during last ca. 4,000 years inferred from a speleothem record: Impact of Indian Summer Monsoon (ISM) and Westerlies. *Quaternary International*, 371, 244–253 doi: 10.1016/j.quaint.2014.10.066.

Kotlia, B.S., Singh, A.K., Zhao, J.-X., Duan, W., Tan, M., Sharma, A.K., and Raza, W. 2017. Stalagmite based high resolution precipitation variability for past four centuries in the Indian Central Himalaya: Chulerasim cave re-visited and data re-interpretation. *Quaternary International*, 444, 35–43, doi: 10.1016/j.quaint.2016.04.007.

Liang, F., Brook, G.A., Kotlia, B.S., Railsback, L.B., Hardt, B., Cheng, H., Edwards, R.L., and Kandasamy, S. 2015. Panigarh cave stalagmite evidence of climate change in the Indian Central Himalaya since AD 1256: Monsoon breaks and winter southern jet depressions. *Quaternary Science Reviews*, 124, 145–161, doi: 10.1016/j.quascirev.2015.07.017.

Lisiecki, L.E. and Raymo, M.E. 2005. A Pliocene-Pleistocene stack of 57 globally distributed benthic δ18O records. *Paleoceanography*, 20(1), PA1003.

Mann, M.E. and Jones, P.D., 2003. Global surface temperatures over the past two millennia, *Geophysical Research Letters*, 30(15), 1820. doi: 10.1029/2003GL017814.

Mann, M.E., Zhang, Z., Rutherford, S., Bradley, R.S., Hughes, M.K., Shindell, D., Ammann, C., Faluvegi, G., and Ni, F. 2009. Global signatures and dynamical origins of the Little Ice Age and Medieval Climate Anomaly. *Science*, 326, 1256–1260. doi: 10.1126/science.1177303.

McDermott, F. 2004. Palaeo-climate reconstruction from stable isotope variations in speleothems: A review. *Quaternary Science Reviews*, 23(7–8), 901–918.

Molnar, P., Boos, W.R., and Battisti, D.S. 2010. Orographic controls on climate and paleoclimate of Asia: Thermal and mechanical roles for the Tibetan Plateau. *Annual Review of Earth and Planetary Sciences*, 38, 77–102.

Mooley, D.A., Parthasarathy, B., Sontakke, N.A., and Munot, A.A. 1981. Annual rain-water over India, its variability and impact on the economy. *Journal of Climatology*, 1(2), 167–186.

Murata, F., Hayashi, T., Matsumoto, J., and Asada, H. 2007. Rainfall on the Meghalaya plateau in north eastern India-one of the rainiest places in the world. *Natural Hazards*, 42, 391–399. doi: 10.1007/s11069-006-9084-z.

Myers, C.G., Oster, J.L., Sharp, W.D., Bennartz, R., Kelley, N.P., Covey, A.K., and Breitenbach, S.F.M. 2015. Northeast Indian stalagmite records Pacific decadal climate change: Implications for moisture transport and drought in India. *Geophysical Research Letters*, 42, 4124–4132. doi: 10.1002/2015GL063826.

Parthasarathy, K. 1960. Some aspects of rainfall in India during the Southwest monsoon season. In: *Proceedings of the Symposium on Monsoon of the World*, India Meteorological Department, New Delhi, India, pp. 185–194.

Roxy, M.K., Ritika, K., Terray, P., Murtugudde, R., Ashok, K., and Goswami, B.N. 2015. Drying of Indian subcontinent by rapid Indian Ocean warming and a weakening land-sea thermal gradient. *Nature Communications*, 6(7423). doi: 10.1038/ncomms8423.

Sanwal, J., Kotlia, B.S., Rajendran, C., Ahmad, S.M., Rajendran, K., Sandiford, M. 2013. Climatic variability in Central Indian Himalaya during the last ~1800 years: Evidence from a high resolution speleothem record. *Quaternary International*, 304, 183–192. doi: 10.1016/j.quaint.2013.03.029.

Singh, D.S., Gupta, A.K., Sangode, S.J., Clemens, S.C., Prakasam, M., Srivastava, P., and Prajapati, S.K. 2015. Multiproxy record of monsoon variability from the Ganga Plain during 400–1200 AD. *Quaternary International*, 371, 157–163. doi: 10.1016/j.quaint.2015.02.040.

Sinha, A., Cannariato, K.G., Stott, L.D., Cheng, H., Edwards, R.L., Yadava, M.G., Ramesh, R., Singh, I.B. 2007. A 900-year (600 to 1500 A.D.) record of the Indian summer monsoon precipitation from the core monsoon zone of India. *Geophysical Research Letters*, 34, L16707. doi: 10.1029/2007GL030431.

Sinha, A., Cannariato, K.G., Stott, L.D., Li, H.-C., You, C.-F., Cheng, H., Edwards, R.L., Singh, I.B. 2005. Variability of Southwest Indian summer monsoon precipitation during the Bølling-Ållerød. *Geology*, 33, 813–816, doi: 10.1130/G21498.1.

Sinha, A., Kathayat, G., Cheng, H., Breitenbach, S.F.M., Berkelhammer, M., Mudelsee, M., Biswas, J., and Edwards, R.L. 2015. Trends and oscillations in the Indian summer monsoon rainfall over the last two millennia. *Nature Communications*, 6, 6309. doi: 10.1038/ncomms7309.

Sinha, N., Gandhi, N., Chakraborty, S., Krishnan, R., Ramesh, R., Yadava, M., Datye, A. 2016. Past rainfall reconstruction using speleothem from Nakarallu cave, Kadapa, Andhra Pradesh, India. In: *Proceedings of the EGU General Assembly*, Vienna, Austria, April 17–22, European Geosciences Union, EPSC2016-811.

Tewari, V.C. 2008. Speleothems from the Himalaya and the Monsoon: A Preliminary study, *Earth Science India*, 1(IV), 231–242.

Tewari, V.C. 2009. Speleothems from the Himalaya, paleoclimate and monsoon, In: S.K. Tandon and A.R. Bhattacaharyya (eds.), *Advances in Earth Science*, Satish Serial Publishing House, New Delhi, India, pp. 91–99.

Tewari, V.C. 2011. Speleothems from Uttarakhand and Meghalaya indicating Holocene Monsoon and Climate. *Journal of the Indian Geological Congress*, 3(1) 87–104.

Tewari, V.C. 2012. Neoproterozoic Blaini glacial diamictite and Ediacaran Krol carbonate sedimentation in the Lesser Himalaya, India. *Geological Society of London, Special Publication*, 366, 265–276. doi: 10.1144/ SP 366.6.

Tewari, V.C. 2013a. Recent devastating natural disaster in the Uttarakhand Himalaya: Causes and sustainable development. In: *National Workshop on Geology and Geoheritage sites of Uttarakhand Himalaya* with special reference to Geoscientific Development of the region, Indian Geological Congress, Uttarakhand, India, November 11–12, pp. 51–53.

Tewari, V.C. 2013b. Speleothems as paleoclimate and paleomonsoon indicator: Evidences from NW Himalaya and the Shillong Plateau, NE India. In: *Proceedings of Selected Topics in Earth System Sciences*, P.K. Verma (ed.) Madhya Pradesh State Council for Science and Technology, (MPCST), Bhopal, MP, pp. 31–39.

Tewari, V.C. 2014. Recent natural disaster in the Uttarakhand Himalaya and future geotechnical remedial measures. *Journal of the Geological Society of India*, 84 (1), 125–126.

Tewari, V.C. 2015. Himalayan Speleothems as proxy for the past climate change and paleomonsoon. *Frontiers of Earth Science*, 2015, 243–250.

Tewari, V.C. 2019. Global cretaceous-tertiary boundary and mass extinction, Deccan volcanism, asteroid impact and paleoclimate change in the northern and southern hemispheres with special reference to the Himalaya. *Journal of the Indian Geological Congress*, 11(1), 23–51.

Tewari, V.C. and Jaiswal, J. 2014. First high-resolution oxygen and carbon isotope variation and paleoclimate change recorded in a Stalagmite from Sahastradhara Cave, Uttarakhand Lesser Himalaya, India. *Journal of the Indian Geological Congress*, 97–98.

Tewari, V.C., Breitenbach, S.F.M., Kausal, N., and Lechleitner, F.A. 2020. The Indian Summer Monsoon from a speleothem $\delta^{18}O$ perspective – A Review of Northeast Indian speleothem Records (abstract). In: *International Conference on Recent Developments in Earth and Environmental Sciences, Natural Resource Management and Climate Change with Special Focus on Eastern Himalayas*, Department of Geology, Sikkim University, Gangtok, India, October 26–27.

Tewari, V.C., Kumar, K., Lokho, K. and Siva, S.N. 2010. Lakadong limestone: Paleocene-Eocene boundary carbonate sedimentation in Meghalaya, northeastern India. *Current Science*, 98(1), 88–94.

Thompson, L.G., Yao, T., Mosley-Thompson, E., Davis, M.E., Henderson, K.A., and Lin, P.N., 2000. A high-resolution millennial record of the South Asian monsoon from Himalayan ice cores. *Science*, 289, 1916–1919. doi: 10.1126/science.289.5486.1916.

Venkateswarlu, B. and Tewari, V.C. 2014. Characterization of strength and durability of highly fractured Krol Limestone near Surabhi Landslide, Mussoorie – Kempty link road. In: *Proceedings Volume of National Seminar on Innovative Practices in Rock Mechanics*, National Institute of Rock Mechanics (NIRM), Bengaluru, India, pp. 329–336.

Wang, Y., Cheng, H., Edwards, R.L., He, Y., Kong, X., An, Z., Wu, J., Kelly, M.J., Dykoski, C.A., and Li, X. 2005. The Holocene Asian monsoon: Links to solar changes and North Atlantic climate. *Science*, 308, 854–857. doi: 10.1126/science.1106296.

Wang, Y.J., Cheng, H., Edwards, R.L., An, Z.S., Wu, J.Y., Shen, C.-C., Dorale, J.A. 2001. A high-resolution absolute-dated Late Pleistocene Monsoon record from Hulu Cave, China. *Science*, 294, 2345–2348. doi: 10.1126/science.1064618.

Yadava, A.K., Bräuning, A., Singh, J., and Yadav, R.R. 2016. Boreal spring precipitation variability in the cold arid western Himalaya during the last millennium, regional linkages, and socio-economic implications. *Quaternary Science Reviews*, 144, 28–43. doi: 10.1016/j.quascirev.2016.05.008.

Yadava, M.G. and Ramesh, R. 2005. Monsoon reconstruction from radiocarbon dated tropical Indian speleothems. *The Holocene*, 15 (1), 48–59.

Yadava, M.G., Ramesh, R., and Pant, G.B. 2004. Past monsoon rainfall variations in peninsular India recorded in a 331 year old speleothem. *Holocene*, 14, 517–524.

Yanai, M. and Wu, G.X. 2006. Effects of the Tibetan plateau. In: B. Wang (ed.), *The Asian Monsoon*, Springer, Berlin, pp. 513–549.

5 A 45 ka Record of Productivity in the Western Bay of Bengal

Implications on the Indian Monsoon and Atlantic Climate

Ishfaq Ahmad Mir
Geological Survey of India

CONTENTS

INTRODUCTION

Ocean productivity involves the uptake of dissolved inorganic carbon and its sequestration into organic compounds by marine primary producers and plays a major role in controlling the partitioning of carbon between the ocean and the atmosphere. Productivity fluctuations can therefore influence the climate by altering the atmospheric concentrations of the greenhouse gas, carbon dioxide (Berger et al., 1989). Ocean paleoproductivity studies document past changes in the biological production of organic matter and skeletal materials. These studies provide an insight into the causes of such fluctuations, the consequences for biogeochemical cycles within the ocean, and their correspondence to Earth's climate. Changes in ocean productivity may partly account for atmospheric CO_2 changes on glacial–interglacial time scales (Sigman and Boyle, 2000). There are a number of studies on glacial–interglacial changes in primary productivity in various oceanic basins of the world. These studies have shown that higher productivity occurred in many low latitude regions during the last glacial maximum (LGM). The Pleistocene records of sediments of the continental slope have shown large fluctuations in the burial rates of organic carbon, which are interpreted as productivity fluctuations. Such fluctuations appear to be in phase with changes in the atmospheric carbon dioxide content (Emerson and Hedges, 1988).

The Bay of Bengal (BoB) is the largest bay in the world. Ganga–Brahmaputra (G–B), Irrawaddy, Godavari, Mahanadi, Krishna, and Kaveri rivers constitute 60% of the total fresh water received; the flux during the southwest monsoon (SWM) coincides with discharge maxima of the G–B rivers (Ittekkot et al., 1991). These discharge supplies nutrients that may enhance primary productivity (Ramaswamy and Nair, 1994), cause strong stratification during the SWM (Ostlund et al., 1980), and also reduce the surface water salinity by 7%. Enormous riverine flux ($2.95 \times 10^{12} m^3/yr$) and excess precipitation over evaporation result in a stable water column (50–80 m), salinity stratification in the BoB, in contrast to other Indian Oceanic regions. The burial of carbon in sediments should vary in response to variation in monsoon winds and rainfall. It is known that the SWM was weak during the LGM but stronger than the present between 6 and 12 ka BP (COHMAP Members, 1988). However, northeast monsoon (NEM) in the northern Indian Ocean becomes more intense during glacial stages and the resulting strong wind could increase the productivity (Duplessy, 1982; Fontugne and Duplessy, 1986). Reconstructions of paleo-productivity in the equatorial Indian Ocean for the last 260,000 years indicate an increase during cold periods (Cayre et al., 1999). In the northern Arabian Sea, the intense cooling of the sea surface during winter is shown to induce convective mixing of the nutrient-enriched thermocline waters into the surface waters resulting in increased productivity (Madhupratap et al., 1996). Observations from northern and western regions of the Arabian Sea indicate that the productivity was higher during the interglacial than the glacial (Naidu and Malmgren, 1995).

A variety of proxies such as the content of $CaCO_3$, abundances of planktonic–benthic foraminifera, organic carbon, total nitrogen, biogenic opal, biogenic barium, and alkenones have been used for estimating marine paleoproductivity by many investigators from various oceanic basins of the world (Pattan et al., 2013; Miret al., 2013). The post-depositional environmental conditions and the process involved within marine sediments can be reconstructed by using multiple proxies such as bioturbation index, sediment lamination, trace and ultra-trace metal concentration, and stable isotope abundances. In the western Bay of Bengal (WBoB), past variations in the input of various sedimentary components and paleoproductivity and nutrient loading studies are not well recorded. In the present study, we report geochemical proxies to understand paleoproductivity and nutrient loading during the past 45 ka.

MATERIALS AND METHODS

SEDIMENT CORE AND CHRONOLOGY

In the present study, a sediment core SK-218/1 collected during 218th cruise of ocean research vessel Sagar Kanya was retrieved from a water depth of 3,307 m (latitude: 14°02′01″N; longitude: 82°00′12″E) off the Krishna–Godavari basin, WBoB (Figure 5.1a). The core site is well below the present-day oxygen minimum zone in the water column (Wyrtki, 1971; Olson et al., 1993; Rao et al., 1994) of BoB. Chronology has been established on eight ^{14}C AMS dates, covering the past 45 ka. The detailed age model for the core is shown in (Figure 5.1b) (Govil and Naidu, 2011).

ELEMENT MEASUREMENT

In this study, a total of 100 sub-samples were analyzed. Samples were thoroughly washed with ultrapure distilled water to make them salt free, dried in an oven, and later powdered. About 50 mg of the sample powder was weighed in a Teflon beaker for geochemical analysis. Ten milliliters of an acid mixture of $HF+HNO_3+HClO_4$ (7:3:1 ratio) was added to the sample powder and the contents were concentrated to a paste by placing the beaker on a hot plate. To this paste, 4 mL of 1:1 HNO_3 was added. After heating for 5 minutes, the material was diluted with ultrapure (18.2 MΩ) water to a final volume of 100 mL. In the same digestion procedure, standard reference materials (MAG-1 and SGR-1) and blanks were also prepared. These solutions were analyzed for a few major, trace, and rare earth elements on an inductively coupled plasma-mass spectrometer (ICP-MS, Thermo X Series 2) at the National Institute of Oceanography (NIO), Goa, India. The accuracy compared to the standard reference material and the precision of the data based on duplicate analysis were better than ±6%.

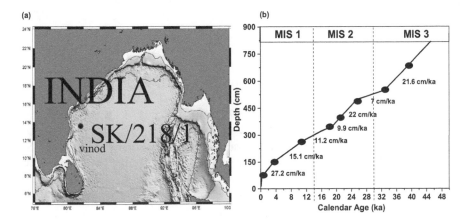

FIGURE 5.1 Map showing the location of sediment core (SK/218/1) in the western Bay of Bengal (a) and plot of depth (cm) versus calendar age (ka) for the sediment core (SK/218/1) derived from Naidu and Govil (2010) (b) and Marine Isotope Stages (MIS 1–3), after Lisiecki and Raymo (2005).

ORGANIC CARBON MEASUREMENT

Aliquots of sediment sub-samples were made carbonate free by treating with 0.1 N HCl. For the analyses of total organic carbon (TOC), carbonate-free sediments (residue) were weighed in tin cups and combusted in a Euro-Vector Elemental Analyzer (EA) coupled with a Delta-V plus stable isotope mass spectrometer (Thermo®) in a continuous flow mode at NIO, Goa, India. Working standard ε-amino-n-caproic acid (ACA) (SMAST, UMASS) and internal sedimentary standard COD were used to check the precision of the measurements (Pattan et al., 2013).

Elemental excess (E_{exc}) or structurally unsupported element content was calculated following the equation of Murray et al. (1993).

$$E_{exc} = E_{total} - \left(Ti_{sample} * E/Al_{PAAS} \right)$$

where PAAS is the Post Archean Australian Shale.

Dry bulk density (ρ_{dry}) was empirically calculated from the $CaCO_3$ content by using the equation of Curry and Lohmann (1986).

$$\rho_{dry} = 0.0066(\%CaCO_3) + 0.29$$

The mass accumulation rate (MAR) of the element or elemental flux (E_{flux}) was estimated by

$$E_{flux} = (E/100) * \rho_{dry} * LSR,$$

where E_{flux} is given in g/cm²/ka, ρ_{dry} is given in g/cm³, and the linear sedimentation rate (LSR) is given in cm/ka.

RESULTS

The barium excess (Ba_{exc}), flux of Ba_{exc}, and organic carbon used as proxies for productivity variability vary from 0.57 to 18.55 with an average of 7.38 (ppm), 0.04–1.01 with an average of 0.44 (g/cm²/ka), and 0.01–0.19 with an average of 0.07 (g/cm²/ka) respectively (Table 5.1). Nutrient proxies of P, Fe, Co, Cu, Ba/Ca, and Zn/Ca vary from 0.03 to 0.08 with an average of 0.05 (wt.%), 1.70–7.38 with an average of 5.28 (wt.%), 9.72–44.19 with an average of 25.07 (ppm), 19.40–119.10 with an average of 74.05 (ppm), 15–608 with an average ratio of 246.15, and 1–186 with an average ratio of 56.66 respectively (Table 5.1). Higher values are observed during MIS 1, MIS 3, and interstadials, whereas lower values are observed during YD, LGM, MIS2, and H events.

DISCUSSION

Organic carbon is a dominant biological component and its measurements in sediments seem to be a logical choice for reconstructing productivity. Sediment trap studies indicate that a general relation between the flux of organic matter to the deep

TABLE 5.1

Descriptive Statistics of Geochemical Data Used in the Present Study as a Proxy for Reconstruction of Paleoproductivity and Nutrient Flux during the Past 45 ka in the WBOB

	Ba_{exc}	Ba_{exc} flux	C_{org} flux	P (%)	Fe (%)	Co (ppm)	Cu (ppm)	Ba/Ca	Zn/Ca
Mean	7.38	0.44	0.07	0.05	5.38	25.07	74.05	246.16	56.66
St. error	0.27	0.02	0.00	0.00	0.09	0.54	1.98	14.03	4.57
Median	7.28	0.37	0.06	0.06	5.33	24.64	73.31	224.00	37.00
Mode	6.02	0.21	0.06	0.06	5.63	27.27	112.20	298.00	34.00
St. Dev.	2.72	0.25	0.04	0.01	0.93	5.40	19.78	140.28	45.67
Kurtosis	2.08	−0.63	0.32	0.65	2.42	1.42	0.15	−0.12	0.25
Skewness	0.90	0.67	0.73	−0.58	−0.74	0.54	0.05	0.77	1.14
Range	17.98	0.97	0.18	0.04	5.68	34.47	99.70	593.00	185.00
Minimum	0.57	0.04	0.01	0.03	1.70	9.72	19.40	15.00	1.00
Maximum	18.55	1.01	0.19	0.08	7.38	44.19	119.10	608.00	186.00

sea and productivity exists. Only a small fraction of the organic matter that arrives to the sediment is ultimately buried. Equations that quantitatively relate the accumulation rate of organic carbon in sediments to overlying productivity (Muller and Suess, 1979) have been suggested. The relationship between organic matter content and productivity has been applied to reconstruct past oceanic productivity in various oceanic basins (Pedersen, 1983; Finney et al., 1988).

Barite ($BaSO_4$) precipitation in the water column is associated with decaying organic matter (Ganeshram et al., 2003). This results in a positive correlation between barite and Ba excess (Ba_{exc}); the Ba fraction is not carried by a terrigenous material, which is not supported by the aluminosilicate structure. Ba_{exc} has been used in many studies, to infer paleoproductivity (Bains et al., 2000). It is determined from the total Ba concentration in the sediment after subtracting the Ba associated with the terrigenous material, which is calculated from the total Al or Ti, and normalization to a constant detrital Ba/Al or Ba/Ti ratio. Barite is a promising paleoproductivity proxy because it is a highly refractory mineral with preservation as high as 30% in oxic sediments (Dymond et al., 1992). Nutrient availability controls the productivity of the surface ocean. Nutrients control oceanic primary productivity, which is believed to be an important control on atmospheric CO_2 on glacial–interglacial timescales. It has been suggested that the concentration of nutrients in the marine sediments provides indirect information about paleoproductivity. Reconstruction of the paleonutrients relies mainly on Ca, Fe, P (as macronutrients), and Cu, Co, Ba, Zn, and their ratios (as micronutrients). In areas where the nutrient pool in the surface ocean is not fully utilized, enrichment of nutrients is recorded by the plankton. This in turn leads to the corresponding changes in the composition of particles that sink in the water column and accumulate in the sediment (Montoya, 1994). Phosphorus (P) is an important nutrient to all living organisms. Marine phytoplankton incorporates P at a relatively constant ratio to their C content. Reconstructing past P accumulation may provide an indirect estimate of C flux and thus productivity (Delaney, 1998).

Iron (Fe) is an essential nutrient for the synthesis of chlorophyll and various algal proteins and is the limiting nutrient in some regions of the ocean such as southern ocean. Divalent trace metals: barium (Ba) and zinc (Zn) have been developed as paleonutrient tracers. Zn is an essential micronutrient for many marine organisms, second only to iron among the biologically important trace metals. Zn/Ca is a very sensitive tracer of past productivity. Ba is incorporated into several taxa of foraminifera, possibly associated with increased productivity. Ba/Ca is also a good tracer of past productivity. Copper (Cu) and cobalt (Co) are bio-limiting elements, are necessary to sustain life, have biological functions, and may exist in low concentrations.

PRODUCTIVITY VARIATION IN WBoB DURING THE LAST GLACIAL AND PRESENT INTERGLACIAL PERIODS

In the present study, the variation of Ba_{exc} (ppm), fluxes (g/cm²/ka) of Ba_{exc} and C_{org}, P (%), Fe (%), Co (ppm), Cu (ppm), and ratios of Ba/Ca and Zn/Ca during the last 45 ka are nearly similar (Figures 5.2–5.4). Highest values were recorded during MIS 1 and MIS 3. Productivity and nutrient flux were highest during the present interglacial than the last glacial period. During LGM, lowest values are noticed, whereas in early MIS 3 and later Holocene, higher values are observed. The physical forcing of SWM was stronger in the interglacial period than it was during the glacial period. There seems to be a clear link of increased productivity with SWM strength. During interglacial periods, intense monsoon precipitation was able to bring more nutrients and fresh water from the nearby landmass, hence increasing productivity. This observation is similar to several previous reports on paleoproductivity from the BoB

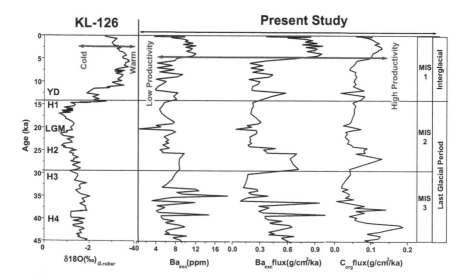

FIGURE 5.2 Correlation of marine $\delta^{18}O_{G.\ ruber}$ (‰) from the northern Bay of Bengal sediment core (KL-126, Kudrass et al., 2001) with productivity variability of the present study (Ba excess, flux of Ba excess, and flux of organic carbon) during the past 45 ka in the western Bay of Bengal. YD is Younger Dryas cooling and LGM is the last glacial maximum.

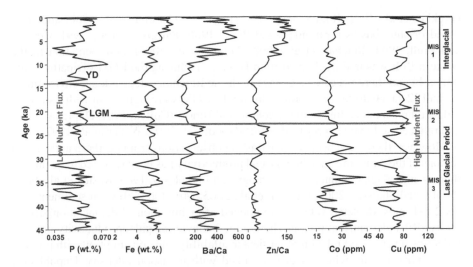

FIGURE 5.3 Reconstruction of nutrient variations during the past 45 ka in the western Bay of Bengal using geochemical nutrient proxies, down the core variability matches well with the paleoproductivity variability. YD is Younger Dryas cooling and LGM is the last glacial maximum.

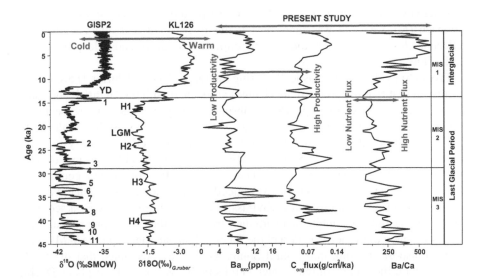

FIGURE 5.4 Paleoproductivity and paleonutrient variability records of the sediment core SK-218 (present study) in comparison with the paleomonsoon record of marine $\delta^{18}O$ values of *Globigerinoides ruber* in the northern Bay of Bengal (126KL, Kudrass et al., 2001) and Greenland ice core $\delta^{18}O$ record (GISP2, Dansgaard et al., 1993). Numbers (1–11) indicate Greenland interstadials; YD is Younger Dryas cooling; H1 to H4 are Heinrich Events and LGM is the last glacial maximum.

(Ittekkot et al., 1991; Ramaswamy and Nair, 1994; Ostlund et al., 1980) and from world oceanic basins (Shimmield and Price, 1986; Paropkari et al., 1991; Howard and Prell, 1994; Reichart et al., 1998; Pattan et al., 2003). Sedimentation in the western BoB was mainly derived from Himalayas during the LGM (Tripathy et al., 2011a, b) and there occurred a change in the sedimentary depositional environment around 12 ka. Hydrological changes during the LGM were significant with a strongly reduced monsoon leading to lower continental runoff (Cullen, 1981; Duplessy, 1982) and reduced sediment supply to the BoB (Goodbred, 2003; Weber et al., 1997).

However, there have been few contrasting observations from northern and eastern regions of the Arabian Sea, where glacial productivity has been reported to be intense than its interglacial counterparts (Fontugne and Duplessy, 1986; Rostek et al., 1997; Prabhu and Shankar, 2005; Banakar et al., 2005; Mir et al., 2013). In sediment core SK-218/1, most depleted values were noticed after initiation of deglaciation around 20 kyr BP. Paleonutrient proxies show same behavior as paleoproductivity proxies, reflecting a very good relation of nutrient availability and consumption with productivity enhancement, which in turn is linked with monsoon intensity. Unpublished data of Mir (2015) show higher terrigenous input to the core location during the interglacial period, which was able to bring more nutrients with discharge of fresh water from the nearby continents by intensification of SWM and increasing surface water productivity.

PRODUCTIVITY VARIATIONS IN THE WBoB DURING STADIALS AND INTERSTADIALS

Productivity was lower during the last glacial period, Younger Dryas (YD), and Heinrich (H) events; it was more during the interglacial and interstadials. This reflects a close link between the Atlantic climate variability and paleoproductivity in the BoB.

Productivity decreased drastically during the LGM, YD, and during H events; our results suggest that the productivity changes associated with the Heinrich and interstadial events, representing episodes of abrupt cooling and warming, respectively, in the North Atlantic extend to the BoB as well. Many studies, Suthof et al. (2001) and Altabet et al. (2002), demonstrate how well the global climate sub-systems are interconnected, and so it is to be expected that the climatic oscillations in the North Atlantic would also affect BoB biogeochemistry. The effect of Atlantic cooling and warming during the past 45 ka is clearly noticed in the excursions of productivity and nutrient proxies used in this study. Kudrass et al. (2001) after studying a sediment core from the BoB in combination with data from the other Asian monsoonal areas suggested changes in snow and dust of the Tibetan Plateau affecting the SWM capacity to transport moisture into central South Asia. They postulated that the SWM initiates, amplifies, and terminates interstadials in the Northern Hemisphere. During warmer interstadials, enrichments in nutrients reflect higher productivity, whereas during colder YD and H events, depletion in values of nutrients reflects decreased productivity. Data from Mir (2015) show higher terrigenous input to the core location during warmer periods, suggested higher terrigenous input to the core location, was able to bring more nutrients with discharge of fresh water from the nearby continents by intensification of SWM, and increased surface water productivity during interglacial and interstadials. Galy et al. (2007) showed that the amount of terrestrial organic carbon buried in the oceanic sediments and its removal in the sea floor are mostly through continentally derived materials.

High precipitation during the SWM and fresh water discharge from the river system of nearby landmass form a salinity gradient in the BoB, ranging from 26 in the north to 34 psu in the south (Levitus and Boyer, 1994). Govil and Naidu (2011) studied the same sediment core and reported that the sea surface temperature (SST) in the BoB was 3.2°C cooler during the LGM than the present, a ~3.5°C rise in SST is documented from 17 to 10 ka. The SST exhibits greater amplitude fluctuations during the MIS 2, which is attributable to the variability of NE monsoon rainfall and associated river discharge into the BoB in association with a strong seasonal temperature contrast. The onset of the strengthening phase of SWM started during Bølling–Allerød as evidenced by the low $\delta^{18}O_{sw}$ values of ~14.7 ka. $\delta^{18}O_{sw}$ values are consistently lower during Holocene with an exception during 5 ka, suggesting the freshening of BoB due to heavy precipitation and river discharge caused by strong SWM. In WBoB sea surface salinity (SSS) during MIS1 to MIS3 varied from 32.5 to 36.5 psu with an average of 34.5 psu. SSS was the highest during colder periods of YD, MIS2, LGM, and Heinrich events (H1 & H2) than Holocene, MIS1, MIS3, and during interstadials and matches very well with the variability of SWM rainfall during MIS 1 to MIS 3.

CONCLUSION

From the present study, the following conclusions can be drawn:

1. During warm periods, productivity was higher compared to colder periods particularly during the present interglacial (MIS1) in the WBoB. This could be due to enhanced SWM intensity, which was able to bring enormous fresh water and nutrients from the nearby landmass.
2. During interglacial and interstadials, intensity of SWM brought more nutrients from **landmass** and at the same time, it formed a strong stratified layer and prevented upwelling of nutrient-rich bottom waters. In such situation, land-derived nutrients increase productivity compared to upwelled bottom water nutrients.
3. Signatures of Younger Dryas, Heinrich events, and interstadials and glacial–interglacial variability of paleoproductivity in the BoB indicate global teleconnections of climate, especially between North Atlantic climate and Indian Ocean.

ACKNOWLEDGMENTS

This work was carried out at the CSIR-National Institute of Oceanography (NIO), Goa during the PhD tenure of the author. The author wishes to thank the Director, NIO, Goa for the facilities provided to carry out this work. The author wishes to thank Dr. P.D. Naidu (Chief Scientist, retired, NIO) for providing the sediment sub-samples. Thanks are also due to Dr. J. N. Pattan (Chief Scientist, retired, NIO) for the PhD mentorship. The author is grateful to Shri M. Sridhar, Director General, GSI and Dr. Ishwar Dan Ashiya, Dy. Director General, GSI, SU: K&G, for providing the facilities to finalize this book chapter. IAM is thankful to the Council for Scientific and Industrial Research (CSIR), New Delhi (India) for the CSIR-SRF/NET fellowship.

REFERENCES

Altabet, M.A., Higginson, M.J., and Murray, D.W. 2002. The effect of millennial-scale changes in Arabian Sea denitrification on atmospheric CO_2. *Nature*, 415, 159–162.

Bains, S., Norris, R., Corfield, R., and Faul, K., 2000. Termination of global warmth at the Palaeocene/Eocene boundary through productivity feedback. *Nature*, 407, 171–174.

Banakar, V.K., Oba, T., Chodabkar, A.R., Kuramoto, T., Yamamoto M., and Minagawa, M. 2005. Monsoon related changes in sea surface productivity and water column denitrification in the Eastern Arabian Sea during last glacial cycle. *Marine Geology*, 219, 99–108.

Berger, W.H., Smetacek, V.S., and Wefer, G., 1989. Ocean productivity and paleoproductivity – An overview. In Berger, W., Smetacek, V.S., and Wefer, G. (eds.), *Productivity in the Oceans: Present and Past*, Wiley, New York, pp. 1–34.

Cayre, O., Beaufort, L., and Vincent, E. 1999. Paleoproductivity in the equatorial Indian Ocean for the last 260,000 yrs: A transfert function based on planktonic foraminifera. *Quaternary Science Reviews*, 18, 839– 857.

COHMAP Members. 1988. Climatic Changes of the last 18 ka; observation and model stimulations. *Science*, 241, 1043–1052.

Cullen, J.L. 1981. Microfossil evidence for changing salinity patterns in thevBay of Bengal over the last 20,000 years. *Palaeogeography, Palaeoclimatology, Palaeoecology*, 35, 315–356.

Curry, W.B. and Lohmann, 1986. Late Quaternary carbonate sedimentation at the Sierra Leone Rise (eastern equatorial Atlantic Ocean). *Marine Geology*, 70, 223–250.

Dansgaard, W., Johnsen, S.J., Clausen, H.B., Dahl-Jensen, D., Gundestrup, N.S., Hammer, C.U., Hvidberg, C.S., Steffensen, J.P., Sveinbjörnsdottir, A.E., Jouzel, J., and Bond, G., 1993. Evidence for general instability of past climate from a 250-kyr ice-core record. *Nature*, 364, 218–220.

Delaney, M.L. 1998. Phosphorous accumulation in marine sediments and the oceanic phosphorous cycle. *Global Biogeochemical Cycles*, 12, 563–572.

Duplessy, J.C. 1982. Glacial–interglacial contrasts in the northern Indian Ocean. *Nature*, 295, 494–498.

Dymond, J., Suess, E., and Lyle, M. 1992. Barium in deep-sea sediment: A geochemical proxy for paleoproductivity. *Paleoceanography*, 7(2), 163–181.

Emerson, S. and Hedges, J.I. 1988. Processes controlling the organic carbon content of open ocean sediments. *Paleoceanography*, 3, 621–634.

Finney, B.P., Lyle, M.W., and Heath, G.R. 1988. Sedimentation at MANOP Site H (eastern equatorial Pacific) over the past 400,000 years: Climatically induced redox variations and their effect on transition metal cycling. *Paleoceanography*, 3, 169–189.

Fontugne, M.R. and Duplessy, J.-C. 1986. Variations of the monsoon regime during the upper Quaternary: evidence from carbon isotopic record of organic matter in north Indian Ocean sediment cores. *Palaeogeography Palaeoclimatology Palaeoecology*, 56, 69–88.

Galy, V., France-Lanord, C., Beyssac, O., Faure, P., Kudrass, H., and Palhol, F. 2007. Efficient organic carbon burial in the Bengal fan sustained by the Himalayan erosional system. *Nature*, 450, 407–4011.

Ganeshram, R., Francois, R., Commeau, J., and Brown-Leger, S. 2003. An experimental investigation of barite formation in seawater. *Geochemica. Cosmochimica Acta*, 67(14), 2599–2605.

Goodbred, S.L. 2003. Response of the Ganges dispersal system to climate change: A source-to-sink view since the last interstade. *Sedimentary Geology*, 162, 83–104.

Govil, P., Naidu, P.D., 2011. Variations of Indian monsoon precipitation during the last 32 kyr reflected in the surface hydrography of the Western Bay of Bengal. *Quaternary Science Reviews*, 30, 3871–3879.

Howard, W.R. and Prell, W.L. 1994. Late quaternary CaCO$_3$ production and preservation in the Southern Ocean: implications for oceanic and atmospheric carbon cycling. *Paleoceanography*, 9, 453–482.

Ittekkot, V., Nair, R.R., Honjo, S., Ramaswamy, V., Bartsch, M., Manganini, S., and Desai, B.N. 1991. Enhanced particle fluxes in Bay of Bengal induced by injection of fresh water. *Nature*, 351, 385–387.

Levitus, S. and Boyer, T. 1994. World ocean atlas 1994. In: *Temperature*, vol. 4, US Department of Commerce, Washington, DC, p. 99.

Kudrass, H.R., Hofmann, A., Doose, H., Emeis, K., Erlenkeuser, H. 2001. Modulation and amplification of climatic changes in the northern hemisphere by the Indian summer monsoon during the past 80 k.y. *Geology*, 29, 63–66.

Lisiecki, L.E. and Raymo, M. E. 2005. A Pliocene-Pleistocene stack of 57 globally distributed benthic δ^{18}O records. *Paleoceanography*, 20, PA1003. doi: 10.1029/2004PA001071.

Madhupratap, M., Prasanna Kumar, S., Bhattathiri, P.M.A., Dileep P Kumar, M., Raghukumar, S., Nair, K.K.C., and Ramaiah, N. 1996. Mechanism of the biological response to winter cooling in the northeastern Arabian Sea. *Nature*, 3(84), 549–552.

Mir, I.A. 2015. Geochemical and organic matter isotopic studies to understand late Quaternary climate in the northern Indian Ocean, PhD Thesis, Goa University, pp. 1–162.

Mir, I.A., Pattan, J.N., Matta, V.M., and Banakar, V.K. 2013. Variation of Paleo-productivity and Terrigenous Input in the Eastern Arabian Sea during the Past 100 ka. *Journal of Geological Society of India*, 81, 647–654.

Montoya, J.P., 1994. Nitrogen isotope fractionation in the modern ocean: Implications for the sedimentary record. In: Zahn, R., Pedersen, T.F., Kaminski, M.A., and Labeyrie, L. (eds.), *Carbon Cycling in the Glacial Ocean: Constraints on the Ocean's Role in Global Change*. NATO ASI Series, 17, Springer, Berlin, pp. 259–279.

Muller, P.J. and Suess, E. 1979. Productivity, sedimentation rate, and sedimentary organic matter in the oceans. I. Organic carbon preservation. *Deep Sea Research Part A*, 26, 1347–1362.

Murray, R.W., Leinen, M., and Isern, A.R. 1993. Biogenic flux of Al to sediment in the equatorial Pacific Ocean: Evidence for increased productivity during glacial periods. *Paleoceanography*, 8, 651–670.

Naidu, P.D. and Govil, P. 2010. A new evidence on sequence of deglacial warming in the tropical Indian Ocean. *Journal of Quaternary Science*. doi: 10.1002/jqs.1392.

Naidu, P.D. and Malmgren, B.A. 1995. A high-resolution record of late Quaternary upwelling along the Oman Margin, Arabian Sea based on planktonic foraminifera. *Paleoceanography*, 11, 129–140.

Olson, D.B., Hitchcock, G.L., Fine, R.S., Warren, B.A. 1993. Maintenance of the low oxygen layer in the central Arabian Sea. *Deep-Sea Research Part II*, 40, 673–685.

Ostlund, H.G., Oleson, R., and Brescher, R. 1980. GEOSECS Indian Ocean radiocarbon and tritium results. Tritium Laboratory Data Report 9. University of Miami, Miami, FL, p. 15.

Paropkari, A.L., Iyer, S.D., Chauhan, O.S., and Prakash Babu, C. 1991. Depositional environments inferred from variations of calcium carbonate, organic carbon and sulfide sulfur: A core from southeastern Arabian Sea. *Geo-Marine Letters*, 11, 96–102.

Pattan, J.N., Masuzawa, T., Naidu, P.D., Parthiban, G., and Yamamoto, M. 2003. Productivity fluctuations in the southeastern Arabian Sea during the last 140 ka. *Palaeogeography, Palaeoclimatology, Palaeoecology*, 193, 575–590.

Pattan, J.N., Mir, I.A., Parthiban, G., Karapurkar, S.G., Matta, V.M., Naidu, P.D., and Naqvi, S.W.A. 2013. Coupling between suboxic condition in sediments of the western Bay of Bengal and southwest monsoon intensification: A geochemical study. *Chemical Geology*, 343, 55–66.

Pedersen, T.F. 1983. Increased productivity in the eastern equatorial Pacific during the last glacial maximum (19,000 to 14,000 yr B.P.). *Geology*, 11, 16–19.

Prabhu, C.N. and Shankar, R. 2005. Palaeoproductivity of the eastern Arabian Sea during the past 200 ka: A multi-proxy investigation. *Deep-Sea Research Part II*, 52, 1994–2002.

Ramaswamy, V. and Nair, R.R. 1994. Fluxes of material in the Arabian Sea and Bay of Bengal sediment trap studies. Proceedings of the Indian Academy of Sciences. *Earth and Planetary Sciences*, 103, 189–210.

Rao, C.K., Naqvi, S.W.A., Kumar, M.D., Varaprasad, S.J.D., Jayakumar, D.A., George, M.D., and Singbal, S.Y.S. 1994. Hydrochemistry of the Bay of Bengal: Possible reasons for a different water-column cycling of carbon and nitrogen from the Arabian Sea. *Marine Chemistry*, 47, 279–290.

Reichart, G.J., Lourens, L.J., and Zachariasse, W.J. 1998. Temporal variability in the northern Arabian Sea OMZ during the last 225,000 years. *Paleoceanography*, 13, 607–621.

Rostek, F., Bard, E., Beafort, L., Sonzogni, C., and Ganssen, G. 1997. Sea surface temperature and productivity record for the past 240 kys in the Arabian Sea. *Deep-Sea Research Part II*, 44, 1461–1480.

Shimmield, G.B. and Price, N.B. 1986. The behaviour of molybdenum and manganese during early diagenesis-offshore Baja California, Mexico. *Marine Chemistry*, 19, 261–280.

Sigman, D.M. and Boyle, E.A. 2000. Glacial/interglacial variations in atmospheric carbon dioxide. *Nature*, 407, 859–869.

Suthof, A., Ittekkot, V., and Gaye-Haake, B., 2001. Millennial-scale oscillanations of denitrification intensity in the Arabian Sea during the late Quaternary and its potential influence on atmospheric N_2O and global climate. *Global Biogeochemical Cycles*, 15, 637–650.

Tripathy, G.R., Singh, S.K., Bhushan, R., and Ramaswamy, V. 2011a. Sr–Nd isotope composition of the Bay of Bengal sediments: impact of climate on erosion in the Himalaya. *Geochemical Journal*, 45, 175–186.

Tripathy, G.R., Singh, S.K., and Krishnaswami, S. 2011b. Sr and Nd isotopes as tracers of chemical and physical erosion. In: Baskaran, M. (ed.), *Handbook of Environmental Isotope Geochemistry*, Springer-Verlag, Berlin, pp. 521–551. http://doi.org/10.1007/978-3-642-10637-8_26.

Weber, M.E., Wiedicke, M.H., Kudrass, H.R., Hübscher, C., and Erlenkeusker, H. 1997. Active growth of the Bengal Fan during sea-level rise and highstand. *Geology*, 25, 315–318.

Wyrtki, K. 1971. *Oceanographic Atlas of the International Indian Ocean Expedition.* National Science Foundation, Washington, DC, p. 531.

6 The Interplay of Mid to Late Holocene Tectonism and Northeast Monsoon on the Landscape Evolution

Palar River Basin, Southern Peninsular India

M.R. Resmi
Banasthali Vidyapith

Hema Achyuthan
Anna University

CONTENTS

INTRODUCTION

Fluvial geomorphic studies reveal the interplay of Tectonics, and Surface Processes operating on the earth's surface. It deals primarily with landforms and stratigraphy to infer the nature, patterns, rates, and history of near-surface processes (Nicoll 2008, Nicoll and Hickin 2010). The river basins are considered as one of the chief repositories of the hydrological and climatic changes since they are the most sensitive

109

elements of the earth's surface, and any shift in tectonic and climatic conditions can change the regime of the fluvial system. These processes can be responsible for the river avulsion/migration, asymmetries of the catchments, and river migration (Cox 1994, Clark et al. 2004, Salvany 2004, Schoenbohm et al. 2004). However, the river hydrodynamic studies are based on the examination of extrinsic control such as tectonic activity and longitudinal cross-sectional morphology of river channels (Slingerland and Smith 1998, Ethridge et al. 1999). Also, the river systems are mainly affected by the changes in sediments and water supply. These are controlled by several factors such as tectonics, climate change (Holbrook and Schumm 1999, Mather 2000, Tooth et al. 2002), and anthropogenic activities (Bernard et al. 2000). Moreover, the river systems respond dynamically to intrinsic changes, such as base level changes (Mather et al. 2002, Stokes et al. 2002).

The rivers are very sensitive to tectonic deformation and the drainage basin records these deformations and stages of its evolution (Gloaguen et al., 2008), which results in the deformation of alluvial landforms (Gao et al. 2007). Geomorphological studies using morphometric analyses (Keller and Pinter 2002) and morphometric indices provide useful information on the tectonic activity in the region (Wells et al. 1988, Burbank and Anderson 2001, D'Alessandro et al. 2008, El Hamdouni et al. 2007, Dehbozorgi et al. 2010, Pérez-Peña et al. 2010, Demoulin 2011, Gao et al. 2013, Joshi et al. 2013). A comprehensive study of landforms and deposits developed or modified by tectonic activity can provide relevant information about the neotectonic activity in the region (Wells et al. 1988, Cox 1994, Bishop 1995, Gupta 1997, Chamyal et al. 2003, Luirei and Bhakuni 2008, John and Rajendran 2008, Mrinalini Devi 2008).

Avulsion and migration are the most important geomorphic markers indicating active tectonism (Jongman 2006). The various geomorphic features and drainage in such areas provide evidence of neotectonic activity in response to movement along faults (Ouchi 1985, Schumm et al. 2000). The impact of vertical movements along faults can be delineated from fluvial geomorphic features and tectonic landforms (Rockwell et al. 1984, Wells et al. 1988, Ascione and Romano 1999, Li et al. 2001). Ground Penetrating Radar (GPR) studies are often used to delineate the characteristics of the main causative fault (Bano et al. 2000, Chow et al. 2001, Rashed et al. 2003, Maurya et al. 2005). Drainage anomalies such as compressed meander and eye-shaped drainage elucidate the structural fabric and tectonic subsidence in the area (Vaidyanadhan 1971, Ramasamy et al. 2011, Singh et al. 1996). Interaction between tectonics and the climate at variable spatial and temporal scales controls weathering and erosion in the catchment areas of major rivers (Burbank et al. 2003, Clift et al. 2010). These also influence the fluvial processes such as river discharge, sediment flux, and channel migration that characterize the landscape in the region (Bookhagen et al. 2005). Both tectonic activities and climatic fluctuations have controlled the catchment evolution of geomorphic landforms and sedimentary processes in the Southern Peninsular Rivers. Furthermore, intensified monsoon phases with corresponding high sediment fluxes during the Late Quaternary put forward a link between millennial-scale climate change and surface processes in the northwest Himalaya (Bookhagen et al. 2005).

PALAR RIVER

Palar River (Figure 6.1) is one of the major rivers draining the Southern Peninsular India and this rhombus-shaped basin is located between latitudes 12°13'36.62"N–13°37'14.87"N and longitudes 78°40'29.40"E–80°08'51.99"E. It rises in the Nandi hills at an elevation of about 900 m above a.m.s.l, Kolar district, Karnataka, traveling 93 km in Karnataka, 30 km in Andhra Pradesh, and 222 km in Tamil Nadu and meets the Bay of Bengal without forming a delta (Figure 6.1). The Palar River flows as an underground river for a long distance and appears near Bethamangala town, from where, gathering water and speed, it flows eastward down the Deccan Plateau. The River drains a total area of 17, 871 km^2 out of which nearly 57% is in Tamil Nadu and the rest is in the states of Karnataka and Andhra Pradesh. In Tamil Nadu, the river enters the Vellore District and after traversing 222 km through the Kanchipuram District meets the Bay of Bengal near Vayalur. Along its course, the Palar River receives two important tributaries, namely the Poiney on the left bank and the Cheyyar on the right bank. These two important tributaries, i.e., the Poiney and the Cheyyar, together account for nearly 25% of the total catchment of the Palar basin. The Poiney River starts in the Chittoor district of Andhra Pradesh at an elevation of 1, 050 m and flows in the easterly and southeasterly direction prior to joining the Palar on its left bank near Walajapet. The Cheyyar, another major tributary, rises in the Javadi hills in the Chengam Taluk of Thiruvannamalai District, Tamil Nadu and flows generally in the northeasterly direction before meeting the

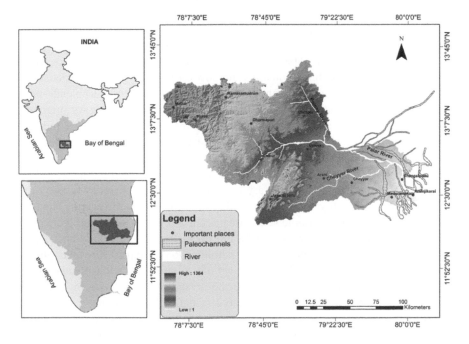

FIGURE 6.1 Digital Elevation Model (DEM) of the Palar River drainage basin with paleochannels and important locations.

Palar near Vayalur. The salient features of this river are discussed below. The Palar River basin can be physiographically divided into three major topographical divisions, namely (1) the hill ranges with an elevation ranging from 300–1, 364 m; this forms a part of the high hill ranges with steep slopes of the Eastern Ghats; (2) The plateau region with an elevation of 300–150 m, which comprises the relict mountains, valleys, etc.; and (3) Lowland zone with an elevation ranging from 150 to 1 m characterized by rolling topography with lateritic ridges, isolated hills, and alluvial valleys, and with an elevation of <1 m. The lowland is also characterized by coastal stretches, alluvial plains, and the coastal plains (150–1 m). The Eastern Ghats comprises Charnockite, Granite, Khondalite, Metamorphic Gneisses, and Quartzite Rock formations. The structure of the Eastern Ghats includes the thrust and strike slip faults all along its range. The highlands are depicted by the hill ranges of Nandi Hills of the Eastern Ghats with elevations >300 m above the a.m.s.l and form an important physiographic province. The average elevation of the Nandi hills is 966 m, where the Palar River originates. The basin has its highest elevation nearly 1, 364 m in the west. The topography is very rugged, and the crest of the mounds and hills are generally very sharp and narrow with very steep slopes. Deep gorges and mountain-fed streams are characteristics of these hill ranges. The middle and lowlands consist of dissected pediplains with an altitudinal range of 1–300 m including the study area, which runs in a NE–SW direction. The plains have a rolling topography with isolated mounds. Floodplains, river terraces, channel and valley fill, colluviums, and isolated mounds and hills are part of the lowlands. The coastal plain in the east where the river meets the sea has an elevation reaching a maximum of 1 m above the mean sea level (a.m.s.l). The Palar River originates from the eastern Dharwar craton and flows through granitoid gneisses (Migmatitic gneiss, Hornblende gneiss, Hornblende-Biotite gneiss). The western part of the study area comprises mainly the migmatites and granite. Migmatites include varieties of gneissic rocks such as the granite gneiss, garnet-biotite gneiss, hornblende gneiss, and hornblende-biotite gneiss. Quartz-mica gneiss is one of the major rock units and shows migmatitic structures. Granites occur as later emplacements along crustal fractures and faults especially in the Chittoor area. Further east, the Tamil Nadu region is traversed by numerous faults and tectonic lineaments. Major lineaments observed in the region include the NW–SE, NE–SW, NNW–SSE, and WNW–ESE trending sets. Some of these faults and lineaments are very old, being of Precambrian, whereas others are comparatively much younger and possibly are of neo-tectonic origin. The upper Paleozoic–Mesozoic rocks in the study area occur mainly in elongated graben-type basins and are grouped under the Gondwana Formation (dated from Upper Permo-Carboniferous to lower Cretaceous) (Krishnan 1968). Charnockite and gneisses occur in the middle reaches of the Palar River and are the most widespread rock type in the study area. It has prevailed over all the crystalline rocks covering more than 50% of the total area in the basin (Figure 6.2). Charnockite and gneiss make up the high hills and steep slopes along the undulating terrain in the northern and southern part of the basin and the lower Gondwana formation followed by coastal alluvium. Within the Palar River basin Gondwana, sediments occur as three distinct patches in Sriperumbudur, Kanchipuram, and Maduranthakam areas. Maduranthakam formation (Talcher sediments of glacial origin) is made up of boulder bed, green shales,

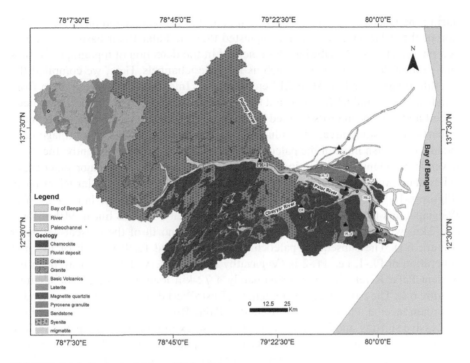

FIGURE 6.2   Geological Map of Palar River.

and overlain by green shale with sandstone partings and ovoid masses of limestone with fine-grained texture (Subramanian and Selvan 2001).

Fluvial sediments and sandstone are confined to the coastal tracts, valleys, and along the river stretch. Based on the environment of deposition, these deposits are further subdivided into different morpho-stratigraphic units such as the marine-fluvial and paleofluvial deposits. A huge thickness of alluvial material is deposited from the middle to lower reaches of the Palar River above the Precambrian terrain. Pebble beds are observed on either in the Palar River or its older channels. These pebble beds are predominantly composed of quartz, but occasionally charnockite gneiss pebbles are also observed. They are well rounded to spherical in shape with a clayey matrix. Around Kanchipuram, close to the Upper Gondwana rocks, an arenaceous formation consisting of pebbles and shingles of quartzite and vein quartz referred to as Conjeevaram gravels (not shown in the map) of probable Pliocene to Lower Pleistocene age is exposed. Laterites associated with reddish-brown ferruginous clayey soil capped over the fluvial sediments are present in the NE side of the study area.

MIGRATION OF THE PALAR RIVER

In the Holocene period, the response of the rivers to tectonic deformation and climate has played a major role in the evolution of the landscape. Narasimhan (1990) indicated a vertical uplift of the land in the area north of the Palar River that occurred during the Quaternary period. Rangaraju et al. (1993) recognized phases of evolution of the Palar River Basin and suggested that the Palar River basin is in a graben phase. The SRTM-DEM analysis helped in the detection of topographic relicts and geomorphological features such as the paleochannels. Here, we compiled the satellite imagery and DEM in GIS environment. Integrating the information from Landsat images and DEM supported by field observations, paleochannels have been traced and have been cross verified with some of the available paleochannel maps in parts of the study area. The enhanced images and the shaded relief image have aided to establish most of the paleochannels in the study area. Presently, the Palar River is preferentially migrating toward south. Therefore, two major paleochannel systems were identified in the northern region of the modern Palar River basin (Figure 6.1). The paleochannel system (PL-1) of the Palar River branches off at Walajapet and terminates in north of the Chennai city. PL-1 is linear, 90 km long, and 2 km wide trending in N–E and occurs 25 km north of the present-day river (Figure 6.1). The subsequent paleochannel is recorded further south of the first generation of PL-1, i.e., PL-2 is (Vegavathy River) observed 17 km away from the present Palar River. This is approximately a 72-km long and 2-km wide channel (Figure 6.1). The paleochannels trend in the East–West direction and when extended upstream meet at a point to the present day Palar River. The paleochannels exhibit successive shifting and in some places, they also exhibit meandering patterns. The Palar River shows a physical continuity with the paleochannel and this is discerned by the channel's morphological characteristics as well as sediment lithology. Both the paleochannels indicate local avulsions or a local fluvial dislocation with a cluster of broken and smaller in width paleochannels. The PL-1 and PL-2 channels differ in their morphology and sedimentation pattern. PL-1 is rather a straight course while PL-2 becomes relatively curved and the degree of sinuosity of the channels is less. PL-1 is linear, curvilinear, and ribbon shaped. However, the paleochannels are covered with vegetation that appears as bands of dark gray tone. Low surface moisture and sparse vegetation along the paleochannels indicate permeable, porous, and coarse-grained materials with high infiltration. However, in few sites, the paleochannel segments were poorly visible as they are deeply buried under the alluvium. Clusters of paleochannels, some of which are pronounced and exhibit long river segments, indicate that the paleochannels have undergone minimal erosion. The path of migration of the Palar River has also been ground checked using GPS-based GCPs (ground control points).

Causes of Migration: Neo-Tectonic Activity

The evaluation of the relative active tectonics in large areas is based on various geomorphic indices. Geomorphic indicators of active tectonics include hypsometric curves, longitudinal river profiles, sinuosity index, stream gradient index, basin shape, asymmetric factor, drainage patterns, and so on. The present study is an attempt to evaluate the neotectonic activity in the five major sub-basins of the Palar River using geomorphic indices. The head reaches and middle reaches of the river are characterized by several faults and lineaments. The major ridge system on the NW side of the

Palar River basin and the seafloor spreading in the Indian Ocean are the main cause of rejuvenating pre-existing faults in the upper reaches of the Palar River.

The SL indices (stream length gradient indices) in the sub-basins II and III are high compared to other sub-basins and the southern part of sub-basin I. The underlying rock group is the Precambrian gneiss and charnockite in the sub-basins I, II, and III; therefore, it can be interpreted that the anomalous value is due to tectonic signals and not due to lithologic control. The low SL values of IV and V indicate that these sub-basins are in the mature stage (Figure 6.3). The hypsometric curve and integral values of the five major sub-basins indicate that fluvial erosion and channel processes control the sub-basins. The hypsometric curve and integral reveal a very low value since the underlying Precambrian gneisses and the granite migmatites are resistant to erosion. However, the impact of the lithology is not enough to explain the relatively lower value of Hi. The tilt trend is predominantly in the southward direction reflecting the MPA (MulkiPulikat Lake Axis) along which the river is flowing south (Figure 6.4).

In sub-basins II and III, Iat values indicate high tectonic activity, which are in the NW–S side of the Palar River. Sub-basin I reveals a moderate Iat value, while sub-basins IV and V reflect low Iat values. The sub-basins II and III show higher SL values, Af values, and VF values, but in these basins, the Hi values are lower, and the hypsometric curves are concave. This is probably due to the strong lithological

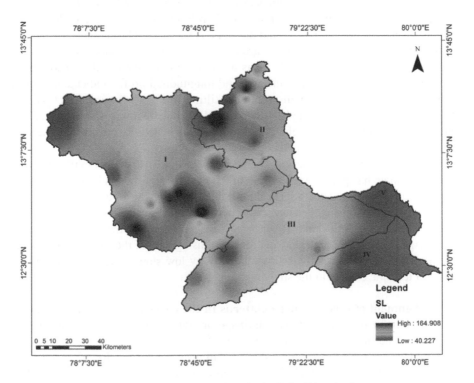

FIGURE 6.3   Spatial distribution of SL values in the Palar River basin.

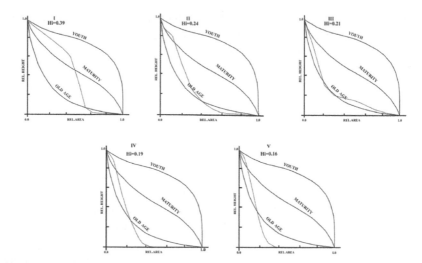

FIGURE 6.4 Hypsometric curves of five major sub-basins (I–V) of the Palar River. Hi is the hypsometric integral calculated and the dotted line indicates the hypsometric curve.

control (Figures 6.5 and 6.6). Hence, sub-basins II and III reflect the role of tectonic uplift. From Figure 6.7, we can delineate several fault zones and lineaments that are mainly present in the sub-basins II and III. Sub-basins IV and V are underlain by alluvial reaches of the Palar River and the slope is very less. The geomorphic indices signify that the sub-basins IV and V are no longer active and hence the patterns of tectonic deformations in the study areas are not similar. Overall, the various sub-basin parameters, morphometric indices, and longitudinal profile indicate the uplift of the sub-basins II and III with prominent tilting in the southward direction. We attribute this to a differential uplift along the MPA under compressive stresses. The data presented show that the highest level of neotectonic activity occurred in the sub-basin II, followed by sub-basins III, I, IV, and V (Figure 6.7).

Since the mid-Holocene period, the response of the rivers to tectonic deformation and climate has played a major role in the evolution of the landscape. However, Narasimhan (1990) indicated a vertical uplift of the land in the area north of the Palar River that occurred during the Quaternary period. Rangaraju et al. (1993) recognized phases of evolution of the Palar River basin and suggested that the Palar River basin is in a graben phase. Geomorphic and remote sensing data of the study area indicate that the Palar River presently drains relatively by low sloping terrain and the evidence of tectonic control in the study area includes:

i. **Alignment of eyed drainage patterns in the current river channel:** The stimuli for the origin of eyed drainage are due to the tectonic subsidence along the lineaments (Figure 6.8). The Palar River channel splits into two and then re-joins further down slope, forming an eye-shaped drainage pattern associated with trending NE–SW lineaments. We identified four eye-shaped drainages in the Palar River. These four eyes-shaped drainage patterns were observed near Vellore, Kanchipuram, and Chengalpattu areas.

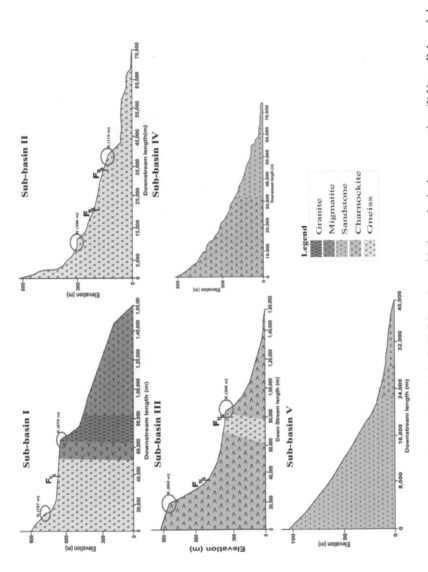

FIGURE 6.5 Longitudinal profiles of the five sub-basin of the Palar River along with the geological cross-section. (I) Upper Palar sub-basin, (II) Poiney sub-basin, (III) Cheyyar sub-basin, (IV) Killiyar sub-basin, and (V) lower Palar sub-basin. The red circle represents the knick points in the profile.

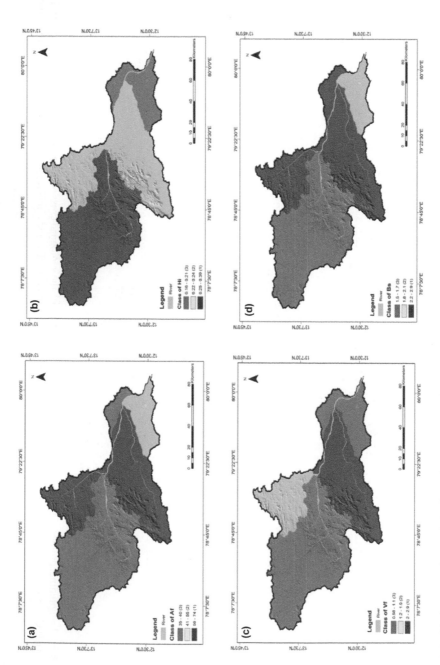

FIGURE 6.6 Various morphometric indices of five major sub-basins of the Palar River. (a) Drainage basin asymmetry (Af), (b) Hypsometry Curve and integral, (c) Ratio of valley floor width to valley height (Vf), and (d)Basin shape index (Bs).

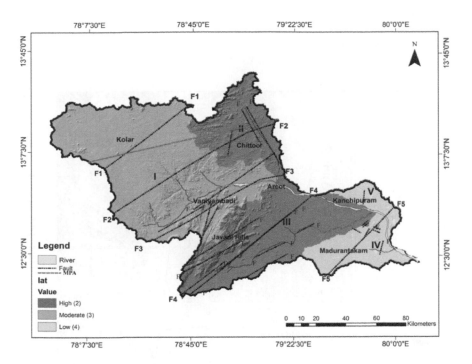

FIGURE 6.7 Distribution of relative tectonic activity classes of the Palar River drainage basin with major (F1–F5) and minor faults (F).

ii. **Distribution of deflected drainage along the active river channel:** Deflected drainages and active lineaments in the fluvial system have been attributed to the tectonic upliftment (Babu 1975, Twidale 2004, Ramasamy et al. 2011). Drainage deflection and eyed drainage associated lineaments also follow the NE–SW trend of the lineaments and thus are interpreted as zones of neotectonic activity. The orientation of the lineaments observed from various drainage anomalies reveals strike slip movements and readjustments of the modern drainage due to fault reactivation (Valdiya 2001).

iii. Presence of many straight and parallel channel segments near the present day Palar River.

iv. **Prevalence of parallel to sub-parallel drainage pattern flowing west to east:** The alignments of the stream channels in the upper, middle, and the lower reaches of the Palar River basin exhibit that this basin is controlled by lineaments and faults in the bedrock (Vemban et al. 1977), which point toward the recent neotectonic activity. Ramasamy et al. (2011) identified several compressed meanders in the river channel at the crossing of lineaments. The fracture swarms, highly dissected and serrated charnockite with granite dykes, also indicate the intensity of neotectonic activity in the region.

The Palar River and its tributaries drain the ancient Archean–Proterozoic rocks. Hence, it was expected that a cratonic river such as the Palar and its major tributaries would display a very mature and senile landscape. However, in the upper reaches

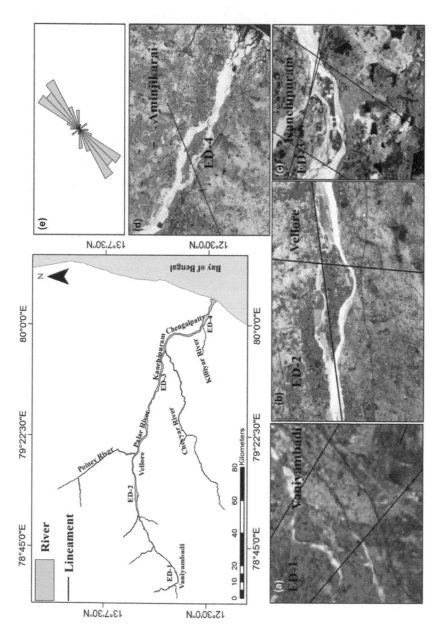

FIGURE 6.8 Drainage anomalies in the Palar River (a–d) and (e) Rose diagram of lineament associated with an eyed drainage pattern, which shows predominantly of the NE–SW direction.

of the Palar River catchment, the regionally extensive, low relief exhibits all the geomorphic evidence of an old, relict landscape, and the longitudinal profile of the Palar river basin displays a relatively smooth curve. The eastern Dharwar craton has undergone denudation episode since the late Miocene period and the prominent anomalies, such as major breaks in the long profiles of the sub-basins I, II, and III, indicate disturbances and disequilibrium conditions associated with active faults. In these sub-basins, knick points are not controlled by either lithology or active tectonism (Figure 6.5).

The major stream courses in the Palar River drainage basin are controlled by the lineaments. The majority of lineaments in all the sub-basins are showing a NE–SW trend. However, in the sub-basin II, the lineaments that are present in fracture swarms are showing an E–W trend. There are five major faults identified in the Palar River basin and a transition zone is also identified between sub-basins II and III, where frequent smaller magnitude earthquake occurs as shown in Figure 6.9. The chronology for the reactivation of the above faults is not well constrained; however, geomorphic evidence shows that peneplain surfaces with lateritic capping of Tertiary age occupy the top in the Eastern Dharwar craton and in various places of the Palar River drainage basin (Radhakrishna 1993). The earthquakes of magnitude M >6:0 for the period 1700–1997 are presented in Figure 6.9. The stress is due to plate boundary forces falls within the fault zones, particularly at the fault intersections (Talwani and Rajendran 1991). Most of the lineaments show a similar trend to the major fault NE–SW trend (Figure 6.9). Furthermore, the sharp turn of the streams near the lower reaches also indicates tectonically controlled structural configuration of the Palar basin.

The major fault F1 occurs in the upper Palar River and is a strike-slip fault following the NNE–SSW trend. In between the faults F2 and F3, a transition zone occurs where frequent minor and major tremors occur and this falls between sub-basins I and II. The MPA also crosses the transition zone. Another major fault F4 traverses sub-basin III, and the Cheyyar River flows parallel to these faults. Upon reactivation of these faults, several lineament anomalies occur in sub-basins II and III. In the transition zone, or immediately next to the zone, near Arcot, the Palar River bifurcates. A rapid shift in the flow direction has resulted due to the movement along these strike-slip faults. The sudden movement along the fault has also opened a new channel and a sudden flux of sediments resulting in stream braiding. That is the reason why the Palar River has enormous quantities of sediment deposited in the riverbed. Hence, the broad channel within the Palar River drainage basin is formed due to the movement along the strike-slip fault in the middle reaches. The presence of a transition zone indicates that the area is under stress. This process has led to the abandonment of older courses of the Palar River while the fault F5 nearer to the east coast signifies a passive tectonic activity regime (Figure 6.9).

The Quaternary sediments give important stratigraphic evidence for delineating the neotectonic activity in the Palar River. The effect of neotectonic tilting of the basin is implicit from the evidence documented in remote sensing studies. The lineament systems of the Palar basin illustrating the connectivity among the zones of squeezed aquifers and the zones of groundwater discharge into the Bay of Bengal. Saravanavel and Ramasamy (2016) further suggested that in areas that have tectonic

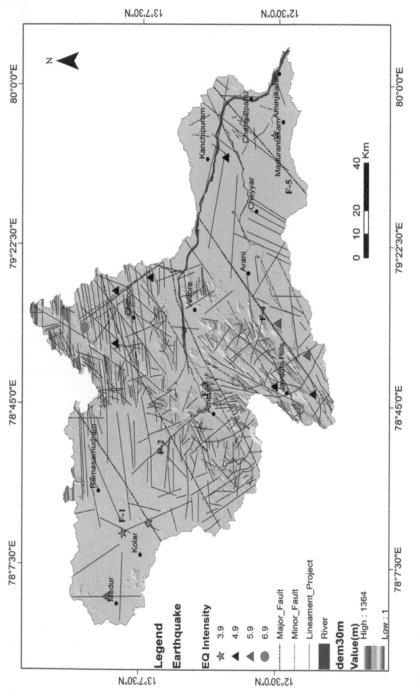

FIGURE 6.9 Distribution map of earthquake epicenters (1700–1997) plotted on the tectonic map of southern Peninsular India.

emergence, the aquifers are squeezed and the freshwater leaks into the ocean through such fractures. The huge amount of sediment dumping pattern and reservoir siltation in the Palar River and into the Bay of Bengal is attributed to the tectonic emergence in the catchment area of the Palar. This is because the tectonically induced reservoir siltation reduces the water holding capacity and as a result, the floodwaters from the river spread to the adjacent plains. The massive floods history of the Palar River corroborates the same.

DEPOSITIONAL ENVIRONMENT AND PALEOCLIMATIC INFERENCES

Paleochannels are one of the major archives to explain paleoclimatic conditions. Two major migration phases of Palar River have been identified between 3.59 and 1.8 ka (Table 6.1). Based on the OSL dates, the initial migratory phase (PL-1) developed subsequent to 3.59±0.63 ka. At PL-1, the fluvial activity spans approximately for a period of 1.24 ka (from 4.83 to 3.59 ka) and ceased since 3.59 ka (Figure 6.10). In PL-1, the association of facies such as Sp (medium to coarse sand with crossbed), St (cross-bedded fine to very coarse with pebbles), and Gp (stratified gravels with trough crossbeds) indicate high-energy depositional environment. Lithofacies Sp has developed mainly as cosets, representing transverse bars documented as the characteristic features of a braided-river environment (Allen 1983, Bristow 1993).

TABLE 6.1
OSL Dates of the Sediments Collected from the Palar River and Its Paleochannels

Sample Name	Depth (cm)	U (ppm)	Th (ppm)	K (%)	Dose rate(Gy/Ka)	ED(Gy)	Age(Ka)
PL-1	90	0.70±0.00	11.49±0.01	2.22±0.03	2.92±0.43	10.48±1.84	3.59±0.63
PL-1	150	0.32±0.00	1.65±0.02	2.76±0.03	2.70±0.04	13.03±2.07	4.83±0.77
PL-2	100	0.52±0.00	2.03±0.01	2.69±0.01	2.72±0.03	6.23±0.78	2.42±0.29
PL-2	240	0.66±0.04	7.24±0.43	2.20±0.04	2.62±0.05	6.70±0.64	2.55±0.49
PL-2	400	0.55±0.01	3.09±0.41	3.10±0.02	3.15±0.04	7.28±0.45	2.31±0.03
PL-2	550	0.75±0.02	13.91±0.07	2.76±0.01	3.58±0.67	8.45±0.89	2.87±1.30
PL-2	600	0.98±0.01	2.11±0.02	2.45±0.06	2.59±0.60	8.44±1.35	3.26±0.98
PL-4	50	0.82±0.00	9.37±0.04	0.63±0.00	1.28±0.02	12.95±2.01	10.09±1.6
PR-1	175	16.63±0.20	0.83±0.01	0.00±3.37	3.62±0.08	5.22±1.05	1.41±0.28
PR-1	290	0.55±0.00	6.14±0.04	3.37±0.05	3.59±0.06	6.76±2.05	1.88±0.57
PR-3	200	0.59±0.01	10±0.14	2.10±0.01	2.69±0.03	0.68±0.51	0.25±0.20
PR-3	300	0.41±0.00	2.98±0.03	1.97±0.00	2.08±0.02	0.78±0.2	0.38±0.27
PR-3	400	0.35±0.00	2.55±0.02	2.42±0.04	2.45±0.41	0.79±0.3	0.39±0.09
PR-3	500	0.35±0.00	1.81±0.01	2.72±0.04	2.68±0.04	2.49±0.69	0.93±0.26
CR	65	0.99±0.02	39.85±0.73	0.81±0.00	3.51±0.08	0.30±0.27	0.09±0.08
CR	115	3.94±0.07	0.29±0.00	0.71±0.01	1.53±0.03	0.17±0.07	0.11±0.04
CR	195	0.24±0.00	3.84±0.01	0.76±0.00	0.99±0.01	0.60±0.14	0.61±0.14

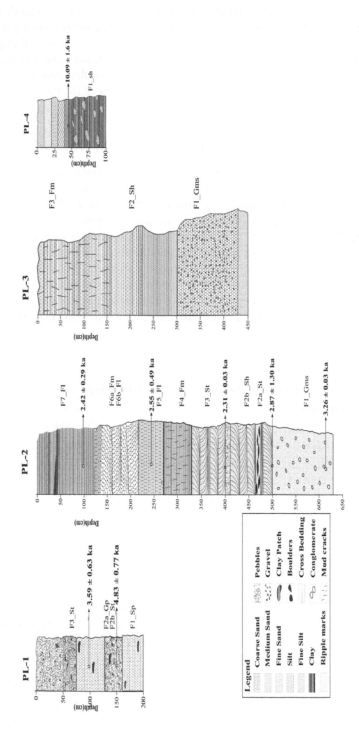

FIGURE 6.10 Exposed lithosection in paleochannels PL-1 (PL-T), PL-2 (PL-V), PL-3, and PL-4(PL-K) on both sides of the Palar River with OSL ages of the sediments.

The abundance of trough crossbed (lithofacies Sp and St) and planar crossbed with gravel (Gp) point to a well-established braided river system and high energy flow with a low rate of sedimentation (0.04–0.03 cm/yr) and this is probably due to the intensified NE monsoon. The mid-Holocene period in the Palar basin was dominated by a high-energy fluvial activity. The subsequent migration phase (PL-2) was identified 7 km away from the site PL-1 with the deposition at PL-2 from 3.26 to 2.42 ka and this phase spans approximately 0.84 ka. At PL-2, the association of facies such as Gms (matrix-supported gravel), St (trough crossbed with medium sand), and Sh (laminated flow) represent debris flow as observed in the lower unit of PL-2 indicating relatively high energy debris channel flow with pebbles (Figures 6.10 and 6.11). The prevalence of sudden gravitational debris flow deposit in the lower unit from PL-2 indicates a rapid uplift of the source area. From Unit 5 (2.55 ± 0.49 ka) association of Fm and Fl, i.e., OF (overbank fines) with ripple marks and desiccation cracks that indicate a slow change in local hydrography, decrease in discharge, a change in the sediment load, and passive meandering deposits are the evidence of channel avulsion

FIGURE 6.11 (a) Overview of the mid-late Holocene section in paleochannel (PL-2). (b) Exposed lithosection in PL-2. (c) Overbank fine (OF) material.

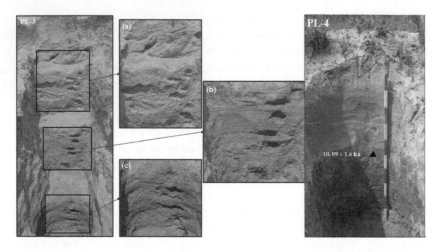

FIGURE 6.12　Exposed section in paleochannel (PL-3). (a) Overbank sediments. (b) Shallow energy deposits. (c) Coarse-grained sediment deposits and PL-4, which is the only paleochannel located in the south of the Palar River.

(Brice 1982, Bridge 2003, Chamyal et al. 2002, Brice 1982, 1983, Thorne 1997). The fine-grained lithofacies associations of PL-2 are attributed to a meandering river system (Klimek 1974). This lithosection also exhibits channel fill by vertical aggradations, fining upward succession, reflecting a progressive abandonment because of rapid upstream avulsion. Over bank, deposits are linked to periods of aggradations, weaker monsoon, higher sediment supply, and a meandering river paleoenvironment suggesting dry environment conditions. The statistical analysis of the entire lithosection also reveals the highly varying energy of the depositional agent.

　The continuous but sporadic migration phases have identified that after PL-2 is PL-3, nearly 12 km away from PL-2 nearer to Kanchipuram (Figure 6.12). The profiles PL-2 and PL-3 are lithological and structurally similar. In these two profiles, the base unit consists of a very coarse sedimentary unit, which indicates a high-energy depositional environment followed by fine lamination and sheet flow (Sh) with fine materials, which suggest a relatively calm depositional environment. From unit 3, we identified the association of Fm and Fl i.e., OF having 2.75 m thickness. This lithosection also exhibits channel fill by vertical aggradations, fining upward succession, reflecting a progressive abandonment because of rapid upstream avulsion. Therefore, we correlated the OF of PL-2 and PL-3, which are formed simultaneously since both are revealing a progressive upstream avulsion. In the PL-3 channel, both the ends are now connected to the present channel. The final migration phase of the Palar River (the current channel) occurred probably between 2.42 and 1.88 ka. In PL-4, we identified a paleochannel, which is situated in south of the Palar drainage basin. This paleochannel is not at all an old course of the Palar River; at the present condition, this channel is in the 4th sub-basin of the Palar River (Figure 6.10). The present channel lithosections (PR-1 and PR-2) exhibit facies assemblage of Sh, Sp, Sl, and Gp with random distribution of pebbles and gravels indicating a transition between high to low energy river flow (Figures 6.13 and 6.14). In the active river system, the

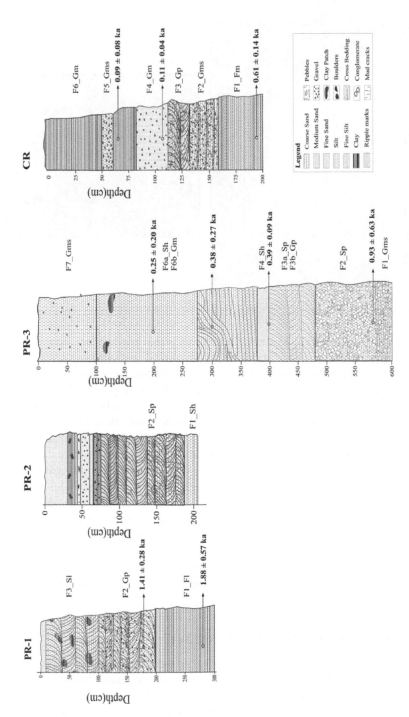

FIGURE 6.13 Exposed lithosection in various places of the Palar River with OSL ages of the sediments.

FIGURE 6.14   The lithosection exposed in Palar River (PR-1) with planar cross-bedding.

prevalence of facies indicates shallow river channels that are developed possibly due to flash floods while the sand lithofacies of Sp and Sl are subordinate. The river system at this stage progressed into several broad, shallow, aggrading, and laterally unstable channels. An increase in the sediment load disproportional to the water discharge may have caused the river to aggrade and excessively widen its channel, thus evolving into shallow channels (Allen and Leeder 1980, Best 1996). The occurrence of cut and fill structures in the active river channel indicates reworking of the sedimentary layers due to small-scale neotectonic activity or due to climatic stimuli resulting in the incision of channels into their own alluvium and subsequent filling. The facies association Gms and Gt point to a high energy flow in the major tributary, the Cheyyar River. Several small magnitude flood deposits were also observed in the Cheyyar River. The slope of the area is decreasing from the north to the south (Figure 6.15). The OSL dates indicate that in PL-1, the rate of sedimentation is low when compared to PL-2 and the present Palar River. However, periods of aggradations are linked to periods of weaker monsoon and higher sediment supply while periods of stronger monsoon are associated with erosion, incision, and reduced sediment supply (Kale 2007).

INFERENCE FROM GEOCHEMICAL PROXY

Climate and rainfall have a major role in chemical weathering processes, which initiate chemical weathering, which increases the intensity of major chemical reactions (Sun et al. 2010). Usually, in warm and humid climate, the rate of weathering is high due to chemical reactions, and the rate of weathering is higher than cold and dry climates (Sun et al. 2010, Achyuthan et al. 2014). In rivers, the materials from various

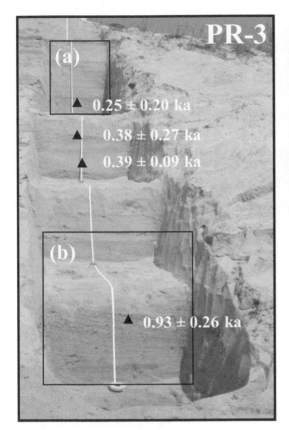
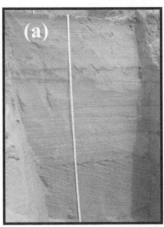

FIGURE 6.15 Lithosection exposed in the Palar River near Vayalur ~11 m thick. The sedimentary structures such as cross-bedding and trough cross-bedding reveal a high energy deposit. This profile was cut ~3 km away from the confluence of Palar River with the Bay of Bengal. (a) This section shows planar bedding followed by sheet flow. (b) Prominent cross and trough cross-bedding sedimentary structures.

sources in the drainage basin as the products of chemical weathering will wash away. However, the present study reveals that a change in discharge and/or sediment load may be due to the decrease in discharge and/or monsoonal fluctuations in the upper catchment (Kale 2007), or due to high channel mobility (Sun et al. 2010), or due to avulsions that are climatically driven (Resmi et al. 2016). However, the ratio of soluble to insoluble chemical elements in a river and paleochannel section sediments can serve as indicators of the chemical weathering conditions in that particular region, especially to precipitation and monsoonal fluctuations (Singh et al. 2016). All the paleochannels are dry and are having well exposed lithosection. Generally, in warm and wet climate conditions, elemental ratios of K/Al, Mg/Al, and Ti/Al in the sediments are high (, Sun et al. 2010, Gayantha et al. 2017) and elements like Fe and Mn are subject to benthic redox conditions (Gayantha et al. 2017). Moreover, the value of Mn/Al and Fe/Al will be also high during the dry periods (Sun et al. 2010).

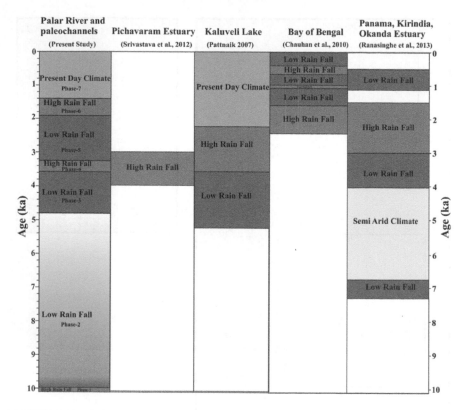

FIGURE 6.16 Schematic diagram of climate variability during the Holocene from records in the region, which experiences NEMR.

In the PL-4 section, prior to 10.09±1.6 ka, an increase in clay content coupled with an increase in Ti/Al, Fe/Al, Mn/Al, Na/Al, Mg/Al, and a decrease in Si/Al, Ca/Al along with CWI is noted (Appendix Table 1). From the higher CIA, Rb/Sr ratio, and decreasing CWI value, we inferred that prior to the 10.09±1.6 ka period, the area experienced an intensified monsoon phase with high weathering. This is further corroborated by the lower values of the major oxides CaO, MgO, TiO_2, and K_2O. Moreover, higher values of CaO/MgO, Na_2O/K_2O, and Na_2O/Al_2O_3 ratios also reflect the intense weathering during that period. Subsequent to 10.09±1.6 ka together with the high values of elemental ratios such as Ti/Al, K/Al, Mg/Al, Ca/Al, Na/Al, K/Al, CWI, and decreasing CIA, values with high sand content is noted. Furthermore, post 10.09±1.6 ka period Si/Al, Ti/Al, Mn/Al, Ca/Al, Na/Al, ratios, and CWI exhibit a gradual increase. These variations point toward an increase in CWI but a decrease in CIA and Rb/Sr ratio, indicating a decrease in rainfall during this period.

The elemental oxide concentrations prior to 4.83±0.77 ka reveal an increase in Si/Al, Fe/Al, Na/Al, K/Al, CaO/MgO, and CWI value is observed. A decrease in Mn/Al, Mg/Al, and CIA values is also noted (Figure 6.16), indicating a reduction in the monsoon precipitation. A significant increase in soluble elemental oxides Al_2O_3,

MgO, CaO, and Na_2O, and lower levels of immobile elements in the lithosection during 4.83 ± 0.77 to 3.59 ± 0.63 ka signify a progressive increase in chemical weathering. This inference is corroborated with the increase in values of Si/Al, Mg/Al, Ti/Al, Fe/Al, Ca/Al, Na/Al, Mn/Al, K/Al, CaO/MgO, and CWI. These findings reveal that chemical weathering in the mid-Holocene period is intense with a decrease in the rate of sedimentation and NEMR. This observation is further supported by low Rb/Sr and Ba/Sr ratios. Post 3.59 ± 0.63 ka, fluctuations in elemental concentrations with high sand percent, K/Al, Ti/Al ratios indicate strong weathering in the catchment. Lower CaO/MgO ratios also point to lower temperature and increase in precipitation in the area, resulting in a short, warm, and humid conditions. This is further supported by low levels of Mn/Al and Fe/Al coupled with increased values of Rb/Sr and Ba/Sr.

In the Vegavathy River (PL-V) section, the upward declining trend of sand, silt, and comparatively low-intensity elemental ratios (K/Al, Mg/Al, and Ti/Al) imply a decrease in rainfall intensity followed by dry conditions. This argument is further demonstrated with the thick OF deposits, mud cracks, and rootlets supported by the relatively high levels of Mn/Al and Fe/Al ratios. The period between 3.26 ± 0.98 and 2.87 ± 1.30 ka reveals a morphological change that occurred in the Vegavathy River, i.e., rapid upstream avulsion followed by fining upward succession with OF and mud cracks. The decreasing trends of elemental ratios K/Al, Mg/Al, Ca/Al, Na/Al, CaO/MgO, Ti/Al, CIA, and CWI are also noticed during this period and this is because of the overall weakening of rainfall. This is further supported by the high Mn/Al and Fe/Al ratios. However, due to upstream avulsion, the grain size, and the rate of sedimentation are high. Furthermore, ameliorated conditions prevailed between 2.87 ± 1.30 and 2.55 ± 0.49 ka due to continued upstream avulsion followed by upstream fining upward succession. A significant decrease in Si/Al, Ti/Al, and Mg/Al and highly varying ratios of Fe/Al, Ca/Al, Na/Al, K/Al, and CWI with an increase in the rate of sedimentation confirm tectonic disturbances in the dry region. However, since 2.55 ± 0.49 to 2.42 ± 0.29 ka, the rate of sedimentation and SiO_2 content are low and this is reflected by high CIA values and Al_2O_3 content. The decreasing trends in elemental ratios (K/Al, Mg/Al, and Ti/Al) with a high CIA value suggest relatively intense chemical weathering in the catchment (Figure 6.16). Post 2.55 ± 0.49 to 2.42 ± 0.29 ka, a prominent increase in Al_2O_3, K_2O, TiO_2, Na_2O, MgO, MnO, CIA, and CWI with a decrease in SiO_2 content and sedimentation rate and low CaO/MgO, Rb/Sr, and Ba/Sr ratios are noted indicating a decrease in precipitation and NEMR intensity.

Furthermore, in the Palar River lithosection, a significant increase in major oxide content is noted between 1.88 ± 0.57 and 1.41 ± 0.28 ka with high ratios of K/Al, Mg/Al, Ti/Al, CIA, Rb/Sr, and Ba/Sr values that indicate an increase in rainfall (Appendix Table 2). A sudden shift in the elemental concentration is noted post 1.41 ± 0.28 ka. However, after 1.41 ± 0.28 ka, the rate of sedimentation decreased, which is reflected by the decrease in K/Al, Mg/Al, and Ti/Al indicating an overall weakening of rainfall. A low value of CaO/MgO is also noted in the entire section, suggesting a decrease in the precipitation and monsoon intensity. Significantly, higher concentrations of all soluble and insoluble elements altogether with high CWI suggest a decrease in weathering and monsoon intensity since the late Holocene period.

PALEOCLIMATIC INFERENCES

In this study, an integration of sedimentology and geochemical data supported by OSL dates reveal the fluctuations in the intensity of NEMR precipitation during the Holocene period in and around the Palar River basin. Based on our data collection, we identify a humid, intensified monsoon phase from the PL-K section prior to ~10 ka marked as phase 1, indicating a high runoff condition because of intense precipitation. Subsequent to phase 1, there is an increase in temperature and dry conditions, and this is marked due to the decrease in NEMR spell, designated as phase 2. Around ~4.83–3.59 ka (phase 3: PL-V), the geochemical and elemental ratio records further point toward a decreasing trend of NEMR. However, prior to phase 3, we observed a dry phase (Figure 6.16) due to the change in elemental concentrations. The geochemical proxy subsequent to ~3.59 ka (phase 4) reveals a sudden rise in NEMR. Resmi et al. (2016) noted that flash flood events also occurred during this period, which might have also led to migration of the Palar River. Phase 5 is marked by a decline in NEMR precipitation resulting in an intense dry phase coupled with rapid upstream avulsion due to neotectonic activity and this was around the period 2.42 ± 0.29 to 3.26 ± 0.98 ka. The variations in elemental concentration prior to 1.88 ka also suggest a dry phase. Hence, from ~ 3.26 ± 0.98 to 1.88 ka, a continuous dry phase prevailed, or the phase 5 climatic conditions continued (Figure 6.16). The geochemical variation of major oxides and their elemental ratio reveals an amelioration in the intensity of NEMR since 1.88–1.44 ka, which has been marked as phase 6. From 1.44 ka to the present, the lowered elemental solubility and mobility reflect periods of weaker monsoon or decline in NEMR during the late Holocene and is marked as phase 7. Hence, it is observed that since the early Holocene period (>10 ka), the intensity of NEMR continuously declined with short periods of intense rainfall and longer periods of weaker NEMR precipitation causing overall lengthier periods of dry conditions.

CONCLUSIONS

The morphological indication of tectonics is supported by sedimentology, mineralogy, and geochemistry reconstruct the evolution of the Palar River drainage basin since the early Holocene period. The drainage network of the Palar River drainage basin is tectonically controlled and the knick points along the longitudinal stream profiles of uniform lithology indicate neotectonic activity along the lineaments. 'V'-shaped valleys with a low Vf ratio have developed in the sub-basins II and III of the Palar River in response to active uplift, while broad U-shaped sub-basins IV and V have formed due to lateral erosion or because of tectonic quiescence. Active strike-slip and reverse faulting in the Palar River drainage basin has produced characteristic assemblages of landforms that are considered as geomorphic markers of active tectonism ultimately resulted in deflected and beheaded streams and river ponding. Presence of these geomorphic markers reveals a strong evidence of neotectonic activity. The differential movements along the faults have resulted in the tilting of the basin that triggered channel avulsion and shifted the river flow from

NE to further south. Frequent occurrences of low-moderate earthquakes, the existence of five major faults, shear zone, and the presence of transition zone within the sub-basins suggest that the area surrounded by the Palar River basin is under stress probably due to continuous stress in MPA and sea floor spreading in the Indian Ocean. Thus, all the tectonic evidence suggests that there is a "prolonged uplift" in the northern parts of the study area, i.e., along the MPA and this would have also triggered during the Quaternary times. This phase was responsible for the migration of the channels on both the sides of MPA. Over thrust, reverse faults and strike-slip faults have been developed in this region due to neotectonic activity. Three major paleochannels were identified in the northern side of the present day Palar River. OSL chronology suggested that the migration started since 3.56 ka and has undergone clockwise migration. Another major paleochannel was identified in the southern side of the present-day channel, which does not belong to the Palar River system. Mid-late Holocene alluvial deposits of the Palar River were characterized by distinct phases of paleoenvironmental changes. From the lithofacies assemblage, we identified that the older Palar River (PL-1) is a braided river system with high-energy flow possibly due to the intensified NE monsoon. In PL-2 and PL-3, we identified debris flow in the lower part of the lithosections and facies assemblage of OF deposit with massive desiccation. These indicate that the debris flow occurred due to rapid avulsion in the upstream and subsequent deposition of the overbank material. The present day Palar basin has also been carved due to the fluctuations in the intensity of the NE monsoons. A detailed chemical analysis of the paleochannels PL-1, PL-2, and PL-3 evidently reveals a mafic source origin while PL-4 has a dominant felsic source. The Palar River sediments have a felsic source as they contain high $SiO_2\%$ and are positively correlated with all the other major oxides. The sediments of the Palar River and the palaeochannels illustrate the depositional history extending from ~10.8 ka to the present. The present study suggests an intense NEMR precipitation and wetter conditions that extend from >10.80 ka (Phase-1), 3.59–3.26 ka (Phase-4), and 1.88–1.44 ka (Phase-6). These phases can be regionally correlated with the wetter phases of the region because of higher NEMR. A decrease in the NEMR precipitation and drier conditions between <10.8 ka (Phase-2), ~4.8–3.59 ka (phase-3), 3.26–1.88 (phase-5), and 1.44 and the present (phase 7) persisted and these phases can be also regionally correlated with the drying trend of the places, which receive NEMR. Our study indicates an antiphase behavior between the NEMR dominated region and SWMR, i.e., an inverse coupling existed between NEMR and SWMR in southern peninsular India.

APPENDIX

Table 1 Major oxide of all Paleochannel (PL-1 to PL-4)

Sl.No.	Depth	SiO2	Al2O3	K2O	CaO	TiO2	Na2O	MgO	P2O5	MnO	Fe2O3	CIA	Cr	Co	Ni	Cu	Zn	Rb	Sr	Y	Zr	Ba	Pb	Th	U	V	Nb	Hf	Sc
PL-1	10	66.72	13.35	1.063	1.626	1.783	1.68	1.176	0.08	0.02	9.31	75.34	175.9	2.9	151.1	47	37.9	68.1	302.6	4.9	89.4	1033	18.8	18.9	3		1.2	1	12.63
PL-1	30	64.58	14.207	1.031	1.562	0.97	1.21	2.163	0.07	0.02	9.27	78.88	161	2.4	172.6	33.6	28.9	70.1	327	3.7	119.7	915	17.9	17.1	2.1	0.9	1.1	0.8	22.5
PL-1	50	65.14	13.063	1.203	1.27	1.1366	1.24	2.108	0.08	0.02	9.11	77.87	168.8	3.6	188.8	52.4	36.1	72.5	286	3.3	50.2	991	17.7	14.7	2.1		1.3	1.1	20.96
PL-1	70	60.52	15.539	1.637	1.029	1.2671	1.07	3.494	0.11	0.03	10.60	80.62	158.1	3	178.2	54.8	32.9	67.6	379.3	4.5	93.2	1102	36.2	16.9	3.4	0.8	1.1	1.4	18.58
PL-1	90	64.28	13.993	0.637	1.24	1.73	0.9	2.977	0.08	0.08	12.51	83.44	154.4	3.2	192.81	45	38.3	100.4	422.6	6.6	86.1	889	22.2	18.7	5	1.9	1.2	1.2	19.28
PL-1	110	60.51	15.469	1.2	1.385	1.907	1.08	3.141	0.06	0.01	14.22	80.85	148.7	2.9	186.7	37.7	31	72.4	306.1	3.9	73.4	992	24.1	18.7	2.2	1.5	1.5	1.4	21.98
PL-1	130	63.49	14.029	1.252	1.525	1.7094	1.95	3.156	0.07	0.02	14.51	74.80	179.8	1.7	179.8	35.7	29.3	75.3	283.8	4.5	68.4	991	18	14.3	2.8	1.2	1.4	1.2	22.85
PL-1	150	64.45	12.719	1.035	1.93	1.93	1.57	3.14	0.07	0.02	13.11	77.21	175.7		201.6	35.7	33.7	70.2	272.1	3.3	46.1	992	22.7	17.9	2.7		1	1	19.95
PL-1	170	63.7	16.692	1.457	1.33	1.6433	1.28	3.149	0.07	0.02	13.34	80.41	149.1	3	181.2	33.4	31.3	74.8	314.9	4.3	58.8	1094	19.4	14.2	2.7	1.3	1	2.1	21.25
PL-2	40	58.78	17.86	2.29	1.376	1.91	1.5	1.66	0.08	0.05	10.50	77.56	185.9	18.8	181.1	34.3	60.5	184.6	419.2	20.5	310.2	730	23	9.5	4.3		8.1	4.8	12.27
PL-2	60	58.42	17.23	2.184	0.739	2.97	2.7	1.778	0.08	0.05	11.89	75.39	196.1	25.6	192.6	34.3	61.1	185.6	479.6	20.4	363.5	663	24.7	20.5	5.5	77	5.9	5.4	18.1
PL-2	90	59.82	16.7	1.412	1.842	3.08	2.73	1.628	0.09	0.05	11.94	77.96	198.8	16.9	188.8	32.5	18.9	177.9	454	7	341.4	659	21.6	22.7	5.1	35.1	7.2	5.1	19.98
PL-2	120	62.22	13.8	1.295	1.565	2.45	3.8	0.434	0.07	0.03	11.94	66.55	168.1	11	198.2	28.6	15.4	179.5	370.7	4.9	106.9	949	16.7	15	3.9		2.8	2.1	18.67
PL-2	180	59.8	11.22	1.289	1.152	2.29	1.38	0.423	0.08	0.02	13.54	72.60	202.81	13.2	202.81	35.1	15.4	179.5	321.9	13	71.4	891	16.7	4.9	0.8	11.9	0.7	2.1	17.9
PL-2	200	60.36	15.5	2.098	2.60	1.38	1.38	0.933	0.08	0.03	12.90	77.00	196.9	15.6	196.9	34.3	37	189	393.2	22.3	293.4	856	23.8	22.9	7.1	16.6	4.3	2.3	13.9
PL-2	240	58.46	15.65	1.351	1.475	2.92	2.46	1.526	0.08	0.04	12.53	74.75	195.8	18.7	189.8	35.8	60.4	193.9	417.4	4.3	369.8	749	24.7	27.7	2.1	23.4	7.9	7.1	18.69
PL-2	280	59.96	16.15	1.562	1.375	3.20	1.64	0.237	0.08	0.02	12.53	77.92	175.7	12.9	201.6	38.2	36.2	72.7	373.7	5	72.7	1110	17.9	7.4	1.8	1.1	1.1	1	19.66
PL-2	300	62.94	15.555	0.54	1.641	2.33	1.5	0.253	0.10	0.02	11.40	80.86	179.1	12.6	195.2	37.5	32.8	156.6	311.3	4.3	126.8	892	14.3	7.9	0.7		1.5	1	18.9
PL-2	330	62.46	13.26	1.155	1.438	3.17	1.48	0.28	0.08	0.02	13.38	76.50	185.1	13	215.4	31.4	32.2	175.3	328.4	3.9	72.9	1027	15.7	6	3	1.2	0.8	2.1	18.66
PL-2	350	58.49	16.53	1.721	1.45	2.17	1.2	0.308	0.09	0.02	13.34	79.21	186.3	12.7	186.3	28.4	35.6	190	328.1	4.4	63	1274	21.2	6.8	3.8	1.3	1	1	19.58
PL-2	375	62.96	15.11	0.73	1.46	3.03	1.12	0.304	0.08	0.02	11.66	82.03	167.3	13.1	210.5	41	37.1	187.8	366.6	5.9	76	1135	19.9	14.3	4.1		1.7	2.1	17.03
PL-2	400	60.62	18.4	1.735	1.581	2.20	1.18	0.286	0.10	0.02	11.53	80.36	189.8	12.8	189.1	26.8	28.1	188.3	387	6.7	77.6	1104	21.2	11.5	3.3	0.9	1	2	15.89
PL-2	430	71.92	9.44	1.299	1.408	3.26	1.43	0.396	0.10	0.02	7.66	69.53	198.5	11	191.7	41	42.4	179.2	337.8	7.4	105.6	1089	16.2	7.8	3.4	1.8	1	2.4	18.56
PL-2	460	70.9	11.656	1.006	1.45	2.43	1.26	0.39	0.11	0.02	9.03	75.74	217.7	12.6	181.4	26.7	28.3	169.8	319.1	6.1	186.2	999	16.6	16.1	3.8	0.8	1	2.2	18.98
PL-2	520	68.2	10.407	1.068	1.464	3.25	1.4	0.315	0.09	0.02	11.34	72.58	196.4	12	197.9	36.4	36.8	173.5	313	7.7	137.2	953	17.4	10.1	3.6	1.3	0.4	2.1	18.96
PL-2	580	69.37	9.782	1.333	0.557	3.32	1.1	0.26	0.08	0.02	9.90	76.59	196.1	11.4	189.1	31.2	36.5	179.5	353.3	5.1	89.5	1026	19.3	19.1	2.6	1.8	3.2	2.2	16.98
PL-2	600	75.84	13.579	0.503	1.371	2.28	0.95	0.205	0.11	0.02	7.59	82.78	185.9	13	185.5	42.3	35.7	180.2	352.4	25.3	81.9	1144	20.3	8.1	2.9	1.6	1.8	8.4	18.74
PL-2	650	73.97	12.4	1.765	1.078	3.88	0.71	0.56	0.11	0.05	9.26	77.73	179.6	11	205.1	37	40	168.7	398.2	8.2	409	909	23.1	49.4	3.5	1.7	2.4	14.58
PL-2		78.72	11.982	1.014	1.797	2.51	0.54	0.324	0.08	0.03	8.01	78.15	175.8	13	199.9	42.3	40	170.8	368.5	10.8	151.7	991	23.3	17.7	3.5	1.7	2.4	2.2	17.58
PL-3	10	64.82	12.28	1.635	1.54	1.80	0.9	0.42	0.09	0.14	12.32	75.09	140.6	11.24	209.9	37	32.4	90.7	339.7	8.2	146.1	1205	23.1	16.4	3.8		1	2.1	22.39
PL-3	60	65.44	14.992	1.126	1.23	1.76	0.12	0.12	0.10	0.02	11.24	85.82	178.6	15.1	184.5	37	27	96.2	328.4	3.3	81.8	1179	21.5	15.1	3	1.2	0.9	2.1	21.98
PL-3	120	63.14	13.87	1.36	1.58	1.90	1.53	0.41	0.13	0.03	12.02	75.64	170.9	9.1	185.6	28.2	30.3	82.6	344.3	6.8	211.3	989	20.3	16.4	2.9	1.2	1.2	2.4	22.05
PL-3	190	61.76	14.04	1.236	1.15	2.71	1.76	0.80	0.11	0.04	13.74	77.20	163.2	15.1	192.1	45.5	49	79.2	412.4	12.4	300.2	858	22.6	13.3	4.6	0.9	3.5	4.4	20.95
PL-3	210	64.84	12.84	1.114	1.11	2.68	1.7	0.61	0.09	0.04	12.75	76.60	112.6	9.1	204.7	37.8	49.1	81.3	419.5	17.4	832	1012	24.4	36.9	4.1	1.5	1.8	14.3	21.09
PL-3	320	63.87	13.705	1.33	1.29	3.16	1.05	0.13	0.12	0.02	11.16	78.89	115.4	8.9	180.2	36	26.6	64.4	412.4	3.8	56.3	871	14	14.9	2.9	1.7	1.3	2.1	19.85
PL-3	410	64.75	12.43	1.289	1.84	2.30	1.86	0.25	0.09	0.02	12.11	71.35	128.6	6	191.6	32.4	29.6	77.3	381	8.6	95.1	1033	18.8	10.7	4.2	0.5	1.1	4.9	17.35
PL-4	10	83.11	12.90	0.78	0.69	0.49	0.61	0.31	0.04	0.03	0.63	86.13	1	3.5	6	4.6	8.1	17.9	283.3	8.2	450.3	474	9	8	0.6		3.1	8.7	17.89
PL-4	50	88	7.29	0.68	1.05	0.57	0.51	0.17	0.07	0.05	1.18	76.44	5.3	3.4	11	47.9	34.5	16.9	125.2	10.4	552.7	334	10.2	4.6	0.7	1.4	3.4	2.2	18.94
PL-4	70	90.68	4.54	0.63	1.05	0.45	0.38	0.36	0.06	0.09	1.61	70.18	17.4	10.2	11	55.8	38	16.3	87.7	9.7	352.4	350	7.2	1	1.2		1	10.3	12.68
PL-4	90	85.44	8.57	0.74	0.10	0.77	1.00	0.36	0.06	0.09	2.02	82.30	15.3	10.08	17.8	46.8	36	18.3	118.7	19.6	715.9	438	12.8	7.3	1.2	10.6	6.5	7.1	11.54
PL-4	100	87.25	8.23	0.84	0.82	0.58	0.53	0.28	0.07	0.05	1.21	79.04	5.4	3.7	10.4	46.9	35.5	19.6	155.7	11.1	478	442	11.6	7.2	1.2	4.2	5.9	19.65
PL-4		82.65	11.02	0.76	0.85	0.75	0.49	0.53	0.05	0.11	2.51	84.01	34.8	10.2	23.8	53.7	45.2	24.8	117.1	18.7	609.8	455	14.1	6	1.5	22.7	5.4	10.97

Table 2 Major oxide of Palar River (PR-1, PR-2, PR-3, CR)

Sl.No.	Depth	SiO_2	Al_2O_3	K_2O	CaO	TiO_2	Na_2O	MgO	P_2O_5	MnO	Fe_2O_3	CIA	Cr	Co	Ni	Cu	Zn	Rb	Sr	Y	Zr	Ba	Pb	Th	U	V	Nb	Hf	Sc
PR1-1	10	79.05	9.58	3.53	1.51	0.2157	3.59	0.29	0.08	0.02	1.51	52.61	66.9	9	8.2	27.5	26.1	87.4	355.4	5.1	88.4	1063	20.8	6.1	4	0.91	0.8	2.1	10.98
PR1-3	30	78.14	9.91	3.85	1.29	0.77	3.46	0.24	0.08	0.02	1.33	53.54	75.7	14.2	10.6	59.1	45.6	94.4	341.1	5	72	1217	21.2	5.4	3.3	0.8	1.1	1.2	8.16
PR1-5	50	77.37	10.22	3.614	1.62	0.41	3.92	0.30	0.08	0.03	1.77	52.76	78.3	3	9.2	25.9	29.9	86.9	380.4	6.2	98.5	1175	22.5	8.4	2.8	1	4.2	1.7	9.95
PR1-7	70	78.58	9.82	3.407	1.60	0.28	3.71	0.36	0.08	0.02	1.57	52.99	61.6	11	12.2	52.4	40.5	82.4	348.3	9.9	82.9	1037	19.3	7.4	3.5	1	1.4	0.9	9.54
PR1-9	90	78.92	9.81	3.244	1.73	0.29	3.62	0.34	0.08	0.03	1.56	53.32	85.6	3	11.7	61	53.9	76.1	357.7	6.8	82.3	1024	17.9	5.1	3.5	1.2	1.7	1	8.15
PR1-11	110	78.83	9.42	3.363	1.43	0.23	3.76	0.26	0.08	0.03	1.51	52.40	71.1	3	7.7	30.9	29	89.1	326.4	4.9	72.4	1032	20.4	7.5	2.4	1.3	1.6	1	9.89
PR1-13	130	80.49	9.12	3.351	1.48	0.15	3.58	0.21	0.07	0.03	1.35	52.03	70.9	3	6.6	33.1	32.2	84.7	343.6	4.1	57.6	996	19.7	4.8	3	1.9	0.8	1.8	9.45
PR1-15	150	78.86	9.58	3.124	1.62	0.20	3.84	0.31	0.08	0.03	1.60	52.74	88.1	3	11.9	42.8	35.2	76.6	379.3	5.7	73.2	1051	18.8	7.9	2.7	0.5	0.6	2.7	8.59
PR1-17	170	77.33	10.22	3.234	2.22	0.44	3.63	0.52	0.10	0.04	2.23	52.95	75.7	3	18.6	58.3	53.8	85	382.9	18.9	158.6	1025	22.1	8.3	3.2	0.4	3.3	2.2	7.91
PR1-19	190	71.31	12.09	2.544	3.32	0.87	4.12	1.08	0.14	0.06	4.17	54.78	60.1	11	21.4	40.2	55	69.5	481.5	21.3	500.7	894	26.6	4.8	3.4	22.2	6.9	7.7	8.94
PR1-21	210	70.51	12.98	2.276	3.55	0.65	3.95	1.35	0.15	0.06	4.18	57.04	81.3	12.2	28.4	58.7	73	69.2	522.9	16.2	344.5	843	23.6	9.8	4.6	1	5.8	2.2	7.98
PR1-23	230	68.56	14.11	2.338	3.23	0.71	4.33	1.45	0.17	0.06	4.19	58.78	84.6	10.2	30	52.5	72.5	67.7	527	17.2	344.7	784	26.7	2.1	5.9	14	6.1	4.9	8.56
PR1-25	250	70.91	12.86	2.471	2.95	0.76	4.53	1.09	0.16	0.05	3.82	56.38	59.7	11	24.3	42.3	64.6	67.2	496	17.3	508.3	814	24.8	2.58	4.4	24.6	6.5	9.2	8.76
PR2-1	10	73.13	12.94	2.36	2.00	0.80	4.28	0.90	0.10	0.05	3.37	59.96	47	10.8	25.8	64.3	64.8	65.5	457.5	13.8	441.8	797	24.3	23.1	7.3	34.6	8.6	5.1	6.31
PR2-2	30	69.53	15.09	2.117	2.13	0.89	2.79	1.41	0.06	0.10	5.75	68.21	64.1	9.9	40.2	49.8	72.4	79.4	357.9	17.6	293.3	751	23.5	19.6	9.3	15.3	7.2	4.3	6.42
PR2-3	50	68.54	17.13	2.388	2.69	0.59	3.92	0.94	0.13	0.05	3.26	65.57	38.6	19.5	25.8	52.1	55.9	67.8	462.9	14.1	255.1	865	20.2	12.5	9.5	13	6	3.5	5.9
PR2-4	70	73.17	12.52	2.405	2.71	0.78	4.44	0.79	0.12	0.04	3.01	56.72	46.5	18.6	23.9	47.2	50.3	63.8	446.2	17.3	540.4	804	20.1	22	6.2	12.8	5.4	6.8	7.8
PR2-5	90	78.68	12.27	2.293	1.23	0.27	2.63	0.20	0.10	0.04	1.42	66.60	9.1	3	16.9	42.3	31.3	56	245.1	4.7	75.3	747	12.5	17	4.3	0.7	5.8	1	8.9
PR2-6	110	75.57	11.77	2.29	2.33	0.84	3.69	0.51	0.11	0.04	2.58	58.61	37.7	11.26	19.6	37.3	49.2	54.1	380	10.6	185	763	14.4	14.2	3.7	1.7	3.3	1.6	7.9
PR2-7	130	74.55	15.14	2.586	1.42	0.56	3.12	0.31	0.10	0.03	1.89	67.99	15.3	11.98	15.7	38.1	32.8	57.8	289.1	6.9	126.9	812	13	14.1	5.2	1.8	2.1	1	8.5
PR2-8	150	80.92	10.34	2.565	1.11	0.54	2.53	0.20	0.06	0.02	1.16	62.49	5.7	11.87	12.7	38.2	35.3	60.2	246.2	3.5	66.9	752	15.2	14.4	5.8	1.8	2.1	3	6.2
PR3-1	10	75.4	11.98	2.185	2.62	0.81	3.97	0.45	0.10	0.04	2.31	57.72	13.3	12.56	13.4	50.3	51.9	56.5	458.4	11	425.5	778	19.2	19.9	4.8	1	4.3	6	7.2
PR3-2	30	74.28	11.09	2.41	2.64	0.78	4.87	0.47	0.11	0.03	2.35	52.78	20.2	12.49	12.3	40.5	46.1	57.5	450.4	12.3	398.2	794	18.5	17.5	5.1	13	6.2	5	7.9
PR3-3	50	76.06	11.07	2.379	2.59	0.68	4.41	0.40	0.12	0.03	2.24	54.14	20.6	11.23	18.8	38.2	49.7	58.7	434.2	11.3	381.1	827	18.7	17.5	5	11.8	5.1	3.6	8.2
PR3-4	70	76.74	9.93	2.315	2.47	0.77	4.35	0.46	0.10	0.04	2.35	52.08	24.3	12.6	15.2	57.2	53.3	57.8	420.1	13.2	385.2	829	19.3	14.1	4.6	32.4	6.4	7	10.23
PR3-5	90	82.78	7.88	2.432	1.53	0.28	3.03	0.19	0.07	0.02	1.41	52.99	6.1	3.5	8.6	33.6	33.2	60.5	320.9	4.4	79	804	16.4	5.6	2.1	13	1.2	1.6	6.8
PR3-6	110	77.3	9.57	2.325	2.30	0.82	4.22	0.41	0.09	0.03	2.33	51.96	30.7	11.9	13.1	55.5	54.2	54	396.9	11.9	402.9	769	18.7	22.9	3.5	34.8	5.8	5.1	8.9
PR3-7	130	76.5	10.61	2.518	2.33	0.60	4.39	0.37	0.08	0.03	1.98	53.45	18.5	3.3	12.9	38.7	39.8	59.6	416.2	9.8	302.3	827	15.9	17.7	4.3	13	4.1	3.3	9.7
PR3-8	150	76.7	12.95	2.613	1.72	0.29	3.41	0.21	0.08	0.02	1.39	62.58	6	3.9	8.7	34.5	38.2	62.9	355.7	5.1	102.8	840	15.7	5.8	2.8	1	1.3	1.8	8.7
PR3-9	170	81.6	8.61	2.35	1.73	0.27	3.36	0.24	0.07	0.02	1.45	53.65	8.2	3.1	9.8	40.3	38.7	57.8	329	5.5	114.1	750	17	9.7	3.9	1	1.6	1.9	10.9
PR3-10	190	71.7	12.74	2.397	3.37	0.65	5.26	0.74	0.12	0.04	2.76	53.61	11.3	12.8	17	58.5	60.4	64.5	560.3	15.4	584	889	23	22.5	7.2	1	5.1	10	6.8
PR3-11	210	80.38	11.07	2.498	1.43	0.20	2.9	0.17	0.06	0.02	1.10	61.85	5.7	3.5	9	45.9	41	57.8	299.5	3.5	58.7	830	14.1	2.8	2.5	1.9	1	1.5	5.9
PR3-12	230	82.62	8.00	2.689	1.34	0.20	2.96	0.12	0.07	0.02	1.20	53.36	5.6	3.2	7	43.1	35.2	62.7	283.2	4	62.5	884	14.2	5.1	2.6	12	1.1	2.7	7.8
PR3-13	250	82.48	8.13	2.619	1.34	0.19	3.21	0.11	0.08	0.02	1.18	53.15	5.4	3.1	9.8	61.5	44.8	60.6	279.1	3.6	73.7	910	15.4	3.3	1.9	1	3.5	2.3	8.5
PR3-14	270	82.05	7.75	2.439	1.34	0.53	2.98	0.21	0.09	0.03	1.62	53.43	8.4	2.9	9.5	38.6	37.5	55.3	266.2	6.2	133.8	870	16.6	20.9	1.7	1.9	1.2	1.7	9.8
PR3-15	290	78.77	9.37	2.513	1.98	0.59	3.51	0.40	0.09	0.04	2.06	53.93	6	2.8	13.7	54.2	47.6	58.3	342.8	9.4	171.7	861	17.1	10.3	2.9	12	3.8	2.9	7.5
PR3-16	310	82.13	8.35	2.404	1.60	0.26	3.57	0.21	0.08	0.02	1.33	52.46	5.5	1.9	10	61.5	42.3	54	294.1	5.2	73.6	802	14.6	4.2	1	25	2.6	2.2	8.4
PR3-17	330	77.38	10.01	2.393	2.11	0.61	4.33	0.37	0.09	0.03	1.88	53.13	19.2	1.45	13.1	46.9	43.5	54.4	362.1	7.7	204.5	797	18.2	18.7	0.9	1.9	10.2	21	9.7
PR3-18	350	81.5	8.44	2.592	1.50	0.33	3.19	0.15	0.09	0.02	1.48	53.67	9.1	2.98	11.2	53.4	42.1	59.3	300.3	5.4	97.2	850	15.6	9.3	2.6	2.3	3.4	2	5.8
PR3-19	370	74.08	10.81	2.39	2.69	1.31	4.34	0.50	0.12	0.05	3.45	53.44	30.5	3.1	16.7	41.3	50	55.4	425.2	22.7	1243	705	22.3	21.1	7.4	11.7	2.5	6.7	7.9
PR3-20	390	78.96	9.16	2.564	1.89	0.50	3.75	0.30	0.10	0.03	1.76	52.76	7.7	11.25	10.9	56.9	50.2	60	344.1	8	191.9	829	15.8	12.2	3.3	12	2.2	1.5	9.5
PR3-21	410	81.7	8.34	2.39	1.58	0.44	3.06	0.27	0.08	0.02	1.48	54.25	16.2	2.78	12.5	39.4	40.2	53.3	295.4	6.6	139.5	853	18.1	12.6	2	13	1.5	1	8.2
PR3-22	430	82.14	8.40	2.446	1.55	0.28	3.2	0.23	0.09	0.02	1.52	53.85	5.9	1.98	10.4	38.3	33.7	54.5	295.8	5.6	111.7	781	14.3	7.1	2.7	1	1.6	2.1	8.6
PR3-23	450	81.48	9.62	2.334	1.37	0.22	2.64	0.26	0.08	0.02	1.31	60.29	5.9	2.79	8	36.9	39.2	56.4	279.5	4.7	82.8	810	12.8	10.6	3.7	12	0.6	1	7.9
PR3-24	470	81.15	9.31	2.485	1.54	0.27	3.1	0.20	0.08	0.02	1.42	56.67	5.3	1.98	10.3	44.3	40.6	55.9	310.8	4.3	146.2	781	15.6	3.5	2.6	1	1.6	2.2	9.5
PR3-25	490	84.49	7.38	2.727	1.09	0.20	2.47	0.16	0.08	0.02	1.22	53.98	7.9	12.25	10.9	37.8	31.1	62.2	257.7	4.2	68.3	848	14.9	5.2	2.7	1	0.6	1	8.7
PR3-26	510	81.86	9.28	2.53	1.33	0.27	2.46	0.16	0.08	0.02	1.43	59.51	7.9	1.96	10.4	33.1	34.9	59.8	281.9	5.7	83.7	861	12.8	6	2.6	32.1	0.7	1.9	6.9
PR3-27	530	83.04	8.58	2.774	1.24	0.19	1.87	0.16	0.07	0.02	1.49	59.31	9.1	12.98	14.7	124.7	74.1	62.5	288	4.8	74.7	972	14.6	5.5	2.8	4.9	2.5	2.1	7.6
PR3-28	550	83.32	7.88	2.563	1.33	0.22	2.78	0.16	0.07	0.02	1.22	54.12	9.3	3.5	7.9	36.8	33.2	57.2	278.2	4.1	62.9	844	15	5.8	2.6	8.6	1.9	2.9	8.2
CR-1	10	71.32	13.74	1.78	3.46	1.38	2.47	1.11	0.15	0.06	4.48	64.06	46.6	16.2	22.1	21.7	46.5	37.1	556.9	12.3	402.9	834	16.8	19.8	4.1	39.9	9.9	5.5	8.5
CR-2	30	75.57	12.58	1.48	3.02	0.87	1.86	0.81	0.10	0.04	3.06	66.43	43.1	11.9	22.4	57.5	59.1	30.9	477.7	8.9	219.7	714	15	16.3	4	29.4	5.5	2.3	7.6
CR-3	50	66.25	11.59	1.35	4.96	2.40	2.43	1.91	0.17	0.12	8.53	57.01	40.8	17.5	32.2	63.5	88	29.1	383.8	20	332.9	620	12.1	16.3	4.7	98	14.8	4	8.4
CR-4	70	69.77	14.20	1.56	3.38	0.85	2.58	1.56	0.12	0.06	4.94	65.38	61.5	12	38	56	72.9	38.8	494.7	14.4	213	754	14.8	18.4	4	57	5.2	2.5	7.9
CR-5	90	80.63	6.98	1.52	1.78	1.64	2.3	0.52	0.11	0.05	3.63	55.46	46	11	20.3	68.6	63.3	32	293.2	8.3	257.4	694	15.1	20.5	6.3	31	9.1	3.7	8
CR-6	110	85.69	6.21	1.31	1.59	0.31	2.38	0.33	0.09	0.04	1.67	54.07	17	10	18.4	74.6	52.1	25.4	261.2	5	83.3	587	10.1	12.7	2	41.23	1.8	1.2	6.27
CR-7	130	83.38	6.57	1.10	1.78	0.72	2.07	0.62	0.11	0.05	2.66	57.02	22.3	10	18.3	53.4	53.7	22.1	254.8	7.9	112.9	547	10.3	18.2	5.2	1.2	4.8	1	8.95
CR-8	150	80.88	6.79	1.25	1.85	1.46	2.74	0.55	0.12	0.05	3.78	53.75	61.7	14.3	18.3	58.2	56.4	23.4	286.1	11.1	219.2	564	11.4	24.4	6.2	90	10	2.2	7.6
CR-9	170	81.16	8.04	1.15	2.00	1.15	2.02	0.63	0.13	0.05	2.99	60.86	41.7	10	19.1	69.2	62.8	22.8	281.7	8.7	148.2	537	12.1	18.1	6.1	44	6.5	2.8	7.5
CR-10	190	84.28	6.04	1.00	1.58	0.54	2.27	0.50	0.11	0.04	2.75	55.43	28	10	16.9	50.1	48.9	19.9	213.7	6.3	110.9	499	9.6	4.3	7.8	2.3	2.6	1	6.2

REFERENCES

Achyuthan, H, Nagasundaram, M, Gourlan, TA, Eastoe, C, Ahmad, SM, & Veena, MP 2014. Mid-Holocene Indian Summer Monsoon variability off the Andaman Islands, Bay of Bengal, *Quaternary International*, vol. 349, pp. 232–244. DOI: 10.1016/j.quaint.2014.07.041.

Allen, JRL 1983. Riverbed forms: progress and problems, In: Collinson, JD & Lewin, J (Eds.), *Modern and Ancient* Fluvial Systems, International Association of Sedimentologists Special Publication. vol. 6, pp. 19–33.

Allen, JRL & Leeder, MR 1980. Criteria for the instability of upper-stage plane beds, *Sedimentology*, vol. 27, pp. 209–217.

Ascione, A & Romano, P 1999. Vertical movements on the eastern margin of the Tyrrhenian extensional basin: New data from Mt. Bulgheria (Southern Apennines, Italy), *Tectonophysics*, vol. 315, pp. 337–356.

Babu, PVLP 1975. Morphological evolution of the Krishna Delta, *Journal of Indian Society of Remote Sensing, Photonirvachak*, vol. 3, pp. 21–27.

Bano, M, Marquis, G, Niviere, B, Maurin, JC & Cushing, M 2000. Investigating alluvial and tectonic features with ground penetrating radar and analyzing diffraction patterns, *Journal of Applied Geophysics*, vol. 43, pp. 33–41.

Bernard, M., Shen-Tu, B, Holt, WE & Davis, D 2000. Kinematics of active deformation in the Sulaiman Lobe and Range, Pakistan, *Journal of Geophysical Research*, vol. 105, pp. 253–279.

Best, JL 1996. The fluid dynamics of small-scale alluvial bedforms, In: Carling, PA & Dawson, MR (Eds.), *Advances in Fluvial Dynamics and Stratigraphy*, pp. 67–125.

Bishop, P 1995. Drainage rearrangement by river capture, beheading and diversion, *Progress in Physical Geography*, vol. 19, pp. 449–473.

Bookhagen, B, Thiede, RC & Strecker, MR 2005. Abnormal monsoon years and their control on erosion and sediment flux in the high, arid northwest Himalaya, *Earth and Planetary Science Letters*, vol. 231, pp. 131–146.

Brice, JC 1982. *Stream Channel Stability Assessment*, Federal Highway Administration, Washington, DC.

Brice, JC 1983. Planform properties of meandering rivers, In: Elliott, CM (Ed.), *River Meandering*, American Society of Civil Engineers, New York, pp. 1–15.

Bridge, JS 2003. *Rivers and Floodplains, Forms, Processes, and Sedimentary Record*, Blackwell Publishing, Oxford.

Bristow, CS 1993. Sedimentology of the rough rock: A Carboniferous braided river sheet sandstone in N England. In: Best, JL & Bristow, CS (Eds.), Geological Society of London Special Publication, Braided Rivers, vol. 75, pp. 291–304.

Burbank, DW & Anderson, RS 2001. *Tectonic Geomorphology*, IInd ed. Blackwell Scientific, Oxford. pp. 270.

Burbank, DW, Blythe, AE, Putkonen, JK, Pratt-Situala, BA, Gabet, EJ, Oskin, ME, Barros, AP & Ohja, TP 2003. Decoupling of erosion and climate in the Himalaya, *Nature*, vol. 426, pp. 652–655.

Chamyal, LS, Maurya, DM & Raj, R 2003. Fluvial systems of dry lands of western India: A synthesis of Late Quaternary palaeoenvironmental and tectonic changes, *Quaternary International*, vol. 104, pp. 69–86.

Chamyal, LS, Maurya, DM, Bhandari, S & Raj, R 2002. Late Quaternary geomorphic evolution of the lower Narmada valley, Western India: Implications for neotectonic activity along the Narmada–Son Fault, *Geomorphology*, vol. 46, pp. 177–202.

Chow, J, Angelier, J, Hua, JJ, Lee, J & Sun, R 2001. Palaeoseismic event and active faulting from ground penetrating radar and high resolution seismic reflection profiles across the Chihshang fault, eastern Taiwan, *Tectonophysics*, vol. 333, pp. 241–259.

Clark, MK, Schoenbohm, LM, Royden, LH, Whipple, KX, Burchfiel, BC, Zhang, X, Tang, W, Wang, E & Chen, L 2004. Surface uplift, tectonics, and erosion of eastern Tibet from large-scale drainage patterns, *Tectonics*, vol. 23, p. TC1006.

Clift, PD, Giosan, L, Carter, A, Garzanti, E, Galy, V, Tabrez, AR & Rabbani, MM 2010. Monsoon control over erosion patterns in the Western Himalaya: Possible feedback into the tectonic evolution, *Geological Society, London, Special Publications*, vol. 342, no. 1, pp. 185–218.

Cox, RT 1994. Analysis of drainage basin asymmetry as a rapid technique to identify areas of possible Quaternary tilt-block tectonics: An example from the Mississippi Embayment, *Geological Society of America Bulletin*, vol. 106, pp. 571–581.

D'Alessandro, L, Miccadei, E & Piacentini, T 2008. Morphotectonic study of the lower Sangro river valley (Abruzzi, central Italy), *Geomorphology*, vol. 102, pp. 145–158.

Dehbozorgi, M, Pourkermani, M, Arian, M, Matkan, AA, Motamedi, H & Hosseiniasl, A 2010. Quantitative analysis of relative tectonic activity in the Sarvestan area, central Zagros, Iran, *Geomorphology*, vol. 30, pp. 329–341. DOI: 10.1016/j.geomorph.2010.05.002.

Demoulin, A 2011. Basin and river profile morphometry: A new index with a high potential for relative dating of tectonic uplift, *Geomorphology*, vol. 126, pp. 97–107.

El Hamdouni, R, Irigaray, C, Fernandez, T, Chacon, J, Keller, EA 2007. Assessment of relative active tectonics, southwest border of Sierra Nevada, southern Spain, *Geomorphology*, vol. 96, pp. 150–173.

Ethridge, FG, Skelly, RL & Bristow, CS 1999. Avulsion and crevassing in the sandy, braided Niobrar River: Complex response to base level rise and aggradation, In: Smith, ND & Rogers, J (Eds.), *Fluvial Sedimentology VI*. Spl. Pub. International Association of Sedimentologist, Blackwell Science, Oxford, UK, vol. 28, pp. 179–191.

Gao, C, Boreham, S, Preece, RC, Gibbard, PL, & Briant, RM 2007. Fluvial response to rapid climate change during the Devensian (Weichselian) Late glacial in the River Great Ouse, southern England, *Sedimentary Geology*, vol. 202, pp. 193–210.

Gao, MX, Zeilinger, G, Xu, XW, Wang, QL & Hao, M 2013. DEM and GIS analysis of geomorphic indices for evaluating recent uplift of the northeastern margin of the Tibetan plateau, *China Geomorphology*, vol. 190, pp. 61–72.

Gayantha K, Routh J & Chandrajith R 2017. A multi-proxy reconstruction of the late Holocene climate evolution in LakeBolgoda, Sri Lanka, *Palaeogeography, Palaeoclimatology, Palaeoecology*, vol. 473, pp. 16–25.

Gloaguen, R, Kaessner A, Wobbe F, Shahzad F, & Mahmood SA 2008. Remote Sensing analysis of crustal deformation using river networks, In: *IEEE International Geoscience and Remote Sensing Symposium*, Boston, MA, pp. IV-1–IV-4.

Gupta, S 1997. Himalayan drainage patterns and the origin of fluvial mega fans in the Ganges foreland basin, *Geology*, vol. 25, pp. 11–14.

Holbrook, J & Schumm, SA 1999. Geomorphic and sedimentary response of rivers of tectonic deformation; a brief review and critique of a tool recognizing subtle epirogenic deformation of modern and ancient settings, *Tectonophysics*, vol. 305, pp. 287–306.

John, B & Rajendran, CP 2008. Geomorphic indicators of Neotectonism from the Precambrian terrain of peninsular India: A study from the Bharathapuzha Basin, Kerala, *Journal of the Geological Society of India*, vol. 71, pp. 827–840.

Jongman, RHG 2006. *Pantanal–Taquari; tools for decision making in integrated water management*, Alterra-rapport 1295, Alterra, Wageningen, p. 215.

Joshi, PN, Maurya, DN & Chamyal, LS 2013. Morphotectonic segmentation and spatial variability of neotectonic activity along the Narmada–Son Fault, Western India- Remote sensing and GIS analysis, *Geomorphology*, vol. 180–181, pp. 292–306.

Kale, VS 2007. Fluvio–sedimentary response of the monsoon-fed Indian rivers to Late Pleistocene–Holocene changes in monsoon strength reconstruction based on existing[14]C dates, *Quaternary Science Reviews*, vol. 26, pp.1610–1620.

Keller, EA & Pinter, N 2002. *Active Tectonics: Earthquakes, Uplift, and Landscape*, 2nd ed., Prentice Hall, Upper Saddle River, NJ.

Klimek, K 1974. The structure and mode of sedimentation in the food-plain deposits in the Wisłoka valley (S. Poland), *Studia Geomorphologic a Carpatho-Balcanica*, vol. 8, pp. 135–151.

Krishnan, MS 1968. *Geology of India and Burma*, Higginbothams Ltd, Madras.

Li, J, Xie, S & Kuang, M 2001. Geomorphic evolution of the Yangtze gorges and the time of their formation, *Geomorphology*, vol. 41, pp. 125–135.

Luirei, K & Bhakuni, SS 2008. Geomorphic imprints of neotectonic activity along the frontal part of eastern Himalaya, Pasighat, East Siang District, Arunachal Pradesh, *Journal of the Geological Society India*, vol. 71, pp. 502–512.

Mather, AE 2000. Adjustment of drainage network to capture induced base-level change: An example from the Sorbas basin, SE Spain, *Geomorphology*, vol. 34, pp. 271–289.

Mather, AE, Stokes, M & Griffiths, JS 2002. Quaternary landscape evolution: A framework for understanding contemporary erosion, SE Spain, *Land Degradation and Management*, vol. 13, pp. 1–21.

Maurya, DM, Patidar, AK, Mulchandani, N, Goyal, B, Thakkar, MG, Bhandari, S, Vaid, SI, Bhatt, NP & Chamyal, LS 2005. Need for initiating ground penetrating radar studies along active faults in India: An example from Kachchh, *Current Science*, vol. 88, pp. 231–240.

Mrinalini Devi, RK 2008. Tectono-geomorphic forcing of the frontal sub-Himalayan streams along the Kimin section in the Arunachal Himalaya, *Journal of the Geological Society of India*, vol. 72, no. 2, pp. 253–262.

Narasimhan, TN 1990. Paleochannels of Palar River west of Madras city possible implication of vertical movement, *Journal of the Geological Society of India*, vol. 136, pp. 471–474.

Nicoll, TJ & Hickin, EJ 2010. Planform geometry and channel migration of confined meandering rivers on the Canadian prairies, *Geomorphology*, vol. 116, pp. 37–47.

Nicoll, TJ 2008. *Planform geometry and kinematics of confined meandering rivers on the Canadian prairies*, M.Sc. Thesis, Simon Fraser University, Burnaby, BC, Canada.

Ouchi, S 1985. Response of alluvial rivers to slow active tectonics movement, *Geological Society of American Bulletin*, vol. 96, pp. 504–515.

Pérez-Peña, JV, Azor, A, Azañón, JM & Keller, EA 2010. Active tectonics in the Sierra Nevada (Betic Cordillera, SE Spain): Insights from geomorphic indexes and drainage pattern analysis, *Geomorphology*, vol. 119, pp. 74–87.

Radhakrishna, BP 1993. Neogene uplift and geomorphic rejuvenation of the Indian Peninsula, *Current Science*, vol. 64, pp. 787–793.

Ramasamy, SM, Kumanan, CJ, Selvakumar, R & Saravanavel, J 2011. Remote sensing revealed drainage anomalies and related tectonics of South India, *Tectonophysics*, vol. 501, pp. 41–51.

Rangaraju, MK, Agrawal, A & Prabhakar, KN 1993. Tectono-stratigraphy, structural styles, evolutionary model and hydrocarbon habitat, Cauvery and Palar basins, In: S.K. Biswas, et al. (Eds.), *Proceedings of Second Seminar on Petroliferous Basins of India*, vol. 1, KDMIPE, ONGC, Dehra Dun , pp. 371–388.

Rashed, M, Kawamura, D, Nemoto, H, Miyata, T & Nakagawa, K 2003. Ground penetrating radar investigations across the Uemachi fault, Osaka, Japan, *Journal of Applied Geophysics*, vol. 53, pp. 63–75.

Resmi, MR, Achyuthan, H & Jaiswal, MK 2016. Middle to late Holocene paleochannels and migration of the Palar River, Tamil Nadu: Implications of neotectonic activity Quaternary International, *Quaternary International*, vol. 443, pp. 211–222. DOI: 10.1016/j.quaint.2016.05.002.

Rockwell, TK, Keller, EA & Johnson, DL 1984. Tectonic geomorphology of alluvial fans and mountain front near Ventura, California, In: Morisawa, M & Hack, JT (Eds.), *Tectonic Geomorphology — Proceedings of the 15th Annual Binghamton Geomorphology Symposium, September 1984*, Allen and Unwin, Boston, MA, pp. 183–208.

Salvany, JM 2004. Tilting neotectonics of the Guadiamar drainage basin, SW Spain, *Earth Surface Processes and Landforms*, vol. 29, pp. 145–160.

Saravanavel, J & Ramasamy, SM 2016. Active tectonics and its impacts over groundwater systems in the parts of Tamil Nadu, India, *Arabian Journal of Geosciences*, vol. 9, p. 429. DOI: 10.1007/s12517-016-2459-x.

Schoenbohm, LM, Whipple, KX, Burchfiel, BC & Chen, L 2004. Geomorphic constraints on surface uplift, exhumation, and plateau growth in the Red River region, Yunnan Province, China, *Geological Society of America Bulletin*, vol. 16, pp. 895–909.

Schumm, SA, Dumont, JF & Holbrook, JM 2000. *Active Tectonics and Alluvial Rivers*, Cambridge University Press, Cambridge, UK, p. 274.

Singh, IB, Ansari, AA, Chandel, RS & Misra, A 1996. Neotectonic control on drainage system in Gangetic plain, Uttar Pradesh, *Journal of the Geological Society of India*, vol. 47, pp. 599–609.

Slingerland, R & Smith, ND 1998. Necessary conditions for meandering-river avulsion, *Geology*, vol. 26, no. 5, pp. 435–438.

Stokes, M, Mather, AE & Harvey, AM 2002. Quantification of river capture induced base-level changes and landscape development, Sorbas Basin, SE Spain In: Jones, S.J. & Frostick, L.E. (Eds.), *Sediment Flux to Basins, Causes, Controls and Consequences*, Geological Society, London, UK, Special Publication, vol. 191, pp. 23–35.

Subramanian, KS and Selvan, TA. 2001. *Geology of Tamil Nadu and Pondicherry*. Geological Society of India, Bangalore, pp. 7–19. &

Sun, Q, Wang, S, Zhou, J, Chen Z, Shen J, Xie X, Wu F, & Chen P. 2010. Sediment geochemistry of Lake Daihai, north-central China: Implications for catchment weathering and climate change during the Holocene, *Journal of Palaeolimnology*, vol. 43, pp. 75–87.

Talwani, P & Rajendran, K 1991. Some Seismological and geometric features of the intraplate earthquakes, *Tectnophys*, vol. 186, pp. 19–41.

Thorne, CR 1997. Channel types and morphological classification, In: Thorne, CR (Ed.), *Applied Fluvial Geomorphology for River Engineering and Management*, Wiley, Chichester, UK, pp. 175–222.

Tooth, S, McCarthy, TS, Brandt, D, Hancox, PJ & Morris, R 2002. Geological controls on the formation of alluvial meanders and floodplain wetlands: The example of the Klip River, Eastern Free State, South Africa, *Earth Surface Processes and Landforms*, vol. 27, pp. 797–815.

Twidale, CR 2004. River patterns and their meaning, *Earth Science Reviews*, vol. 67, pp. 159–218.

Vaidyanadhan, R 1971. Evolution of the drainage of Cauvery in South India, *Journal of the Geeological Society of India*, vol. 12, pp. 14–23.

Valdiya, KS 2001. River response to continuing movements and the scarp development in central Sahyadri and adjoining coastal belt, *Journal of the Geological Society of India*, vol. 51, pp. 139–166.

Vemban, NA, Subramanian, KS, Gopalakrishnan, K & VenkataRao, V 1977. Major faults/dislocations/lineaments of Tamil Nadu Miscellaneous Publication, *Geological Survey of India*, vol. 31, pp. 53–56.

Wells, S, Bullard, T, Menges, T, Drake, P, Karas, P, Kelson, K, Ritter, J & Wesling, J 1988. Regional variations in tectonic geomorphology along segmented convergent plate boundary, *Pacific Costa Rica, Geomorphology*, vol. 1, pp. 239–265.

7 Importance of the Geochemistry of Northern Indian Ocean Sediments for Assessing the Quaternary Climate Change and Future Directions

Busnur Rachotappa Manjunatha
Mangalore University

Keshava Balakrishna
Manipal Institute of Technology, Manipal
Academy of Higher Education

Jithin Jose and A. Naveen Kumar
Mangalore University

CONTENTS

INTRODUCTION

The Cenozoic Era is not only the youngest and most important in the geologic history in terms of shaping the landscape of the world but also responsible for the initiation of the global monsoon and evolution of life (Davies et al. 1975; Kolla and Kidd 1982; Molnar et al. 1993; Prell et al. 1991; Zachos et al. 1999; Erwin 2009; Wang et al. 2014, 2019; Saupe et al. 2020). In order to get more insights into the origin of ocean basins, and their evolution, paleoclimate, and paleoceanography, they are investigated through the collection of long cores that often involve drilling from the international agencies, such as Deep-Sea Drilling Project (DSDP), Ocean Drilling Program (ODP), and Integrated Ocean Drilling Program (IODP). The important drilled sites in the Northern India Ocean are shown in Figure 7.1. During such expeditions, detailed investigations have been made on isotopes, clay minerals, particle size, mineral magnetic parameters, and micropaleontological proxies to (a) know the prominence of insolation changes on climate change, (b) fluctuations in the velocity

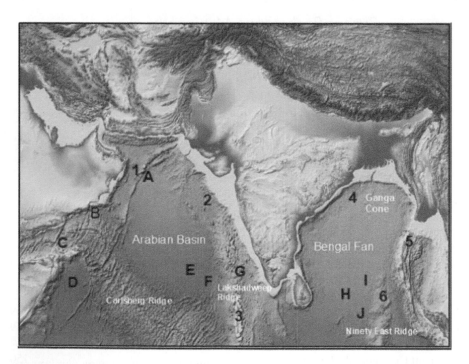

FIGURE 7.1 Physiography of South Asia and sea-floor topography of Northern Indian Ocean (adopted from https://maps.ngdc.noaa.gov/viewers/marine_geology) with DSDP (A: 222, 223 & 224; B: 233; C: 231; D: 234 & 235; E: 221; F: 220; G: 219; H: 218; I: 217; and J: 216; Roonwal 1986), ODP (1 Leg 117), and IODP (2: 355, 3: 259, 4: 353) drilled sites (Clemens et al. 2016).

of monsoon winds (c) type of vegetation existed on the mainland (d) influence of river input to the ocean system, and (e) rate of upwelling and primary productivity, denitrification, and evolution of oxygen minimum zone in the marine environment (Roonwal 1986; Prell et al. 1992; Bloemendal et al. 1993; Naidu and Malmgren 1996; Gupta et al. 2015; Pande et al. 2016; Wang and An 2008; Wang et al. 2017, 2018, 2019).

In the Cenozoic Era, the Quaternary is the most recent period for understanding climate change, glaciation, sea-level fluctuations from glacial to interglacial periods (120 m), an increase in atmospheric CO_2 level, and evolution of human beings (Denton et al. 2010; Anderson et al. 2013; Singh et al. 2020). The Quaternary Period begins with the Marine Isotopic Stage (MIS) 103 of the Gelasian Stage demarcated at 2.58 Ma to distinguish from the older Piacenzian Stage of the Neogene Period (Cohen et al. 2013; Head and Gibbard 2015; Gibbard and Lewin 2016; Head 2019). The Quaternary Period has been subclassified into the older Pleistocene and the younger Holocene epochs with demarcation at 11.7 ka. The Quaternary Period has been witnessed by the well-demarcated glacial stages separated by interglacial ones. Mulitin Milankovitch's first dominant type of climatic cycles – the precession or eccentricity that explains seasonal changes in the long-term climatic conditions from warm to cold, and vice versa – is due to changes in the incoming solar radiation at time scales from 80 to 120 ka, with a maximum at 400 ka. However, time scales are at an average of 100 ka (Denton et al. 2010; Wang et al. 2019). In the Mediterranean Sea, a significant increase in the faunal density and sapropel accumulation were noticed from Mio-Pliocene – Pliocene to the Quaternary Period due to an increase in the ocean circulation as well as ventilation of the deep water (Thunnell et al. 1995).

The cyclic changes in glacial and interglacial climate can be understood by the study of marine sediments as well as polar ice cores. The polar ice caps are indeed better-suited archive for the reconstruction of high-resolution climate change over a million-year time scale (Ma; Luthi et al. 2008; Masson-Delmotte et al. 2008). Although marine sediments and oceanic rocks provide paleoclimatic, paleoceanographic (Molnar and England 1990; Prell et al. 1991; Wang and An 2008; Erwin 2009; Wang et al. 2019; Saupe et al. 2020), and paleotectonic information since the Jurassic Period, the resolution of climate change derived from the drilled cores is much lower than that from the ice cores (100–1,000 years; Wang et al. 2019). Therefore, the polar ice caps are indispensable for deriving the high-resolution climatic fluctuations. Their age range is, however, restricted from a few hundred thousand years to nearly 1 million years (Grootes et al. 1993; Luthi et al. 2008; Higgins et al. 2015). Although these ice cores are useful for understanding the human's impact on the earth's environment, particularly in tracing the historical variations in the concentration of greenhouse gases, they are insufficient to cover the entire time span of the Quaternary Period (Luthi et al. 2008; Masson-Delmotte et al. 2008). In order to fill up this gap, Voosen (2017) has retrieved the blue ice core of about 2.7 Ma old, from the northwest of Dome C Site (0.8 Ma old core) in Antarctica. The results of this core are yet to become available.

Although there are many terrestrial paleoclimatic archives, such as speleothems, marsh/peatland, lake sediments, Himalayan ice caps, river flood plains, sand dunes, tree rings, and paleosols, they fail to retain continuous records of the climate change, owing to the postdepositional disturbances in the sedimentary column (physical processes – erosion and vertical mixing, bioturbation, lapses in sedimentation; hiatus). In this regard, marine sediments seem to be better for retrieving paleoclimatic and paleoceanographic changes over a longer time scale.

SCOPE AND OBJECTIVES

In India, the Pliocene and Pleistocene deposits are quite widely distributed in foothills of the Himalaya, Vale of Kashmir, Andaman and Nicobar Islands, and Saurashtra, Cambay, Kerala and Tamil Nadu, Andhra Pradesh and West Bengal coastal regions (Sastry 1997). In this chapter, the geochemical importance of studying marine sediments of the Northern Indian Ocean has been outlined to infer paleoclimatic and paleoceanographic changes occurred during the Quaternary Period. Further, suggestions have been made to improve the resolution of reconstruction to progress the climate forecasting models.

GEOCHEMICAL PROXIES FOR UNDERSTANDING THE CLIMATE CHANGE

Geochemistry is an interdisciplinary science of geology and chemistry that concerns about the distribution and migration and governing laws of elements in different layers of the earth in space and time (Mason and Moore 1982; White 2018). There are many related subjects of this discipline, for instance, cosmochemistry that deals with the origin, synthesis, and abundance of elements in the universe as well as in other heavenly bodies (Mason and Moore 1982). Like genera and species in biology, elements and isotopes in geochemistry are useful in determining not only sources of materials, such as minerals, rocks, sediments, and water (including the air mass), but also their ages. Isotopes are different atomic masses of the same element, which can be further categorized into radioactive and stable isotopes depending upon the radioactive disintegration or fractionation of isotopes due to variations in physicochemical conditions in the earth system, respectively (Mason and Moore 1982; White 2018). Another group of isotopes are formed due to the interaction of atmospheric gases with cosmic rays in the upper atmosphere, which are commonly known as cosmogenic isotopes (Krishnaswami and Lal 1978). A certain group of radioactive isotopes are liberated as fission products during the usage of nuclear fuel and nuclear weapon testing. Such nuclides are known as fission products or artificial nuclides or fallout nuclides (Krishnaswami and Lal 1978). Many long-lived radioactive isotopes (U-238/Pb-206, Th-230/234U, Pb-210, K-40, C-14, Be-10) are useful in determining the age of minerals, rocks, and fossils; stable isotopes provide information about the physicochemical and biological conditions existed in the geological history (Table 7.1). Hence, such isotopes are useful for the paleoclimatic and paleoceanographic studies (Krishnaswami and Lal 1978; Ivanovich and Harmon 1992; Walker 2005; Lal and Baskaran 2011; Baskaran 2012; and chapters and references therein). Therefore, isotopes disciplines can be segregated as isotope geology/isotope geochemistry or geochronology. Geochemistry becomes an indispensable branch of geology in solving

TABLE 7.1
Important Radioactive Isotopes and Stable Nuclides Used Widely in Earth Science

Isotope	Half-Life (Years)	Dating Range (Years)	Datable Material/Applications
Primordial Nuclides			
^{87}Rb/^{87}Sr	4.88×10^{10}	Precambrian rocks and minerals	Whole rock, feldspar, and micaceous minerals
^{40}K/^{40}Ar	1.19×10^{10}	Precambrian rocks and minerals	Whole rock, feldspar, and micaceous minerals
^{147}Sm/^{143}Nd	1.06×10^{11}	Precambrian rocks and minerals	Ferro-magnesium minerals and whole rock
^{238}U/^{206}Pb	4.47×10^{9}	Precambrian rocks and minerals	Any material containing uranium, zircon, uraninite, monazite, pitchblende, and whole rock
U-Th Decay Series Nuclides			
^{230}Th	7.53×10^{4}	350×10^{3}	Deep-sea sediments, speleothems, fossils, corals, and polymetallic nodules
^{234}Th	24 days	~6 months	High rate of sedimentation and sediment mixing rates
^{231}Pa	3.43×10^{4}	Up to 200×10^{3}	Marine sediments
^{234}U	2.48×10^{5}	Up to 13×10^{5}	Fossils, corals
^{210}Pb	22.3	Up to 150	Accumulation rates of snow fall and sedimentation
Cosmogenic			
^{3}H			
^{14}C	5,730	$40–45 \times 10^{-4}$	Any material containing carbon and calcium carbonate
^{10}Be	1.5×10^{6}	9.0×10^{6}	Oceanic sediments, rock encrustation, and polymetallic nodules
^{7}Be	53 days	1	High rate of sedimentation and sediment mixing rates
Nuclear Fission Products			
^{134}Cs	~2	10	Trace contamination at nuclear power stations
^{137}Cs	30	150	Land erosion and validated ^{210}Pb dating
^{14}C	5,730	75^{6}	Mixing of the upper ocean water
Stable Isotopes			
δ^{18}O	-	-	Paleo-precipitation, temperature, and determine marine isotopic stages
δ^{15}N	-	-	Denitrification rates in the oceans
δ^{13}C	-	-	Sources of organic matter in the ocean sediments
δ^{34}S	-	-	Redox processes in the sediments

Source: Krishnaswami and Lal (1978); Mason and Moore (1982); Faure (1986); Ivanovich and Harmon (1992); Baskaran (2012).

most of the problems, including the assessment of the anthropogenic impact on earth (Siegel 1979; Brocker and Peng 1982; Byrne et al. 1996; Turekian and Holland 2013; White 2018). Among isotopes, long-lived radiogenic isotopes are useful for understanding the composition and evolution of rocks and sediments, and fingerprinting their sources, including their temporal information (White 2018). Based on half-lives and their measurable radioactivity in the material to be dated, radionuclides can be classified into three types (Krishnaswami and Lal 197):

1. Primordial radionuclides are not only having long half-lives but also present during the formation of the earth and will continue to be present over a long time (U-Pb, Sm/Nd, Rb-Sr, K-Ar, Re-Os). These nuclides are suitable for the determination of ages of old rocks (White 2018).
2. Cosmogenic nuclides are produced by the interaction of cosmic rays with atmospheric gases in the upper atmosphere (e.g., Be-10, C-14, H-3, Al-27, Cl-36). These isotopes are used to determine ages of carbonaceous and calcareous sediments, polymetallic nodules as well as deep-sea sediments.
3. Fission products or artificial or manmade/anthropogenic or nuclear bomb-produced nuclides (Cs-137, Sr-90, Pu-239, etc.), which are produced during the nuclear fission reactions. The maximum fallout nuclides from the atmosphere during 1963 can be used as a stratigraphic marker in the lake and marine sedimentary columns from which average sedimentation can be estimated. Fission products like Cs-137 are useful for estimating the changes in erosion rates of the river basin from which the human's impact on deforestation can be studied.

Furthermore, depending upon the applications, several branches of geochemistry have been emerging along with overlapping subjects (White 2018):

Exploration geochemistry (concerns about the fingerprinting of economically important mineral/metal deposits), biogeochemistry (interaction between geological, biological, aqueous, and atmosphere, and budgeting of carbon, nitrogen, phosphorus, and sulfur in the earth, atmosphere, and ocean systems that are also related to global warming), pure and theoretical geochemistry (modeling geochemical processes based on experimental data), Precambrian geochemistry (the study of ancient rocks and their geochemical environment of formation), organic geochemistry (the study of the interaction of organic molecules with nonliving organic matter in the earth), and low-temperature aqueous geochemistry and high-temperature geochemistry (related to geochemical processes operating in the earth systems). Some branches, like environmental geochemistry, concern about the health of the entire earth and ocean, including the biota; nutrient deficiencies in living organisms and biota can be determined: iodine deficiency in the human being causes goiter and cretinism, a severe form of mental retardation; selenium deficiency causes heart diseases and cancer; and fluorine enrichment causes fluorosis. Similarly, aluminum, arsenic, chromium, and lead pollution can affect the health of not only human beings but also animals (Paul et al. 2015; Appleton et al. 1996; Tepanosyan et al. 2017; Modabberi et al. 2018).

Likewise, many related branches of geochemistry can be categorized: petroleum geochemistry, sedimentary geochemistry, marine geochemistry, medical geology, and archaeological geochemistry (to trace the origin and age of artifacts used in the ancient world). Such reconstructions are also useful in determining the paleoclimate and its role in the civilization/societal collapse as well as in tracing routes of human migration (White 2018).

ARCHIVES AND PROXIES FOR UNDERSTANDING PALEOCLIMATE AND PALEOCEANOGRAPHY

Marine sediment is the composite of clastic sediment transported from continents through the fluvial and aeolian pathways, together with the *in situ* biogenic (calcareous and siliceous) and authigenic components (Fe-Mn nodules, metal precipitates) (Calvert 1976; Calvert and Price 1983). Hence, based on mineral abundance, these sediments can be classified into cosmogenous or extraterrestrial, biogenous, lithogenous, and hydrogenous types (Li and Schoomaker 2003). In this classification, there are many common elements dominantly found to occur in two or more sediment types. For instance, Fe is dominant in the lithogenous, hydrogenous, and cosmogenous types of sediments. Similarly, silica is dominant in lithogenous, hydrogenous, and biogenous types of sediments. In the above-mentioned classification, aeolian sediment is predominant in the temperate regions bordering the deserts. Geochemical cycling of elements principally occurs through geologic processes that begin with the magmatic differentiation and associated processes leading to the formation of a wide variety of igneous rocks; their weathering by physical/mechanical and chemical processes forming weathered products; their transportation by geological agents, such as wind, water, and ice; and their deposition in the lakes and oceans forming sedimentary rocks; and then conversion to metamorphic rocks without undergoing the stage of melting upon the deep burial of rocks under the influence of temperature, pressure, and fluids (Goldberg 1954; Mason and Moore 1982; White 2018).

Virtually, all elements and isotopes can be used as proxies because each one of these will provide information about the processes that are operating in different compartments of the earth. In the marine environment, proxies are similar to tracers, which can be classified into the following types (Chester 1990): (1) conservative or nonreactive and (2) nonconservative or reactive types. In the former type, elements or isotopes are hydrophilic, which are generally allochthonous; however, they tend to remain in the water for a long time. Such proxies are useful for tracing sources of water in the oceans, their fluxes, and water column circulation processes (e.g., temperature, electrical conductivity, trace metals, CFCs, stable and radioactive isotopes, and radiogenic nuclides, including some of those liberated during the nuclear fission processes (Ra-228, Kr-85, He-3 Sr-90, Cs-137, Ar-39, C-14 and U-234 with half-lives of 5.8, 10.7,12.4, 28.6, 30.2, 270, 5,730, and 245,500 years, respectively)). The nonconservative or reactive type of tracers is hydrophobic, in the sense that such elements or isotopes are removed from the ocean water column and added on to the ambient particles settling in the ocean water column. Examples for this type are Th-234, Po-210, Th-228, Pb-210, Th-230, and Be-10 with half-lives of 0.07, 0.38, 1.9, 22.3, 75,000 years, and 1.5 Ma, respectively. Since the particle reactive types of isotopes vary in their half-lives, they can be applied to determine the rate of sedimentation in different oceanographic basins, starting from nearshore regions to the deep ocean (all particle reactive isotopes except for Th-230 and Be-10 for the former and latter cases, respectively).

In contrast to radionuclides, stable isotopes do not emit radiation; however, their fractionation in the marine sediments serves as tracers for determining paleotemperature and precipitation (Fischer and Wefer 1999; e.g., $\delta^{18}O$), denitrification ($\delta^{15}N$), sources of organic matter ($\delta^{13}C$), redox processes ($\delta^{34}S$), and particle fluxes (^{230}Th and ^{210}Pb), and for determining sedimentation rates in the oceans (Borole 1988; Manjunatha and Shankar 1992; Somayajulu et al. 1999).

There are another group of isotopes that are produced due to the radioactive decay of long-lived primordial nuclides; for e.g., ^{87}Rb decays to ^{87}Sr (Tripathy et al. 2011; Kleine 2015). The abundance of radiogenic isotopes expressed to their same stable elements, for instance, $^{87}Sr/^{86}Sr$, $^{143}Nd/^{144}Nd$, $^{206}Pb/^{204}Pb$, $^{207}Pb/^{204}Pb$, $^{208}Pb/^{204}Pb$, and $^{40}Ar/^{36}Ar$. The fractionations of radiogenic isotopes during the geological processes, such as weathering and erosion, and geochemical differentiation processes are not only applied to understand the current earth processes, but also in the past (Tripathy et al. 2011; Goswami et al. 2012). They are often used to determine the time span of geological events (such as absolute ages of rocks and sediments) as well as to quantitatively find out sources of pollutants in the environment (Stille and Shields 1997). Therefore, the radiogenic isotope geochemistry has become increasingly important in measuring rates of weathering; tracing particle sources in the atmosphere, paleosols, and marine sediments; and determining water masses in rivers and carbon transport through rivers to the ocean (Degens et al. 1991; Stille and Shields 1997; Zachos et al. 2001; Banner 2004; Das et al. 2006; Tripathy et al. 2011; Goswami et al. 2012; Kleine 2015; Liu et al. 2020). Among many radiogenic isotopes, Sr and Nd are the most widely used isotopes in the determination of ages as well as paleohydrological/weathering conditions in the terrestrial system (Das et al. 2006; Tripathy et al. 2011; Goswami et al. 2012).

Despite the above-mentioned isotopes, several elements are useful in one way or the other to understand the paleoclimate and paleoceanographic processes (Chester 1990; Calvert and Pedersen 1993; Fischer and Wefer 1999; Sirocko et al. 2000). Elements that are inferred to trace the terrigenous material in the marine sediments are Al, Fe, Si, Ca, Mg, and some of the trace elements and rare earth elements. Similarly, Al, Ti, Zr, Mg, and Ca are particularly used to determine aeolian sediment proportion in the marine sediment. Several biogeochemical processes are operating in the ocean water column (with an average depth of 3.8 km). In the ocean water column, the presence of iodine and Mn indicates oxygenation, and enrichment of Cr, Mn, Mo, Re, U, and vindicate anoxic conditions, while higher concentrations of Cu, Cd, Ni, and Zn represent sulfidic environment (Calvert and Pedersen 1993; Pattan et al. 2017). In the deep-sea oxygenated bottom water conditions, precipitation of Fe-Mn occurs and acts as a scavenging agent. It scavenges dissolved elements in the seawater column to form polymetallic nodules and Fe-Mn encrustations (Goldberg 1954; Goldberg et al. 1988; Chester 1990; Frank et al. 1999). Biogenic silica, C_{org}, Ca, Mg, Sr, Cd, Ba, Ni, and Zn are used to infer autochthonous (biogenic) proportion in the marine sediment. Apart from elements, compounds such as clay minerals, pollen, phytoliths, mineral magnetic parameters, and organic tracers/lipids are also used to derive the paleoclimate of the continents bordering the oceans (Fischer and Wefer 1999).

PHYSIOGRAPHY, SEA-BOTTOM TOPOGRAPHY, AND SEDIMENTS OF THE NORTHERN INDIAN OCEAN

The Northern Indian Ocean is unique in the global ocean system as having two northern limbs, i.e., the eastern Bay of Bengal and the western Arabian Sea. The limbs are closed in the northern portion by land dominated by the Himalayan Mountain chain, resulting in blocking the atmospheric and ocean circulation. The semi-annual differences in meteorological and oceanographic parameters create summer monsoon (June–September) and winter monsoon (December–February) with two intermediate phases of fall intermonsoon (October and November) and spring intermonsoon (March–May). Though Arabian Sea and Bay of Bengal latitudinal coordinates are quite similar, there are contrasting differences in the atmospheric and oceanographic parameters. In contrast to the Bay of Bengal, the greater part of the Arabian Sea is bordered by deserts, and hence, there is more evaporation of seawater than the input of fresh water from precipitation/river discharge. The currents in the Northern Indian Ocean reverse semi-annually due to the changes in wind direction as well as the intrusion of the low-saline Bay of Bengal water into the Arabian Sea (Wyrtki 1971; Schott and McCreary 2001; Schott et al. 2009). The West Indian Coastal Current (WICC) flows along the eastern Arabian Sea in the northerly direction as a result of the intrusion of low-saline water from the Bay of Bengal. However, during the summer, the reversal of flow takes place due to the influence of southwest monsoon winds.

CONTINENTAL MARGINS OF THE INDIAN MAINLAND

Indian subcontinent is blessed with a fair amount of water resource due to monsoon rainfall with an annual average of 119 cm/yr (Kaur and Purohit 2014). About 75% of annual rainfall generally occurs over the major part of India during the summer monsoon (Tyagi et al. 2012). Although the west coast and the adjacent Western Ghats of India receive high rainfall (200 to 400–600 cm/yr), rivers draining this terrain are generally small (<100 km long), drainage area is <5,000 km^2 and thereby forming a marine-dominated coast (Manjunatha and Shankar 1992; Shankar and Manjunatha 1997). Conversely, the east-flowing rivers are long with large drainage areas (>20,000 km^2; Rao 1979; Manjunatha and Shankar 1992), and river mouths are dominated by the formation of large deltas. Therefore, the east coast is commonly known as the deltaic coast, where the marine activities on the coastal regions are pacified by the rivers. There are three important river systems which discharge water and sediments to the Northern Indian Ocean: Himalayan Rivers, Peninsular Rivers, and rivers of the Myanmar. The sediment input from rivers draining the Indian subcontinent accounts (~1.2 billion tons; Subramanian 1993; Sebastian et al. 2019) about 10% of the global riverine sediment input to the global ocean. Of this, only about 20 million tons of sediment is discharged into the Arabian Sea, while the rest is dumped into the Bay of Bengal. Owing to such contrasting difference in sediment input, the sediment thickness in the continental marginal region of the Bay of Bengal is about four to five times higher than that for the Arabian Sea (Straume et al. 2019).

The east coast and west coast are the two important coastlines bordering the mainland of India, and cover distances of 2,686 and 3,341 km, with continental shelf areas of 0.31×10^6 and $0.13 \times 10^6 \text{km}^2$, respectively (Figure 7.1; Rao and Wagle 1997; Manjunatha and Shankar 1992; Nayak 2005; Sebastian et al. 2019). A brief review of the nature of estuaries along the west coast of India has been provided by various researchers (Ahmed 1972; Bruckner 1989; Shankar and Manjunatha 1997). Compared to the deltaic east coast, the west coast is dominated by ria-type of estuaries, marshes, lagoons, tidal flats and mudflats, spits, and bars (Ahmed 1972). Ria-type of estuaries are common along the Konkan Coast where a macrotidal effect is observed, leading to the intrusion of seawater up to 30–40 km inland and river mouths characterized by broad creeks. Estuaries of the Karnataka Coast are shallow, narrow river mouths with meso- to microtidal range where the seawater intrusion rarely exceeds 15–20 km into mainland (Shankar and Manjunatha 1997). The Kerala Coast is unique in terms of the dominance of backwaters with the microtidal estuaries and lagoons (Ahmed 1972).

The width of both continental shelves is broader towards the north, but narrow in the south. The width of the eastern continental shelf is minimum in the southern part, off Karaikal and maximum off Ganga–Brahmaputra Delta (16 and 210 km, respectively; Sebastian et al. 2019). Similarly, the width of the western continental shelf is minimum in the southern part (60 km off Kochi) but maximum in the northern part (350 km in the Gulf of Khambhat; Manjunatha and Shankar 1992; Sebastian et al. 2019).

GEOCHEMICAL STUDIES OF THE INDIAN SUBCONTINENT AND THE ADJACENT CONTINENTAL MARGIN

The geochemical studies carried out on the Indian subcontinent and the adjacent ocean margin can be classified into the following categories.

RIVERINE STUDIES

The geochemical studies of rivers are useful for understanding the weathering processes and sequestration of atmospheric CO_2, and estimating fluxes of dissolved and sediment load, including carbon transport to the ocean (Ramesh and Subramanian 1988; Sarin et al. 1989; Ramesh 1999a, b; Ittekkot 1988; Ittekkot and Laane 1991; Ittekkot et al. 1985; Manjunatha and Shankar 1994; Shankar and Manjunatha 1994; Manjunatha et al. 1996, 2001; Gupta et al. 1997; Humborg et al. 2000; Subramanian 2000; Balakrishna et al. 2001; Balakrishna and Probst 2005; Chakrapani and Veizer 2005; Sarma et al. 2009, 2011; Tripathy et al., 2011; Gurumuthy et al. 2012, 2015, 2017; Singh and Kumar 2017; Singh et al. 2017; Krishna et al. 2016, 2018; Banerjee et al. 2020; Li et al. 2020). These studies are the most important for tracing biogenic and lithogenous sources through major and trace elemental measurements, radiogenic isotopes as well as clay minerals, and determining their temporal and spatial variability in response to changes in the sedimentation in the ocean basin. The geochemical characteristics of rivers depend upon the prevalence of the climate over the river basins. Therefore, continuous monitoring of rivers is useful for accessing the human's impact.

Many cities of India are developed on river banks since the ancient civilization. Further, industrial and urban developments are planned based primarily on the availability of water resources for many purposes, including the easy discharge of solid wastes and sewage/industrial effluents. In India, the urban population has increased tenfold over a span of 50 years with an annual growth rate of 2.31%. This has increased pollutant levels in rivers and coastal regions (Kroeze and Seitzinger 1998; Mukhopadhyay et al. 2002; Milliman et al. 2008; Prasad et al. 2013; Ramesh et al. 2015; Balakrishna et al. 2017; Bharathi and Sarma 2019; Joshua et al. 2020). Due to population increase, and urban and industrial development, there has been a substantial increase in the pollutant load of rivers that will affect the water quality and human health in the coastal region (Sharma et al. 1994; Magesh et al. 2013; Samanta et al. 2015; Bhaumik et al. 2016; Chakraborty et al. 2014, 2019). Such studies are needed to prevent human activities on the riverine and coastal environment.

ESTUARINE AND NEARSHORE SEDIMENTS

Estuaries are ephemeral features existing between rivers and the sea. Estuaries are constantly subjected to change with regard to tidal, diurnal, seasonal, and glacial–interglacial cycles along with sediments transported from rivers and the sea, including anthropogenic sources (Gorsline 1967; Meade 1969, 1982; Borole et al. 1982a, b; Kempe 1988; Subramanian et al. 1985, 1987a, b, 1988; Subramanian and Jha 1988; Sarin et al. 1985, 1989, 1990; Manjunatha and Shankar 2000; Shankar and Manjunatha 1997; Manjunatha and Balakrishna 1999; Ramesh et al. 1999a, b; Jha et al. 2009; Fernandes et al. 2011; Araujo et al. 2018; Badesab et al. 2018; Nayak et al. 2018; Naik et al. 2019; Mazumdar 2020). About 90% of river-transported sediment is deposited in estuaries and the adjacent continental margin regions due to drastic changes in physicochemical properties of water from the riverine to the marine environment (Schubel and Kennedy 1984; Songlin 1987; Chester 1990; Chester and Jickells 2012). Estuarine studies are carried out to quantify the removal of dissolved and particulate elements during the mixing of the river with the seawater (Schubel and Kennedy 1984; Songlin 1987; Chester 1990; Chester and Jickells 2012). Therefore, it appears that chemical processes operating in the estuaries are most complicated in the earth system due to continuous interactions between water and solid particles, adsorption and desorption processes, biological uptake and regeneration of nutrients and associated elements, flocculation of trace elements, Fe-Mn oxides, clay minerals, and rapid sedimentation and resuspension due to tidal activity and digenetic/redox remobilization of elements from the deposited sedimentary column. All these processes change the gross river input of elements to the oceans (Chester 1990). There a number of studies carried out in India over four decades to understand the estuarine removal processes (Borole et al. 1977, 1982a, b; Balakrishna et al. 1997, 2001; Qasim 2003; Kessarkar et al. 2009, 2013a, b; Rao et al. 2011; Shynu et al. 2012; Pradhan et al. 2014a, b; Sarma et al. 2012, 2014; Priya et al. 2015; Suja et al. 2016, 2017;

Mitra et al. 2020; and references cited therein). In order to understand the behavior of dissolved elements during the mixing, laboratory studies have been conducted on some tropical river water with the Arabian Seawater (Shankar and Manjunatha 1994; Balakrishna et al. 1997). The results of the laboratory mixing experiments on Kali River water with Arabian Seawater indicate that the rapid removal (7%–60%) of dissolved metals (Cu, Pb, Zn, Ni, Co, Mn, Cr, and Fe) takes place within half an hour of mixing, vis-à-vis at 24 hours; the removal rates are generally increasing with salinity, implying about 90% of removal of dissolved components during estuarine mixing (Balakrishna et al. 1997). From the above studies, it is possible to determine whether nearshore/continental margin sediments are the source or sink for elements to the ocean water column. Generally, reducing conditions are prevailing in the continental margin areas where postdepositional remobilization processes are operating.

The elements removed from the nearshore environments are transported to the open ocean through seawater circulation and precipitate wherever oxygenated conditions exist at the sediment–water interface (Goldberg 1954; Goldberg et al. 1988; Chester 1990). Owing to this process, there is a contrasting geochemical difference between the continental margin sediments and deep-sea sediments that are enriched with redox-sensitive elements, such as Mn, Co, Co, Ni, and many of rare earth elements (Goldberg 1954; Chester 1990). These estuarine and nearshore sediments are known as hydrogenous elements to signify the origin from the ocean water column (Chester 1990).

CONTINENTAL MARGIN SEDIMENTS

The eastern and western continental margins of India are typical examples for the passive type of continental margin, which have originated from geodynamic processes of the breakup and drifting of the eastern Gondwana Supercontinent (Kolla and Kidd 1982; Gombos et al. 1995; Pandey et al. 2019; Yatheesh 2020). Therefore, sediment depositing on the vast continental margins of India is ideal for retrieving paleoclimatic information since the Cenozoic Era. The sediments of the Indian Ocean are classified into terrigenous, biogenic, brown clays, ferro-manganese nodules, volcanogenic, and cosmogenic (Kolla and Kidd 1982). The main factors governing the distribution are sources of sediment and their distribution by the ocean circulation. The terrigenous sediments are mainly restricted to the continental marginal areas, with the exception that Ganga and Brahmaputra riverine sediments are traced throughout the meridian off their mouths in the Bay of Bengal and extending beyond the equator. The terrigenous sediments deposited on the eastern continental margin are much thicker than that on the western continental margin of India (5–20 and 1–5 km, respectively; Straume et al. 2019). Compared to the eastern continental margin, the western continental margin is broad with offshore ridges (Kolla and Kidd 1982). Therefore, sediments depositing on Lakshadweep Ridge and Pratap Ridge are particularly useful for understanding the detailed paleoceanography. The continental margins of India have been investigated for various purposes, including the exploration for fossil fuel and mineral resources (Roonwal 1986; Bastia and Radhakrishna 2012). The geochemistry of surficial and core pelagic sediments of the Arabian Sea has been studied in detail (Shankar et al. 1987; Sirocko et al. 2000) with the major objectives of determining sources of sediments and associated geochemical changes and the response of the Arabian Sea with regard to the deglacial changes in climate since 24 ka.

A number of studies have been carried out on the continental margins of India, particularly related to the scope of this chapter (Borole et al. 1982a, b; Paropkari 1990; Rao and Murty 1990; Rao and Rao 1995; Rao and Wagle 1997; Paropkari et al. 1999; Kurian et al. 2013; Sebastian et al. 2019). The surficial distributions of major, minor, trace, and rare earth elements; clay minerals; and their interpretation to trace sources of sediments; geochemical changes from riverine to nearshore sediment; and their distribution and controlling factors have been studied in detail, since five decades (Murty et al. 1969, 1970; Chakraborty et al. 1985, 2014, 2018; Mascarenhas et al. 1985; Paropkari 1990; Rao et al. 1988; Rao and Murty 1990; Paropkari et al. 1992, 1993; Rao and Rao 1995; Manjunatha and Shankar 1997; Padmalal et al. 1997; Thamban et al. 1997, 2001; Ramesh et al. 1999a, b; Agnihotri et al. 2003; Kessarkar et al. 2003; Selvaraj et al. 2004; Alagarsamy and Zhang 2005, 2010; Laluraj and Nair 2006; Sangode et al. 2007; Jacob et al. 2008; Singh et al. 2011; Kurian et al. 2013; Acharya et al. 2015; Acharya and Panigrahi 2016; Naik et al. 2017; Sebastian et al. 2019; Pandey et al. 2020 and references cited therein). These studies are also helpful in determining the transport and the pattern of sediments as well as postdepositional remobilization processes of sediments deposited on the continental margin and the abyssal plains (Manjunatha and Shankar 1996, 1997; Nath 2001; Nath et al. 1997;). These studies indicate the seaward decrease of terrigenous sediments, lithological changes in the sources of riverine sediments, and redox processes. However, a study conducted off the Netravati–Gurpur Rivers, near Mangaluru, the west coast of India, shows the relative importance of rivers rather than the bottom water currents controlling the terrigenous sediment distribution (Manjunatha 1991). One of the tracers used to determine the inverse of terrigenous sediment in the marine sediment is the calcium, which is very low in tropical rivers (Manjunatha 1991; Manjunatha and Shankar 1997). The calcium contents in the sediments of the western continental margin of India generally range between 5% and 60% (Paropkari 1990; Rao and Murty 1990). Therefore, increasing the density of sampling of sediments and their analyses would help us to resolve this problem. Compared to surficial sediments, there are not many investigations, especially on the geochemistry of sediment cores (Sarin et al. 1979; Borole et al. 1977, 1982a, b; Manjunatha 1991; Manjunatha and Shankar 1996; Somayajulu et al. 1999; Sarkar et al. 2000; Karbassi et al. 2001; Bhushan et al. 2001; Agnihotri et al. 2003; Pattan and Pearce 2009; Pattan et al. 2001, 2003, 2005; Staubwasser et al. 2002; Banakar et al. 2010; Govil and Naidu 2011; Sruthi et al. 2014; Kumar et al. 2015; Tiwari et al. 2015; Nagoji and Tiwari; 2017; Gourlan et al. 2020; and references cited therein).

Since climate change is one of the thrust areas of the earth system science as well as the scope of this book, there are several papers related to the Late Quaternary Period spanning the Marine Isotopic Stages (MIS)-1 to not more than MIS-5 (Davies et al. 1975; Sarin et al. 1979; Shankar and Manjuatha 1995; Von Rad et al. 1999a, b; Sarkar et al. 2000; Sirocko et al. 1993, 2000; Thamban et al. 2001, 2002; Staubwasser et al. 2002, 2003; Staubwasser and Weiss 2006; Rao et al. 2010; Rashid et al. 2011; Kessarkar et al. 2013a, b; Saraswat et al. 2013; Singh et al. 2016, 2017; Chandana et al.

2018; Saravanan et al. 2019; Banerji et al. 2020; Singh et al. 2020; references cited therein). Nevertheless, there are some very significant studies carried out in the continental margins around India beyond the MIS-5 (Kolla and Kidd 1982; Shimmield et al. 1990; Bloemendal et al. 1992, 1993; Naidu and Malmgren 1996; Reichart et al. 1997; Rostek et al. 1997; Prabhu et al. 2004; Deplazes et al. 2013, 2014; Khim et al. 2018; Alonso-Garcia et al. 2019; Yu et al. 2019; and references cited therein). Though the Quaternary Period has a time span until 2.58 Ma, there is a long gap of information from the base of the Quaternary Period to the beginning of the last interglacial period (MIS-103 to MIS-5). Nevertheless, some studies have been carried out by the international group of scientists, including many Indians on samples, collected during DSDP, ODP, and IODP expeditions and international cruises (Prell et al. 1992; Shimmield et al. 1990; Clemens and Prell 2003; Clemens et al. 1996, 2010; Reichart et al. 1997; Rostek et al. 1997; Schulz et al. 1998; von Rad et al. 1999a, b; Gupta et al. 2015; Kunkelova et al. 2018; Clift and Webb 2019; Clift et al. 2019; Saraswat et al. 2019; Cai et al. 2020).

A recent review on the Neogene climate of the Indian subcontinent and adjoining ocean indicates (Singh et al. 2020) that (1) the initiation of monsoon in the Himalayan region can be dated back to 11.6 Ma, which seems to be younger than the monsoonal record inferred from the Maldives carbonate platform attributed to the impact of global tectonics; (2) an increase in the wind intensity is evidenced by the increase of ventilation in the water column, primary productivity, and denitrification in the Arabian Sea from ~2.8 to 3.2 Ma; (3) the carbon isotope record of pedogenic nodules in the Siwalik succession suggests a major shift in the vegetation from ~2.8 to 1.1 Ma due to the intensification of monsoonal precipitation; and (4) strengthened winter monsoon between ~1.65 and 1.85 Ma due to the increase in ocean circulation, as a result of the east–west sea surface temperature gradient in the Pacific Ocean.

The drilled cores from Maldives Island provide evidence of the deposition of sediments from the physical oceanographic processes, such as erosion and often redistribution of sediments by currents due to the initiation of South Asian Monsoon around 21 Ma (Late Miocene), with further intensification ~10–13 Ma (Middle Miocene to Upper Miocene; Rea 1994; Clift et al. 2008, 2010; Gupta et al. 2015). This indicates the gradual intensification of South Asian Monsoon with the tectonic rise of the Himalayan Mountain. Ocean drilling records in the Northern Indian Ocean indicate that the initiation of monsoon was accompanied by the increase in the sediment input from the Himalayan Rivers from 16 to 10 Ma (Rea 1992; Clift 2008), beginning of upwelling the northwestern Arabian Sea ~8.5 Ma (Kroon et al. 1991), later a dry phase marked by the abundance of hematite over goethite ~7.7 Ma, followed by the humid phase from 6.3 to 5.95 Ma (Clift et al. 2019).

Recently, the mineralogical study on drilled cores of the Laxmi Basin, northeastern Eastern Arabian Sea (Pandey et al. 2016; Routledge et al. 2019), indicates a prominent change in the sediment sources between the Deccan Trap rocks and Indus River (smectite/(illite + chlorite)), suggesting the intensification of the Indian summer monsoon noticed around 3.7 Ma (Cai et al. 2020).

The northwestern Arabian Sea, off Oman, is one of the high primary productive upwelling regions in the global ocean. Recently, Bialik et al. (2020) have probed sediment core collected at Ocean Drilling Program Site 722B to trace the

initiation of the southwest monsoon winds beyond the Neogene Period. The depletion of redox-sensitive element – Mn noticed ~14.5 Ma – marks the intense reducing
conditions caused by increased primary productivity and the deposition of organic
matter-enriched sediments, attributed to the intensification of the summer monsoon.
Similarly, a significant increase in the flux of biogenic silica from ~12.5 to 11 Ma and
high rate of denitrification between 11 and 9.5 Ma in the northwestern Arabian Sea
are the compound effects of the intensification of the summer monsoon. In a nutshell,
the commencement of globally high upwelling in the northwestern Arabian Sea has
been dated back to 14.8 Ma (Zhuang et al. 2017) due to the uplift of Tibetan Plateau
(~25 Ma), followed by the closure of Tethyan Seagate (20 Ma) and progressive glaciations over Antarctica since the Miocene Period (Bialik et al. 2020).

CONCLUSIONS AND FUTURE DIRECTIONS

Based on the scope of this chapter, and an overview of the literature review on geochemical and paleoclimatic studies carried out in the continental margins of India, the
following conclusions are drawn along with suggestions for the future work. Though
the available information is quite sufficient to understand the paleoclimate/paleocean-
ography over the past 100 ka, there is a need to cover a long temporal scale from
100 ka till the beginning of the Quaternary Period. The water depths selected by the
previous investigators for a sampling of sediment cores in the ocean basins were quite
deep (700–4,000 m), where the sedimentation is low. Hence, the resolution of paleoclimatic reconstruction was much lower ranging from hundreds to thousands of years.
These studies have been carried out in some selected sections of sediment cores by
measuring parameters like (a) carbon, nitrogen, and oxygen isotopes, (b) microfossils,
(c) organic matter, (d) major and trace elements, (e) clay minerals, and (f) mineral
magnetic parameters along with accelerator mass spectrometric (AMS) ^{14}C dating
to derive paleoclimatic and paleoceanographic information. Some workers have provided bits and pieces of information to draw paleoceanographic changes; however,
their results there are often inconsistent with those of other proxies.

There are many efforts to minimize the temporal resolution between multidecadal
and centennial scales, but such studies are restricted to the deeper continental mar-
gins/deep-ocean basins, especially the transition from the tropical to temperate
regions (Anderson et al. 2002; Gupta et al. 2015; Deplazes et al. 2013, 2014). There
are not many high-resolution studies in the core and trough of the monsoon regions
in India. Since India has quite a broad coastline and the adjacent continental shelves,
such regions need to be explored in detail by collecting long sediment cores (either
by giant piston coring or by drilling) covering the entire Quaternary Period. It is
suggested that sediment cores so collected must be studied in detail to identify climatic anomalies and their cyclicities in the past to determine tele-connecting forces.
Further, the ideally suitable conditions for such high-resolution studies are the monsoon core and monsoon trough regions of India and neighboring seas (18°N–28°N
and 65°E and 88°E; and north of 20°N– and 75°E–85°E respectively; Figure 7.2;
Sikka and Narasimha 1995; Rajeevan et al. 2010). In addition, the nearshore regions,
such as river flood plains, deltas, estuaries, backwater, and marshes, are probable regions to carry out high-resolution study; however, one has to make sure of

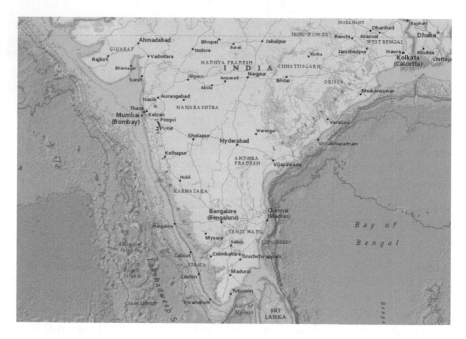

FIGURE 7.2 Bathymetric map of the continental margins of the Indian mainland (https://maps.ngdc.noaa.gov/viewers/fishmaps/) showing the broad western continental margin and the adjacent ridges ideally suited for a detailed study.

collecting undisturbed sediment cores. The selection of sampling sites and proxies to be measured depends upon the objectives of the work.

Therefore, there is a need to improve the temporal variations of the paleoclimate by retrieving the long and undisturbed sediment cores from the region of high sedimentation, to increase the temporal efficiency of sampling between centennial to millennial and decadal to annual scales, to measure as many paleoclimatic proxies as possible to develop a high-resolution multiproxy's marine stratigraphic record of India during the Quaternary Period. The data generated from such work will be useful for improving futuristic climatic models. It is difficult for an individual group to take up this kind of work, and should involve international organizations (IODP) or a national network of research organizations to work on Quaternary of India, covering both terrestrial and marine domains. In the marine scenario, based on the available information, sites for the collection of long cores can be demarcated based on sedimentation rates and multichannel seismic profiles. Long cores from undisturbed sediment sequences should be subjected to (a) X-radiography, (b) high-resolution X-ray fluorescence (XRF) core scanning, (c) sub-sampling at minimum intervals, and (d) measure as many multi-proxies as possible along with radioisotope dating. This should provide precise information on paleoclimatic and paleoceanographic changes in the area of investigation during the Quaternary Period. Although there are some efforts done by measuring some of the above-mentioned parameters at a centennial time scale (Anderson et al. 2002; Gupta et al.

2015; Deplazes et al. 2014; Von Rad et al. 1999a, b; Tiwari et al. 2015; Saravanan et al. 2019), this needs to be extended for a longer temporal scale.

ACKNOWLEDGMENTS

The lead author (BRM) is thankful to Dr. Neloy Khare for the invitation to write this chapter. He is grateful to the Ministry of Earth Sciences (GEOTRACES and SIBER) and Ministry of Science and Technology (FIST), Government of India, for funding.

REFERENCES

Acharya, S. S., M. K. Panigrahi, A. K. Gupta, and S. Tripathy. 2015. Response of trace metal redox proxies in continental shelf environment: The Eastern Arabian Sea scenario. *Continental Shelf Research*. 106: 70–84.

Acharya, S. S., and M. K. Panigrahi. 2016. Evaluation of factors controlling the distribution of organic matter and phosphorus in the Eastern Arabian Shelf: A geostatistical reappraisal. *Continental Shelf Research*. 126: 79–88.

Ahmed, E. 1972. *Coastal Geomorphology of India*, Orient Longman, Bombay.

Appleton, J. D., R. Fuge, and G. J. H. McCall. 1996. Environmental Geochemistry and Health with Special Reference to Developing Countries, Geological Society Special Publication No 113, London.

Agnihotri, R., S. K. Bhattacharya, M. M. Sarin, and B. L. K. Somayajulu. 2003. Changes in surface productivity and subsurface denitrification during the Holocene: A multiproxy study from the eastern Arabian Sea. *The Holocene*. 13: 701–713.

Alagarsamy, R., and J. Zhang. 2005. Comparative studies on trace metal geochemistry in Indian and Chinese rivers. *Current Science*. 89: 299–309.

Alagarsamy, R., and J. Zhang. 2010. Geochemical characterization of major and trace elements in the coastal sediments of India. *Environmental Monitoring and Assessment*. 161: 161–176.

Alonso-Garcia, M., T. Rodrigues, F. Abrantes, M. Padilha, C.A. Alvarez-Zarikian, T. Kunkelova, J.D. Wright, and C. Betzler. 2019. Sea-surface temperature, productivity and hydrological changes in the Northern Indian Ocean (Maldives) during the interval~575-175 ka (MIS 14 to 7). *Palaeogeography, Palaeoclimatology, Palaeoecology*. 536: 109376–12.

Anderson, D. M., J. T. Overpeck, and A. K. Gupta. 2002. Increase in the Asian SW monsoon during the past four centuries. *Science*. 297: 596–599.

Araujo, J., S. W. A. Naqvi, H. Naik, and R. Naik. 2018. Biogeochemistry of methane in a tropical monsoonal estuarine system along the west coast of India. *Estuarine, Coastal and Shelf Science*. 207: 435–443.

Badesab, F., V. Gaikwad, T. R. Gireeshkumar, O. Naikgaonkar, K. Deenadayalan, S. V. Samiksha, P. D. Kumar, V. J. Loveson, S. D. Iyer, A. Khan, P. B. Udayakrishnan, and A. Sardar. 2018. Magnetic tracing of sediment dynamics of mudbanks off southwest coast of India. *Environmental Earth Sciences*. 77: 625–641.

Bialik, O. M., G. Auer, N. O. Ogawa, D. Kroon, N. D. Waldmann, and N. Ohkouchi. 2020. Monsoons, upwelling, and the deoxygenation of the northwestern Indian Ocean in response to middle to late Miocene global climatic shifts. *Paleoceanography and Paleoclimatology*. 35: e2019PA003762.

Balakrishna, K., B. R. Manjunatha, and R. Shankar. 1997. A laboratory study of the flocculation of dissolved heavy metals in Kali River during estuarine mixing, west coast of India. *Journal of the Geological Society of India*. 50: 753–758.

Balakrishna, K., R. Shankar, M. M. Sarin, and B. R. Manjunatha. 2001. Distribution of U–Th nuclides in the riverine and coastal environments of the tropical southwest coast of India. *Journal of Environmental Radioactivity*. 57: 21–33.

Balakrishna, K., and J. L. Probst. 2005. Organic carbon transport and C/N ratio variations in a large tropical river: Godavari as a case study, India. *Biogeochemistry*. 73: 457–473.

Balakrishna, K., A. Rath, Y. Praveenkumar Reddy, K. S. Guruge, and B. Subedi. 2017. A review of the occurrence of pharmaceuticals and personal care products in Indian water bodies. *Ecotoxicology and Environmental Safety*. 137: 113–120.

Banakar, V. K., B. S. Mahesh, G. Burr, and A. R. Chodankar. 2010. Climatology of the Eastern Arabian Sea during the last glacial cycle reconstructed from paired measurement of foraminiferal $\delta^{18}O$ and Mg/Ca. *Quaternary Research*. 73: 535–540.

Banerjee, S., P. Ghosh, R. Nagendra, B. Bhattacharya, B. Desai, and A. K. Srivastava. 2020. Marine and fluvial sedimentation including erosion and sediment flux in Peninsular Indian phanerozoic basins. *Proceedings of the Indian National Science Academy*. 86: 351–363.

Banerji, U. S., P. Arulbalaji, and D. Padmalal. 2020. Holocene climate variability and Indian Summer Monsoon: An overview. *The Holocene*. 30: 744–773.

Banner, J. 2004. Radiogenic isotopes: Systematics and applications to earth surface processes and chemical stratigraphy. *Earth-Science Reviews*. 65: 141–194.

Baskaran, M. 2012. *Handbook of Environmental Isotope Geochemistry*, Advances in Isotope Geochemistry, Springer, Heidelberg.

Bastia, R., and M. Radhakrishna. 2012. *Basin Evolution and Petroleum Prospectivity of the Continental Margins of India*, Elsevier, New York.

Bharathi, M. D., and V. V. S. S. Sarma. 2019. Impact of monsoon-induced discharge on phytoplankton community structure in the tropical Indian estuaries. *Regional Studies in Marine Science*. 31: 100795. doi:10.1016/j.rsma.2019.100795.

Bhaumik, U., M., K. Mukhopadhyay, N. P. Srivasta, and A. P. Sharma. 2016. *Indian Estuaries*. Hindustan Publishing Corporation, India.

Bhushan, R., K. Dutta, and B. L. K. Somayajulu. 2001. Concentrations and burial fluxes of organic and inorganic carbon on the eastern margins of the Arabian Sea. *Marine Geology*. 178: 95–113.

Bloemendal, J., J. W., King, F. R. Hall, and S.-H. Doh. 1992. Rock magnetism of late Neogene and Pleistocene deep-sea sediments: Relationship to sediment source, diagenetic processes, and sediment lithology. *Journal of Geophysical Research*. 97: 4361–4375.

Bloemendal, J., J. W. King, A. Hunt., P. B. Demenocal, and A. Hayashida. 1993. Origin of the sedimentary magnetic record at Ocean Drilling Program Sites on the Owen Ridge, western Arabian Sea. *Journal of Geophysical Research*. 98: 4199–4219.

Borole, D. V., S. Krishnaswami, and B. L. K. Somayajulu. 1977. Investigations on dissolved uranium, silicon and on particulate trace elements in estuaries. *Estuarine and Coastal Marine Science*. 5: 743–754.

Borole, D. V., S. Krishnaswami, and B. L. K. Somayajulu. 1982a. Uranium isotopes in rivers, estuaries and adjacent coastal sediments of western India-their weathering, transport and oceanic budget. *Geochimica et Cosmochimica Acta*. 46: 125–137.

Borole, D.V., M. M. Sarin, and B. L. K. Somayajulu. 1982b. Composition of Narbada and Tapti estuarine particles and adjacent Arabian Sea sediments. *Indian Journal of Geo-Marine Sciences*. 11: 51–62.

Borole, D. 1988. Clay sediment accumulation rates on the monsoon-dominated western continental shelf and slope region of India. *Marine Geology*. 82: 285–291.

Brocker, W. S., and T.-H. Peng. 1982. Tracers in the Sea. Columbia University, Palisades, NY.

Bruckner, H. 1989. Late Quaternary shorelines in India. In: *Late Quaternary Correlation and Applications*, eds. D. B. Scott, P. A. Pirazzoli, and C. A. Honing, pp. 169–194. Kluwer, Maine.

Byrne, R. H., and E. R. Sholkovitz. 1996. Marine chemistry and geochemistry of the lanthanides. In: Handbook on the Physics and Chemistry of Rare Earths, ed. J. K. A. Gschneidner and L. Eyring, pp. 498–593, Elsevier, New York.

Cai, M., Z. Xu, P. D. Clift, B-K Khim, D. Lim, Z. Yu, D. K. Kulhanek, and T. Li. 2020. Long-term history of sediment inputs to the eastern Arabian Sea and its implications for the evolution of the Indian summer monsoon since 3.7 Ma. *Geological Magazine.* 157: 908–919.

Calvert, S. E. 1976. The mineralogy and geochemistry of nearshore sediments. In *Chemical Oceanography*, eds. J. P. Riley, and R. Chester, pp. 87–280. Academic Press, New York.

Calvert, S. E., and N. B. Price. 1983. Geochemistry of Namibian shelf sediments. In *Coastal Upwelling*, eds. E. Suess, and J. Thiede, pp. 337–376. Plenum, New York.

Calvert, S. E., and T. F. Pedersen. 1993. Geochemistry of recent oxic and anoxic marine sediments: Implications for the geological record. *Marine Geology.* 113: 67–88.

Chakraborty, P., A. Mascarenhas, K. L. Paropakari, and P. S. N. Murty. 1985. Geochemistry of sediments of the eastern continental shelf. *Mahasagar Bulletin-National Institute of Oceanography.* 18: 1–25.

Chakraborty, P., D. Ramteke, S. Chakraborty, and B. N. Nath. 2014. Changes in metal contamination levels in estuarine sediments aroud India - An assessment. *Marine Pollution Bulletin.* 78: 15–25.

Chakraborty, S., P. Chakraborty, A. Sarkar, A. Kazip, M. B. L. Mascarenhas-Pereira, and B. N. Nath. 2018. Distribution and geochemical fractionation of lead in the continental shelf sediments around India. *Geological Journal.* 54: 1190–11204.

Chakraborty, P., S. Jayachandran, J. Lekshmy, P. Padalkar, L. Sitlhou, K. Chennuri, K. Shetye, A. Sardar, and R. Khandeparker. 2019. Seawater intrusion and resuspension of surface sediment control mercury (Hg) distribution and its bioavailability in water column of a monsoonal estuarine system. *Science of the Total Environment.* 660: 1441–1448.

Chakrapani, G. J., and J. Veizer. 2005. Dissolved inorganic carbon isotopic compositions in the Upstream Ganga River in the Himalayas, *Current Science.* 89: 553–556.

Chandana K. R., U. S. Banerji, and R. Bhushan. 2018. Review on Indian summer monsoon (ISM) reconstruction since LGM from Northern Indian Ocean. *Earth Science India.* 11:71–84.

Chester, R. 1990. *Marine Geochemistry.* Unwin Hyman, London.

Chester, R., and T. Jickells. 2012. Marine Geochemistry. John Wiley & Sons, New York.

Clemens, S. C., D. W. Murray, and W. L. Prell. 1996. Nonstationary phase of the Plio-Pleistocene Asian monsoon. *Science.* 274: 943–948.

Clemens, S. C., and W. L. Prell. 2003. A 350,000 year summer-monsoon multi-proxy stack from the owen ridge, northern Arabian Sea. *Marine Geology.* 201: 35–51.

Clemens, S. C., W. L. Prell, and Y. Sun. 2010. Orbital-scale timing and mechanisms driving Late Pleistocene Indo-Asian summer monsoons: Reinterpreting cave speleothem $\delta^{18}O$. *Paleoceanography.* 25: PA4207. doi:4210.1029/2010PA001926.

Clemens, S. C., W. Kuhnt, L. J. LeVay, and the Expedition 353 Scientists. 2016. Expedition 353 summary. *Proceedings of the International Ocean Discovery Program Volume 353*, College Station, TX. doi:10.14379/iodp.proc.353.101.2016.

Clift, P. D., K. V. Hodges, D. Heslop, R. Hannigan, H. Van Long, and G. Calves. 2008. Correlation of Himalayan exhumation rates and Asian monsoon intensity. *Nature Geoscience.* 1: 875–880.

Clift, P. D., L. Giaosan, A. Carter, E. Garzanti, V. Galy, A. R. Tabrez, M. Pringle, I. H. Campbell, C. France-Lanord, J. Blusztajn, and C. Allen. 2010. Monsoon control over erosion patterns in the western Himalaya: Possible feed-back into the tectonic evolution. In: *Monsoon Evolution and Tectonics-Climate Linkage in Asia*, eds. P. D. Clift, R. Tada, and H. Zheng, pp. 185–218. Geological Society, London, Special Publication 342.

Clift, P. D., and A. G. Webb. 2019. A history of the Asian monsoon and its interactions with solid Earth tectonics in Cenozoic South Asia. In *Himalayan Tectonics: A Modern Synthesis*, eds. M. P. Searle, and P. J. Treloar, pp. 631–652. Geological Society, London, Special Publication no. 483.

Clift, P. D., D. K. Kulhanek, P. Zhou, M. G. Bowen, S. M. Vincent, M. Lyle, and A. Hahn. 2019. Chemical weathering and erosion responses to changing monsoon climate in the Late Miocene of Southwest Asia. *Geological Magazine*. 157; 939–955. doi: 10.1017/S0016756819000608.

Cohen, K. M., S. C. Finney, P. L. Gibbard, and J.-X. Fan. 2013. The ICS international chronostratigraphic chart. *Episodes*. 36: 199–204.

Das, A., S. Krishnaswami, and A. Kumar. 2006. Sr and 87Sr/86Sr in rivers draining the Deccan Traps (India): Implications to weathering, Sr fluxes, and the marine Sr/Sr record around K/T. *Geochemistry, Geophysics, Geosystems*. 7: Q06014. doi:10.1029/2005GC001081.

Davies, T. A., O. E. Weser, B. P. Luyendyk, and R. B. Kidd. 1975. Unconformities in the sediments of the Indian Ocean. *Nature*. 253: 15–19.

Degens, E. T., S. Kempe, and J. E. Richey. 1991. Summary: Biogeochemistry of major world river. In: *Biogeochemistry of Major World Rivers*, eds. E. T. Degens, S. Kempe, and J. E. Richey, pp. 323–347. SCOPE 42. John Wiley & Sons, New York.

Denton, G. H., R. F. Anderson, J. R. Toggweiler, R. L. Edwards, J. M. Schaefer, and A. E. Putnam. 2010. The last glacial termination. *Science*. 328: 1652–1656.

Deplazes, G., A. Lückge, L. C. Peterson, A. Timmermann, Y. Hamann, K. A. Hughen, U. Röhl, C. Laj, M. A. Cane, and D. M. Sigman. 2013. Links between tropical rainfall and North Atlantic climate during the last glacial period. *Nature Geoscience*. 6: 213–217.

Deplazes, G., A. Lückge, J.-B. W. Stuut, J. Pätzold, H. Kuhlmann, D. Husson, M. Fant, and G. H. Haug. 2014. Weakening and strengthening of the Indian monsoon during Heinrich events and Dansgaard-Oeschger oscillations. *Paleoceanography*. 29: 99–114.

Erwin, D. H. 2009. Climate as a driver of evolutionary change. *Current Biology*. 19: R575–R583.

Faure, G. 1986. *Principles of Isotope Geology*, 2nd edn. John Wiley & Sons, New York.

Fernandes, L., G. N. Nayak, D. Ilangovan, and D. V. Borole. 2011. Accumulation of sediment, organic matter and trace metals with space and time, in a creek along Mumbai coast, India. *Estuarine, Coastal and Shelf Science*. 91: 388–399.

Fischer, G., and G. Wefer. 1999. *Use of Proxies in Paleoceanography*. Springer, Berlin, Heidelberg.

Frank, M., R. O'Nions, J. Hein, and V. Banakar. 1999. 60 Myr records of major elements and Pb-Nd isotopes from hydrogenous ferromanganese crusts: Reconstruction of seawater paleochemistry. *Geochimica et Cosmochimica Acta*. 63: 1689–1708.

Gibbard, P. L., and J. Lewin. 2016. Partitioning the quaternary. *Quaternary Science Reviews*. 151: 127–139.

Goldberg, E. D. 1954. Marine geochemistry 1. Chemical scavengers of the sea. *Journal of Geology*. 62: 249–265.

Goldberg, E. D., M. Koide, K. Bertinc, V. Hodge, M. Stallard, D. Martincic, N. Mikac, M. Branica, and J. K. Abaychi. 1988. Marine geochemistry 2. Scavenging redux, *Applied Geochemistry*. 3: 561–571.

Gombos, A. M., W. G. Powell, and I. O. Norton. 1995. The tectonic evolution of western India and its impact on hydrocarbon occurrences: An overview. *Sedimentary Geology*. 96: 119–129.

Gorsline, D. S. 1967. Contrasts in coastal bay sediments on the Gulf and Pacific coasts. In: *Estuaries*, ed. G. H. Lauff, pp. 219–225. American Association for the Advancement of Science, Washington, DC.

Goswami, V., S. K. Singh, R. Bhushan, and V. K. Rai. 2012. Temporal variations in 87Sr/86Sr and εNd in sediments of the southeastern Arabian Sea: Impact of monsoon and surface water circulation. *Geochemistry, Geophysics, Geosystems.* 13: Q01001, doi:10.1029/2011GC003802.

Gourlan, A. T., F. Albarede, H. Achyuthan, and S. Campillo. 2020. The marine record of the onset of farming around the Arabian Sea at the dawn of the Bronze Age. *The Holocene.* 30: 878–887. doi:10.1177/0959683620902218.

Govil, P., and P. D. Naidu. 2011. Variations of Indian monsoon precipitation during the last 32 kyr reflected in the surface hydrography of the Western Bay of Bengal. *Quaternary Science Reviews.* 30: 3871–3879.

Grootes, P. M., M. Stuiver, J. W. C. White, S. Johnsen, and J. Jouzel. 1993. Comparison of oxygen isotope records from the GISP2 and GRIP Greenland ice cores. *Nature.* 366: 552–554.

Gupta, L. P., V. Subramanian, and V. Ittekkot. 1997. Biogeochemistry of particulate organic matter transported by the Godavari River, India. *Biogeochemistry.* 38: 103–128.

Gupta, A. K., A. Yuvaraja, M. Prakasam, S. C. Clemens, and A. Velu. 2015. Evolution of the South Asian monsoon wind system since the late Middle Miocene. *Palaeogeography, Palaeoclimatology, Palaeoecology.* 438: 160–167.

Gurumurthy, G. P., K. Balakrishna, J. Riotte, J. J. Braun, S. Audry, H. N. Udaya Shankar, and B. R. Manjunatha. 2012. Controls on intense silicate weathering in a tropical river, southwestern India. *Chemical Geology.* 300–301: 61–69.

Gurumurthy, G. P., K. Balakrishna, M. Tripti, J. Riotte, S. Audry, J. J. Braun, and H. N. Udaya Shankar. 2015. Use of Sr isotopes as a tool to decipher the soil weathering processes in a tropical river catchment, southwestern India. *Applied Geochemistry.* 63: 498–506 doi:10.1016/j.apgeochem.2015.03.005.

Gurumurthy, G. P., M. Tripti, J. Riotte, R. Prakyath, and K. Balakrishna. 2017. Impact of water-particle interactions on molybdenum budget in humid tropical rivers and estuaries: Insights from Nethravati, Gurupur and Mandovi river systems. *Chemical Geology.* 450: 44–58.

Head, M. J., and P. L. Gibbard. 2015. Formal subdivision of the quaternary system/period: Past, present, and future. *Quaternary International.* 383: 4–35.

Head, M. J. 2019. Formal subdivision of the Quaternary System/Period: Present status and future directions. *Quaternary International.* 500: 32–51. doi:10.1016/j.quaint.2019.05.018.

Higgins, J. A., A. V. Kurbatov, N. E. Spaulding, E. Brook, D. S. Introne, L. M. Chimiak, Y. Yan, P. A. Mayewski, and M. L. Bender. 2015. Atmospheric composition 1 million years ago from blue ice in the Allan Hills, Antarctica. *Proceedings of the National Academy of Sciences.* 112: 6887–6891.

Humborg, C., D. J. Conley, L. Rahm, F. Wulff, A. Cociasu, and V. Ittekkot. 2000. Silicon retention in river basins: Far-reaching effects on biogeochemistry and aquatic food webs in coastal marine environments. *Ambio.* 29: 45–50.

Ivanovich, M., and R. S. Harmon. 1992. *Uranium-Series Disequilibrium: Applications to Earth, Marine, and Environmental Sciences,* 2nd edn. Clarendon Press, Oxford.

Ittekkot, V., S. Safiullah, B. Mycke, and R. Seifert. 1985. Seasonal variability and geochemical significance of organic matter in the River Ganges, Bangladesh. *Nature.* 317: 800–802.

Ittekkot, V. 1988. Global trends in the nature of organic matter in river suspensions. *Nature.* 332: 436–438.

Ittekkot, V., and R. W. P. M. Laane. 1991. Fate of riverine particulate organic matter. In: *Biogeochemistry of Major World Rivers,* eds. E. T. Degens, S. Kemp, and J. E. Richey, pp. 233–243. John Wiley & Sons, New York.

Jacob, J., N. C. Kumar, K. A. Jayaraj, T. V. Raveendran, K. K. Balachandran, T. Joseph, M. Nair, C. T. Achuthankutty, Nair, K.K.C., R. George, and Z. P. Ravi. 2008. Biogeochemistry of the surfacial sediments of western and Eastern Continental Shelves of India. *Journal of Coastal Research*. 24: 1240–1248.

Jha P. K., J. Tiwari, U. K. Singh, M. Kumar, and V. Subramanian. 2009. Chemical weathering and associated CO_2 consumption in the Godavari river basin, India. *Chemical Geology*. 264: 364–374.

Joshua, D. I., Y. Praveenkumar Reddy, V. P. Prabhasankar, A. P. D'Souza, N. Yamashita, and K. Balakrishna. 2020. First report of pharmaceuticals and personal care products in two tropical rivers of southwestern India. *Environmental Monitoring and Assessment*. 192: 529.

Karbassi, A. R., R. Shankar, and B. R. Manjunatha. 2001. Geochemistry of shelf sediments off Mulki, Southwest coast of India and its paleoenvironmental significance. *Journal of the Geological Society of India*. 58: 37–44.

Kaur, S., and M. K. Purohit. 2014. *Rainfall Statistics of India – 2013*. India Meteorological Department (Ministry of Earthsciences) Report no. ESSO/IMD/HS/R.F.REPORT/02(2014)/18, 99p.

Kempe, S. 1988. Estuaries-their natural and anthropogenic changes. In: *Scales and Global Change,* eds. T. Rosswal, R. G. Woodman, and P. G. Risser, pp. 251–285. SCOPE. Wiley, New York.

Kessarkar, P. M., V. Purnachadra Rao, S. M. Ahmad, and G. A. Babu. 2003. Clay minerals and Sr–Nd isotopes of the sediments along the western margin of India and their implication for sediment provenance. *Marine Geology*. 202: 55–69.

Kessarkar, P. M., V. P. Rao, R. Shynu, P. Mehra, and B. E. Viegas. 2009. The nature and distribution of particulate matter in the Mandovi estuary, central west coast of India. *Estuaries and Coasts*. 33: 30–44.

Kessarkar, P. M., R. Shynu, V. P. Rao, F. Chong, T. Narvekar, and J. Zhang. 2013a. Geochemistry of the suspended sediment in the estuaries of Mandovi and Zuari rivers, central west coast of India. *Environmental Monitoring and Assessment*. 185: 4461–4480.

Kessarkar, P. M., V. Purnachadra Rao, S. W. A Naqvi, et al. 2013b. Variation in the Indian summer monsoon intensity during the Bølling-Ållerød and Holocene. *Paleoceanography and Paleoclimatology*. 28: 413–425.

Khim, B.-K., K. Horikawa, Y. Asahara, J. E. Kim, and M. Ikehara. 2018. Detrital Sr-Nd isotopes, sediment provenances, and depositional processes in the Laxmi Basin of the Arabian Sea during the last 800 kyrs. *Geological Magazine*, 157, 895–907. doi:10.1017/S0016756818000596.

Kleine, T. 2015. Radiogenic isotopes. In *Encyclopedia of Astrobiology*, eds. R. J. Amils, C. Quintanilla, H. J. Cleaves II, W. M. Irvine, D. L. Pinti, and V. Michel. Springer, Berlin, Heidelberg.

Krishna, M. S., M. H. K. Prasad, D. B. Rao, R. Viswanadham, V. V. S. S. Sarma, and N. P. C. Reddy. 2016. Export of dissolved inorganic nutrients to the northern Indian Ocean from the Indian monsoonal rivers during discharge period. *Geochimica et Cosmochimica Acta*. 172: 430–443.

Krishna, M. S., R. Viswanadham, M. H. K. Prasad, V. R. Kumari, and V. V. S. S. Sarma. 2018. Export fluxes of dissolved inorganic carbon to the northern Indian Ocean from the Indian monsoonal rivers. *Biogeosciences Discussions*. 16: 505–519.

Krishnaswami, S., and Lal, D. 1978. Radionuclide limnochronology. In: *Lakes: Chemistry, Geology, Physics*, ed. A. Lerman, pp. 153–177. Springer, New York.

Kroeze, C., and Seitzinger, S. P. 1998. Nitrogen inputs to rivers, estuaries and continental shelves and related nitrous oxide emissions in 1990 and 2050: A global model. *Nutrient Cycling in Agroecosystems*. 52: 195–212.

Kroon, D., T. N. F. Steens, and S. R. R. Troelstra. 1991. Onset of monsoonal related upwelling in the western Arabian Sea as revealed by planktonic foraminifers. *Proceedings of the Ocean Drilling Program, Scientific Results.* 117: 257–263.

Kumar, A., B. R. Manjunatha, and P. J. Kurian. 2015. Glacial-interglacial productivity contrasts along the eastern Arabian Sea: Dominance of convective mixing over upwelling. *Geoscience Frontiers.* 6: 913–925.

Kunkelova, T., S. J. A. Jung, E. S. de Leau, N. Odling, A. L. Thomas, C. Betzler, G. P. Eberli, C. A. Alvarez-Zarikian, M. Alonso-García, O. M. Bialik, C. L. Blättler, J. A. Guo, S. Haffen, S. Horozal, A. L. H Mee, M. Inoue, L. Jovane, L. Lanci, J. C. Laya, T. Lüdmann, N. N. Bejugam, M. Nakakuni, K. Niino, L. M Petruny, S. D. Pratiwi, J. J. G. Reijmer, J. Reolid, A. L. Slagle, C. R. Sloss, X. Su, P. K Swart, J. D Wright, Z. Yao, J. R. Young, S. Lindhorst, S. Stainbank, A. Rueggeberg, S. Spezzaferri, I. Carrasqueira, S. Hu, and D. Kroon. 2018. A two million year record of low-latitude aridity linked to continental weathering from the Maldives. *Progress in Earth and Planetary Sciences.* 5: 86.

Kurian, S., B. N. Nath, N. C. Kumar, and K. K. C. Nair. 2013. Geochemical and isotopic signatures of surficial sediments from the western continental shelf of India: Inferring provenance, weathering, and the nature of organic matter geochemical and isotopic signatures of sediments from the Indian west coast, *Journal of Sedimentary Research.* 83: 427–442,

Lal, D., and M. Baskaran. 2011. Applications of cosmogenic-isotopes as atmospheric tracers. In: *Advances in Isotope Geochemistry*, ed. M. Baskaran, pp. 575–589. Springer, Heidelberg.

Laluraj, C. M., and S. M. Nair. 2006. Geochemical index of trace metals in the surficial sediments from the western continental shelf of India, Arabian Sea. *Environmental Geochemistry and Health.* 28: 509–518.

Li, Y.-H., and J. E. Schoomaker. 2003. Chemical composition and mineralogy of marine sediments. In: *Treatise on Geochemistry*, eds. H. D. Holland, and K. K. Turekian, Vol. 7, pp. 1–35. Elsevier, NY.

Liu, J., J. Zhong, H. Ding, F. J. Yue, C. Li, S. Xu, and S-L. Li. 2020. Hydrological regulation of chemical weathering and dissolved inorganic carbon biogeochemical processes in a monsoonal river. *Hydrological Processes.* 34: 2780–2792.

Luthi, D., M. Le Floch, B. Bereiter, T. Blunier, J.M. Barnola, U. Siegenthaler, D. Raynaud, J. Jouzel, H. Fischer, K. Kawamura, and T. F. Stocker. 2008. High-resolution carbon dioxide concentration record 650, 000–800, 000 years before present. *Nature.* 453: 379–382.

Magesh, N. S., N. Chandrasekar, S. K. Kumar, and M. Glory. 2013. Trace element contamination in the estuarine sediments along Tuticorin Coaste Gulf of Mannar, southeast coast of India. *Marine Pollution Bulletin.* 73: 355–361.

Manjunatha, B. R. 1991. *Geochemistry and Magnetic Susceptibility of Riverine, Estuarine and Marine Environments around Mangalore, West Coast of India.* PhD dissertation, Mangalore University India.

Manjunatha, B. R., and R. Shankar. 1992. A note on the factors controlling the rate of sedimentation along western continental shelf of India. *Marine Geology.* 104: 219–224.

Manjunatha, B. R., and R. Shankar. 1994. Magnetic and sedimentological studies of Netravati and Gurpur river- bed sediments, west coast of India. *Journal of the Geological Society of India.* 44: 413–426.

Manjunatha, B. R., K. Balakrishna, R. Shankar, A. Thiruvengadasami, R. K. Prabhu, T. R. Mahalingam, and M. A. R. Iyengar. 1996. The transport of elements from soils around Kaiga to the Kali river, southwestern coast of India. *Science of the Total Environment.* 191: 109–118.

Manjunatha, B. R., and R. Shankar. 1996. Signature of non-steady-state diagenesis in continental shelf sediments. *Estuarine, Coastal and Shelf Science.* 42: 361–369.

Manjunatha, B. R., and R. Shankar. 1997. The influence of rivers on the geochemistry of shelf sediments, southwestern coast of India. *Environmental Geology.* 31: 107–116.

Manjunatha, B. R., and K. Balakrishna. 1999. An insight into the depositional history of Late Quaternary sediments around Mangalore, west coast of India. *Indian Journal of Marine Sciences.* 28: 449–454.

Manjunatha B. R., and R. Shankar. 2000. Behaviour and distribution patterns of particulate metals in estuarine and coastal surface waters near Mangalore, southwest coast of India. *Journal of the Geological Society of India.* 55: 157–166.

Manjunatha, B. R., K. Balakrishna, and R. Shanakar. 2001. Geochemistry and assessment of metal pollution in soils and river components of a monsoon-dominated environment near Karwar, southwest coast of India. *Environmental Geology.* 40: 1462–1470.

Mascarenhas, A., A. L. Paropakari, and P. S. N. Murty. 1985. Geochemistry of sediments of the eastern continental shelf. *Mahasagar-Bulletin of the National Institute of Oceanography.* 18: 1–25.

Mason, B., and C. B. Moore. 1982. Principles of Geochemistry. John Wiley and sons, New York.

Masson-Delmotte, V., S. Hou, A. Ekaykin, J. Jouzel, A. Aristarain, R. T. Bernardo, D. Bromwich, O. Cattani, M. Delmotte, S. Falourd, M. Frezotti, H. Gallee, L. Genoni, E. Isakkson, A. Landais, M. M. Helsen, G. Hoffmann, J. Lopez, V. Morgan, H. Motoyama, D. Noone, H. Oerter, J. R. Petit, A. Poyer, R. Uemura, G. A. Schmidt, E. Schlosser, J. Simões, E. J. Steig, B. Stenni, M. Stievenard, M. R. van den Broeke, R. S. W. van de Wal, W. J. van de Berg, F. Vimeux, and J. W. C. White. 2008. A review of Antarctic surface snow isotopic composition: Observations, atmospheric circulation, and isotopic modeling. *Journal of Climate.* 21: 3359–3387.

Mazumdar, A. 2020. Recent contributions to the geochemistry and sedimentology of estuaries, mangroves, and mudbanks along the Indian Coast: A status report. *Proceedings of the Indian National Science Academy.* 86: 343–350.

Meade, R. H. 1969. Landward transport of bottom sediments in estuaries of the Atlantic coastal plain. *Journal of Sedimentary Petrology.* 39: 222–234.

Meade, R. H. 1982. Sources, sinks, and storage of river sediment in the Atlantic drainage of the United States. *Journal of Geology.* 90: 235–252.

Milliman, J. D., K. L. Farnsworth, P. D. Jones, K. H. Xu, and L. C. Smith. 2008. Climatic and anthropogenic factors affecting river discharge to the global ocean, 1951–2000. *Global and Planetary Change.* 62: 187–194.

Mitra, S., M. Sudarshan, M. P. Jonathan, S. K. Sarkar, and S. Thakur. 2020. Spatial and seasonal distribution of multi-elements in suspended particulate matter (SPM) in tidally dominated Hooghly river estuary and their ecotoxicological relevance. *Environmental Science and Pollution Research.* 27: 12658–12672.

Modabberi, S., M. Tashakor, N. Sharifi Soltani, and A. S. Hursthouse. 2018. Potentially toxic elements in urban soils: Source apportionment and contamination assessment. *Environmental Monitoring and Assessment.* 190: 715.

Molnar, P., P. England, and J. Martinod. 1993. Mantle dynamics, uplift of the Tibetan Plateau, and the Indian monsoon. *Reviews of Geophysics.* 31: 357–396. doi: 10.1029/93RG02030.

Mukhopadhyay, S. K., H. Biswas, T. K. De, S. Sen, and T. K. Jana. 2002. Seasonal effects on the air–water carbon dioxide exchange in the Hooghly estuary, NE coast of Bay of Bengal, India. *Journal of Environmental Monitoring.* 4: 549–552.

Murty, P. S. N., C. V. G. Reddy, and V. V. R. Varadachari. 1969. Distribution of organic matter in the marine sediments off the west coast of India. *Proceedings of the National Academy of Sciences, India.* 35B: 167.

Murty, P. S. N., C. M. Rao, and C. V. G. Reddy. 1970. Distribution of nickel in the marine sediments off the west coast of India. *Current Science.* 39: 30–32.

Nagoji, S. S., and M. Tiwari. 2017. Organic carbon preservation in Southeastern Arabian Sea sediments since mid-Holocene: Implications to South Asian Summer Monsoon variability, *Geochemistry, Geophysics, Geosystems.* 18: 3438–3451. doi:10.1002/2017GC006804.

Naidu, P. D., and B. Malmgren. 1996. A high-resolution record of late Quaternary upwelling along the Oman Margin, Arabian Sea based on planktonic foraminifera. *Paleoceanography and Paleoclimatology.* 11: 129–140.

Naik, D. K., R. Saraswat, D.W. Lea, S. R. Kurtarkar, and A. Mackensen. 2017. Last glacial-interglacial productivity and associated changes in the eastern Arabian Sea. *Palaeogeography, Palaeoclimatology, Palaeoecology.* 483: 147–156.

Naik, R., J. Araujo, A. Pratihary, S. Kurian, and S. W. A. Naqvi. 2019. Sedimentary sulphate reduction and organic matter mineralization across salinity gradient of the Mandovi Estuary, West coast of India. *Estuarine, Coastal and Shelf Science.* 221: 21–29.

Nath, B. N., M. Bau, B. Ramalingeswara Rao, and Ch. M. Rao. 1997. Trace and rare earth elemental variation in Arabian Sea sediments through a transect across the oxygen minimum zone. *Geochimica et Cosmochimica Acta.* 61: 2375–2388.

Nath, B. N. 2001. Geochemistry of sediments. In: *The Indian Ocean : A Perspective*, eds. R. Sen Gupta, and E. Desa, pp. 645–690. Oxford & IBH Publishing, New Delhi.

Nayak, G. N. 2005. Indian Ocean coasts – coastal geomorphology. In: *Encyclopedia of Coastal Science*, ed. M. L. Schwartz, pp. 554–557. Springer, Dordrecht, The Netherlands.

Nayak, G. N., S. P. Volvoikar, and T. Hoskatta. 2018. Changing depositional environment and factors controlling the growth of mudflat in a tropical estuary, west coast of India. *Environmental Earth Sciences.* 77: 741–761.

Padmalal, D., K. Maya, and P. Seralathan. 1997. Geochemistry of Cu, Co, Ni, Zn Cd and Cr in the surficial sediments of a tropical estuary, southwest coast of India: A granulometric approach. Environ*mental* Geo*logy.* 31: 85–93.

Pandey, D. K., P. D. Clift, and D. K. Kulhanek, and Expedition 355 Scientists. 2016. Arabian Sea Monsoon. *Proceedings of the International Ocean Discovery Program 355.* International Ocean Discovery Program, College Station, TX. doi:10.14379/iodp. proc.355.2016.

Pandey, D. K., A. Pandey, and S. A. Whattam. 2019. Relict subduction initiation along a passive margin in the northwest Indian Ocean. *Nature Communications.* 10: 2248. doi: 10.1038/s41467-019-10227-8.

Pandey, D. K., R. Nair, and A. Kumar. 2020. The Western Continental Margin of India: Indian Scientific Contributions (2016-2018). *Proceedings of the Indian National Science Academy.* 86: 331–341.

Paropkari, A. L. 1990. Geochemistry of sediments from the Mangalore-Cochin shelf and upper slope off southwest India: Geological and environmental factors controlling dispersal of elements. *Chemical Geology.* 81: 99–119.

Paropkari, A. L., C. Prakash Babu, and A. Mascarenhas. 1992. A critical evaluation of depositional parameters controlling the variability of organic carbon in Arabian Sea sediment. *Marine Geology.* 107: 213–226.

Paropkari, A. L., C. Prakash Babu, and A. Mascarenhas. 1993. New evidence for enhanced preservation of organic carbon in contact with oxygen minimum zone on the western continental slope of India. *Marine Geology.* 111: 7–13.

Paropkari, A. L., P. V. Shirodkar, R. Alagarsamy, S. Kaisary, and A. Mesquita. 1999. Trace elements in near shore sediments along the east and west coast of India. *Proceedings of an International Workshop, Japan, 1999*, pp. 300–305. http://drs.nio.org/drs/handle/2264/1755.

Pattan, J. N., P. Shane, N. J. G. Pearce, V. K. Banakar, G. Parthiban. 2001. An occurrence of the w 74 ka Youngest Toba tephra from the western Continental margin of India. *Current Science.* 80: 1322–1326.

Pattan, J. N., T. Masuzawa, D. P. Naidu, G. Parthiban, and M. Yamamoto. 2003. Productivity fluctuations in the southeastern Arabian Sea during the last 140 ka. *Palaeogeography, Palaeoclimatology, Palaeoecology.* 193: 575–590.

Pattan, J. N., T. Masuzawa, and M. Yamamoto. 2005. Variations in terrigenous sediment discharge in a sediment core from southeastern Arabian Sea during the last 140 ka. *Current Science.* 89: 1421–1425.

Pattan, J. N., and N. J. G. Pearce. 2009. Bottom water oxygenation history in southeastern Arabian Sea during the past 140 ka: Results from redox-sensitive elements. *Palaeogeography, Palaeoclimatology, Palaeoecology.* 280: 396–405.

Pattan, J. N., G. Parthiban, A. Garg, and N. R. C. Moraes. 2017. Intense reducing conditions during the last deglaciation and Heinrich events (H1, H2, H3) in sediments from the oxygen minimum zone off Goa, eastern Arabian Sea. *Marine and Petroleum Geology.* 84: 243–256. doi:10.1016/j.marpetgeo.2017.03.034.

Paul, D., B. Choudhary, T. Gupta, and M. T. Jose. 2015. Spatial distribution and the extent of heavy metal and hexavalent chromium pollution in agricultural soils from Jajmau, India. *Environmental Earth Sciences.* 73: 3565–3577.

Prabhu, C. N., R. Shankar, K. Anupama, M. Taieb, R. Bonnefille, L. Vidal, and S. Prasad. 2004. A 200-ka pollen and oxygen-isotopic record from two sediment cores from the eastern Arabian Sea. *Palaeogeography, Palaeoclimatology, Palaeoecology.* 214: 309–321.

Pradhan, U. K., Y. Wu, P. V. Shirodkar, J. Zhang, and G. Zhang. 2014a. Sources and distribution of organic matter in thirty five tropical estuaries along the west coast of India-a preliminary assessment. *Estuarine, Coastal and Shelf Science.* 151: 21–33.

Pradhan, U. K., Y. Wu, P. V. Shirodkar, J. Zhang, and G. Zhang. 2014b. Multi-proxy evidence for compositional change of organic matter in the largest tropical (peninsular) river basin of India. *Journal of Hydrology.* 519: 999–1009.

Prasad, M. H. K., V. V. S. S. Sarma, V. V. Sarma, M. S. Krishna, and N. P. C. Reddy. 2013. Carbon dioxide emissions from the tropical Dowleiswaram reservoir on the Godavari river, southeast of India. *Journal of Water Resource and Protection.* 5: 534–545.

Prell, W. L., and N. Niitsuma, eds. 1991. *Proceedings of the International Ocean Discovery Program: Scientific Results*, Vol. 117. College Station, TX. doi:10.2973/odp.proc.sr.117.1991.

Prell, W. L., D. W. Murray, S. C. Clemens, and D. M. Anderson. 1992. Evolution and variability of the Indian Ocean Summer Monsoon: Evidence from the western Arabian Sea drilling program. In *Synthesis of Results from Scientific Drilling in the Indian Ocean*, eds. R. A. Duncan, D. K. Rea, R.B. Kidd, U. von Rad, and J. K. Weissel, pp. 447–469. American Geophysical Union, Washington, DC. Geophysical Monograph no. 70.

Priya, K. L., P. Jegathambal, and E. J. James. 2015. Seasonal dynamics of turbidity maximum in the Muthupet estuary, India. *Journal of Ocean University of China.* 14: 765–777.

Qasim, S. Z. 2003. *Indian Estuaries.* Allied Publishers, New Delhi.

Rajeevan, M., S. Gadgil, and B. Jyoti. 2010. Active and break spells of the Indian Summer Monsoon. *Journal of Earth System Science.* 119: 229–247.

Ramesh, R., and V. Subramanian. 1988. Temporal, spatial and size variation in sediment transport in the Krishna River Basin, India. *Journal of Hydrology.* 98: 53–65.

Ramesh, R., A. L. Ramanathan, R. Arthur James, V. Subramanian, S. B. Jacobsen, and H. D. Holland. 1999a. Rare earth elements and heavy metal distribution in estuarine sediments of east coast of India. *Hydrobiologia.* 57: 1–11.

Ramesh, R., A. L. Ramanathan, R. Arthur James, V. Subramanian, S. B. Jacobsen, and H. D. Holland. 1999b. Rare earth elements and heavy metal distribution in estuarine sediments of east coast of India. *Hydrobiologia.* 397, 89–99.

Ramesh, R., R. S. Robin, and R. Purvaja. 2015. An inventory on the phosphorus flux of major Indian rivers. *Current Science.* 108: 1294–1299.

Rao, K. L. 1979. *India's Water Wealth: Its Assessment, Uses and Projections*. Orient Longman, New Delhi.

Rao, V. P., N. P. Reddy, and Ch. M. Rao. 1988. Clay mineral distribution in the shelf sediments off the northern part of the east coast of India. *Continental Shelf Research*. 8: 145–151.

Rao, Ch. M., and P. S. N. Murty. 1990. Geochemistry of continental margin sediments of the central west coast of India. *Journal of the Geological Society of India*. 35: 19–37.

Rao, V. P., and B. R. Rao. 1995. Provenance and distribution of clay minerals in the sediments of the western continental shelf and slope of India. *Continental Shelf Research*. 15: 1757–1771.

Rao, P., and Wagle, B. G. 1997. Geomorphology and surficial geology of the western continental shelf and slope of India: A review. *Current Science*. 73: 330–350.

Rao, V. P., P. M. Kessarkar, M. Thamban, and S. K. Patil. 2010. Paleoclimatic and diagenetic history of the late quaternary sediments in a core from the Southeastern Arabian Sea: Geochemical and magnetic signals. *Journal of Oceanography*. 66: 133–146.

Rao, V. P., R. Shynu, P. M. Kessarkar, D. Sundar, G. S. Michael, T. Narvekar, V. Blossom, and P. Mehra. 2011. Suspended sediment dynamics on a seasonal scale in the Mandovi and Zuari estuaries, central west coast of India. *Estuarine, Coastal and Shelf Science*. 91: 78–86.

Rashid, H., E. England, L. Thompson, and L. Polyak. 2011. Late glacial to Holocene Indian summer monsoon variability based upon sediment records taken from the Bay of Bengal. *Terrestrial, Atmospheric and Oceanic Sciences*. 22: 215–228.

Rea, D. K. 1994. The paleoclimatic record provided by eolian deposition in the deep sea: The geologic history of wind. *Reviews of Geophysics*. 32: 159–195.

Reichart, G. J., M. den Dulk, H. J. Visser, C. H. van der Weijden, and W. J. Zachariasse. 1997. A 225 kyr record of dust supply, paleoproductivity and the oxygen minimum zone from the Murray Ridge (northern Arabian Sea). *Palaeogeography, Palaeoclimatology, Palaeoecology*. 134: 149–169.

Roonwal, G. S. 1986. *The Indian Ocean: Exploitable Mineral and Petroleum Resources*. Springer, Berlin, Germany.

Rostek, F., E. Bard, L. Beaufort, C. Sonzogni, and G. Ganssen. 1997. Sea surface temperature and productivity records for the past 240 kyr in the Arabian Sea. *Deep-Sea Research. Part II*. 44: 1461–1480.

Routledge, C. M., D. K. Kulhanek, L. Tauxe, G. Scardia, A. D. Singh, S. Steinke, E. M Griffith, and R. Saraswat. 2019. A revised chronostratigraphic framework for International Ocean Discovery Program Expedition 355 sites in Laxmi Basin, Eastern Arabian Sea. *Geological Magazine*. 156: 1–18.

Samanta, S., T. K. Dalai, J. K. Pattanaik, S. K. Rai, and A. Mazumdar. 2015. Dissolved inorganic carbon (DIC) and its δ ^{13}C in the Ganga (Hooghly) River estuary, India: Evidence of DIC generation via organic carbon degradation and carbonate dissolution. *Geochimica et Cosmochimica Acta*. 165: 226–248.

Sangode, S. J., R. Sinha, B. Phartiyal, O. S. Chauhan, R. K. Mazari, T. N. Bagati, N. Suresh, S. Mishra, R. Kumar, and P. Bhattacharjee. 2007. Environmental magnetic: Studies on some Quaternary sediments of varied depositional settings in the Indian sub-continent. *Quaternary International*. 159: 102–118.

Saraswat, R., D. W. Lea, R. Nigam, A. Mackensenand, and D. K. Naik. 2013. Deglaciation in the tropical Indian Ocean driven by interplay between the regional monsoon and global teleconnections. *Earth and Planetary Science Letters*. 375: 166–75.

Saraswat, R., S. R. Kurtarkar, R. Yadav, A. Mackensen, D. P. Singh, S. Bhadra, A. D. Singh, M. Tiwari, S. P. Prabhukeluskar, S. R. Bandodkar, D. K. Pandey, P. D. Clift, D. K. Kulhanek, K. Bhishekar, and S. Nair. 2019. Inconsistent change in surface hydrography of the eastern Arabian Sea during the last four glacial–interglacial intervals. *Geological Magazine*. 157: 989–1000. doi: 10.1017/S0016756819001122.

Saravanan, P., A. K. Gupta, H. Zheng, M. K. Panigrahi, and M. Prakasam. 2019. Late Holocene long arid phase in the Indian subcontinent as seen in shallow sediments of the eastern Arabian Sea. *Journal of Asian Earth Sciences*. 181: 103915.

Sarin, M., D. Borole, and S. Krishnaswami. 1979. Geochemistry and geochronology of sediments from the Bay of Bengal and the equatorial Indian Ocean. *Proceedings of the Indian Academy of Sciences—A*. 88: 131–154.

Sarin, M. M., K. S. Rao, S. K. Bhattacharya, R. Ramesh, and B. L. K. Somayajula. 1985. Geochemical studies of the river estuarine systems of Krishna and Godavari. *Mahasagar: Bulletin of the National Institute of Oceanography*. 18: 129–143.

Sarin, M. M., S. Krishnaswami, K. Dilli, B. L. K. Somayajulu, and W. S. Moore. 1989. Major ion chemistry of the Ganga-Brahmaputra river system: Weathering processes and fluxes to the Bay of Bengal. *Geochimica et Cosmochimica Acta*. 53: 997–1009.

Sarin, M. M., K. Krishnaswami, B. L. K Somayajulu, and W. S. Moore. 1990. Chemistry of uranium-thorium and radium isotopes in the Ganga-Brahmaputra river system: Weathering processes and fluxes to the Bay of Bengal. *Geochimica et Cosmochimica Acta*. 54: 1387–1396.

Sarkar, A., R. Ramesh, B. L. K. Somayajulu, R. Agnihotri, A. J. T. Jull, and G. S. Burr. 2000. High-resolution Holocene monsoon record from the eastern Arabian Sea. *Earth and Planetary Science Letters*. 177: 209–218.

Sarma, V. V. S. S., S. N. M. Gupta, P. V. R. Babu, T. Acharya, N. Harikrishnachari, K. Vishnuvardhan, N. S. Rao, N. P. C. Reddy, V. V. Sarma, Y. Sadhuram, T. V. R. Murty, and M. D. Kumar. 2009. Influence of river discharge on plankton metabolic rates in the tropical monsoon driven Godavari estuary, India. *Estuarine, Coastal and Shelf Science*. 85: 515–524.

Sarma, V. V. S. S., N. A. Kumar, V. R. Prasad, V. Venkataramana, S. Appalanaidu, B. Sridevi, B. S. K. Kumar, M. D. Bharati, C. V. Subbaiah, T. Acharyya, G. D. Rao, R. Viswanadham, L. Gawade, D. T. Manjary, P. P. Kumar, K. Rajeev, N. P. C. Reddy, V. V. Sarma, M. D. Kumar, Y. Sadhuram, and T. V. R. Murty. 2011. High CO_2 emissions from the tropical Godavari estuary (India) associated with monsoon river discharges. *Geophysical Research Letters*. 38: L08601.

Sarma, V. V. S. S., R. Viswanadham, G. D. Rao, V. R. Prasad, B. S. K. Kumar, S. A. Naidu, N. A. Kumar, D. B. Rao, T. Sridevi, M. S. R. Krishna, N. P. C. Reddy, Y. Sadhuram, and T. V. R. Murty. 2012. Carbon dioxide emissions from Indian monsoonal estuaries. *Geophysical Research Letters*. 39: L03602.

Sarma, V. V. S. S., M. S. Krishna, V. R. Prasad, B. S. K. Kumar, S. A. Naidu, G. D. Rao, and N. P. C. Reddy. 2014. Distribution and sources of particulate organic matter in the Indian monsoonal estuaries during monsoon. *Journal of Geophysical Research Biogeosciences*. 119: 2095–2111.

Sastry, M. V. A. 1997. The Pliocene-Pleistocene boundary in the Indian subcontinent. In: The Pleistocene Boundary and the Beginning of the Quaternary, ed. J. A. Van Couvering, pp. 232–238. Cambridge University Press, Cambridge.

Saupe, E. E., H. Qiao, Y. Donnadieu, A. Farnsworth, A. T. Kennedy-Asser, J.-B. Ladant, D. J. Lunt, A. Pohl, P. Valdes, and S. Finnegan. 2020. Extinction intensity during Ordovician and Cenozoic glaciations explained by cooling and palaeogeography. *Nature Geoscience*. 13: 65–70.

Schott, F. A., and J. P. McCreary. 2001. The monsoon circulation of the Indian Ocean. *Progress in Oceanography*. 51: 1–123.

Schott, F. A., S. P. Xie, and J. P. McCreary. 2009. Indian Ocean circulation and climate variability. *Reviews of Geophysics*. 47: 1–46.

Schubel, J. R., and V. S. Kennedy. 1984. The estuary as a filter: An introduction. In The Estuary as a Filter, ed. V. S. Kennedy, pp. 1–11, Academic Press, New York.

Schulz, H., U. von Rad, and H. Erlenkeuser. 1998. Correlation between Arabian Sea and Greenland climate oscillations of the past 110,000 years. *Nature*. 393: 54–57.

Sebastian, T., B. N. Nath, M. Venkateshwarlu, P. Miriyala, A. Prakash, P. Linsy, M. Kocherla, A. Kazip, and A. V. Sijinkumar .2019. Impact of the Indian summer monsoon variability on the source area weathering in the Indo-Burman ranges during the last 21 kyr—A sediment record from the Andaman Sea. *Palaeogeography, Palaeoclimatology, Palaeoecology*. 516: 22–34.

Selvaraj, K., V. Ram Mohan, and P. Szefer. 2004. Evaluation of metal contamination in coastal sediments of the Bay of Bengal, India: Geochemical and statistical approaches. *Marine Pollution Bulletin*. 49: 174–185.

Shankar, R., K. V. Subbarao, and V. Kolla. 1987. Geochemistry of surface sediments from the Arabian Sea. *Marine Geology*. 76: 253–279.

Shankar, R., and B. R. Manjunatha. 1994. Elemental composition and particulate metal fluxes from Netravati and Gurpur rivers to the coastal Arabian Sea. *Journal of the Geological Society of India*. 43: 255–265.

Shankar, R., and B. R. Manjunatha. 1995. Late-Quaternary sedimentation history in the Laccadive trough, southeastern Arabian Sea. *Journal of the Geological Society of India*. 45: 689–694.

Shankar, R., and B. R. Manjunatha. 1997. Onshore transport of shelf sediments into the Netravati-Gurpur estuary, west coast of India: Geochemical evidence and implications. *Journal of Coastal Research*. 13: 331–340.

Sharma, P., D. V. Borole, and M. D. Zingde. 1994. Pb based trace element fluxes in the nearshore and estuarine sediments off Bombay. *Marine Chemistry*. 47: 227–241.

Shimmield, G. B., S. R. Mowbray, and G. P. Weedon. 1990. A 350 ka history of the Indian Southwest Monsoon-evidence from deep-sea cores, northwest Arabian Sea. *Transactions of the Royal Society of Edinburgh Earth Sciences*. 81: 289–299.

Shynu, R., V. P. Rao, P. M. Kessarkar, and T. G. Rao. 2012. Temporal and spatial variability of trace metals in suspended matter of the Mandovi estuary, central west coast of India. *Environmental Earth Sciences*. 65: 725–739.

Siegel, F. R. 1979. *Review of Research on Modern Problems in Geochemistry*. Earth Sciences 16: United Nations Educational, Scientific and Cultural Organisation. Paris.

Sikka, D. R., and R. Narasimha. 1995. Genesis of the monsoon trough boundary layer experiment (MONTBLEX). *Proceedings of the Indian Academy of Sciences - Earth & Planetary Sciences*. 104: 157–187.

Singh, A. D., S. J. A. Jung, K. Darling, R. Ganeshram, T. Ivanochko, and D. Kroon. 2011. Productivity collapses in the Arabian Sea during glacial cold phases. *Paleoceanography*. 26: PA3210.

Singh, A. D., P. D. Naidu, and R. Saraswat. 2016. Indian contributions to marine micropalaeontology (2010-2015). *Proceedings of the Indian National Science Academy*. 82: 663–673.

Singh, D. P., R. Saraswat, and D. K. Naik. 2017. Does glacial-interglacial transition affect sediment accumulation in monsoon-dominated regions? *Acta Geologica Sinica (English Edition)*. 91: 1079–1094.

Singh, U. K., and B. Kumar. 2017. Pathways of heavy metals contamination and associated human health risk in Ajay River basin, India. *Chemosphere*. 174: 183–199.

Singh, A. D., A.K. Ghosh, R.C. Mehrotra, R. Patnaik, and M. Tiwari. 2020. Recent Advances in Understanding Neogene Climatic Evolution: Indian Perspective. *Proceedings of the Indian National Science Academy*. 86: 379–388. doi: 10.16943/pinta/2020/49776.

Sirocko, F., M. Sarnthein, H. Erlenkeuser, H. Lange, M. Arnold, and J. C. Duplessy. 1993. Century scale events in monsoonal climate over the past 24,000 years. *Nature*. 364: 322–324.

Sirocko, F., D. Garbe-Schönberg, and C. Devey. 2000. Processes controlling trace element geochemistry of Arabian Sea sediments during the last 25,000 years. *Global and Planetary Change.* 26: 217–303.

Songlin, H. 1987. On the aggregation of suspended fine-grained matter and behaviour of trace metals in estuaries: A review. In Transport of Car-bon and Minerals in Major World Rivers, eds. E.T. Degens, S. Kempe, and G. Webin, pp. 195–206, Geologisch Palaeontolo-gisches Institut, University of Hamburg, SCOPE/UNEP, Sonderbad.

Somayajulu, B. L. K., R. Bhushan, A. Sarkar, G. S. Burr, and A. J. T. Jull. 1999. Sediment deposition rates on the Continental margins of eastern Arabian Sea using 210-Pb, 137-Cs and 14-C. *Science of the Total Environment.* 237–238: 429–439.

Sruthi, K. V., P. J. Kurian, and P. R. Rajani. 2014. Distribution of major and trace elements of a sediment core from the eastern Arabian Sea and its environmental significance. *Current Science.* 107: 1161–1167.

Staubwasser, M., F. Sirocko, P. M. Grootes, and H. Erlenkeuser. 2002. South Asian monsoon climate change and radiocarbon in the Arabian Sea during early and middle Holocene. *Paleoceanography and Paleoclimatology.* 17(4): 1063.

Staubwasser, M., F. Sirocko, P. M. Grootes, and H. Erlenkeuser. 2003. Climate change at the 42 ka BP termination of the Indus valley civilization and Holocene South Asian mon-soon variability. *Geophysical Research Letters.* 30: 1425.

Staubwasser, M., and H. Weiss. 2006. Holocene climate and cultural evolution in late prehistoric–early historic West Asia. *Quaternary Research.* 66: 372–387.

Stille, P., and G. Shields. 1997. *Radiogenic Isotope Geochemistry of Sedimentary and Aquatic Systems (Lecture Notes in Earth Sciences)* 68, p. 0217. Springer, Berlin.

Straume, E. O., C. Gaina, S. Medvedev, K. Hochmuth, K. Gohl, J. M. Whittaker, R. Abdul Fattah, J. C. Doornenbal, and J. R. Hopper. 2019. GlobSed: Updated total sediment thick-ness in the world's oceans. *Geochemistry, Geophysics, Geosystems.* 20: 1756–1772.

Subramanian, V., L. Van't Dack, and R. Van Grieken. 1985. Chemical composition of river sediments from the Indian subcontinent. *Chemical Geology.* 48: 271–279.

Subramanian, V., G. Biksham, and R. Ramesh. 1987a. Environmental geology of peninsular river basins of India. *Geological Society of India.* 30: 393401.

Subramanian, V., L. Van't Dack, and R. Van Grieken. 1987b. Heavy metal distribution in the sediments of Ganges and Brahmaputra river. *Environmental Geology and Water Sciences.* 9: 93–103.

Subramanian, V., and P. K. Jha. 1988. Geochemical studies on the Hoogly (Ganges) estu-ary. *Mitteilungen aus dem Geologisch-Paläontologischen Institut der Universität Hamburg,* SCOPE/UNEP Sonderb. 66: 267–288.

Subramanian, V., P. K. Jha, and R. Van Grieken. 1988. Heavy metals in the Ganges estuary. *Marine Pollution Bulletin.* 19: 290–293.

Subramanian, V. 1993. Sediment load of Indian Rivers. *Current Science.* 64: 928–930.

Subramanian, V. 2000. *Water: Quantity-Quality Perspective in South Asia.* Kingston International Publishers, Surrey, UK.

Suja, S., P. M. Kessarkar, R. Shynu, V. P. Rao, and L. L. Fernandes. 2016. Spatial distribution of suspended particulate matter in the Mandovi and Zuari estuaries: Inferences on the estuarine turbidity maximum. *Current Science.* 110: 1165–1168.

Suja, S., L. L. Fernandes, and V. P. Rao. 2017. Distribution and fractionation of REE and Y in suspended and bottom sediments of the Kali estuary, western India. *Environmental Earth Sciences.* 76: 174.

Tepanosyan, G., L. Sahakyan, O. Belyaeva, N. Maghakyan, and A. Saghatelyan. 2017. Human health risk assessment and riskiest heavy metal origin identification in urban soils of Yerevan, Armenia. *Chemosphere.* 184: 1230–1240.

Thamban, M., V. P. Rao, and S. V. Raju. 1997. Controls on organic carbon distribution in sedi-ments from the eastern Arabian Sea Margin. *Geo-Marine Letters.* 17: 220–227.

Thamban, M., V. P. Rao, R. R. Schneider, and P. M. Grootes. 2001. Glacial to Holocene fluctuations in hydrography and productivity along the southwestern continental margin of India, *Palaeogeography, Palaeoclimatology, Palaeoecology.* 165: 113–127.

Thamban, M., V. P. Rao, and R. R. Schneider. 2002. Reconstruction of late Quaternary monsoon oscillations based on clay mineral proxies using sediment cores from the western margin of India. *Marine Geology.* 186: 527–539.

Tiwari, M., S. S. Nagoji, and R. Ganeshram. 2015. Multi-centennial scale SST and Indian summer monsoon precipitation variability since mid-Holocene and its nonlinear response to solar activity, *The Holocene.* 25: 1–10.

Tripathy, G. R., S. K. Singh, and S. Krishnaswami. 2011. Sr and Nd isotopes as tracers of chemical and physical erosion. In: *Handbook of Environmental Isotope Geochemistry,* ed. M. Baskaran, pp. 521–551. Springer-Verlag. doi:10.1007/978-3-642-10637-8_26.

Turekian, K. K. and H. D. Holland. 2013. *Treatise on Geochemistry.* Elsevier. ISBN-13-978-0080959757.

Thunnell, R. C., E. Tappa, and D. M. Anderson. 1995. Sediment fluxes and varve formation in Santa Barbara Basin, offshore California. Geology. 23: 1083–1086.

Tyagi, A., P. Asnani, U. De, H. Hatwar, and A. Mazumdar. 2012. *The monsoon monograph (Vol. 2).* Indian Meteorological Department. Ministry of Earth Sciences, Government of India, New Delhi.

Von Rad, U., M. Schaaf, K. H. Michels, H. Schulz, W.H. Berger, and F. Sirocko. 1999a. A 5000-yr record of climate change in varved sediments from the oxygen minimum zone off Pakistan, Northeastern Arabian Sea. *Quaternary Research.* 51: 39–53.

Von Rad, U., H. Schulz, V. Riech, M. den Dulk, U. Berner, and F. Sirocko. 1999b. Multiple monsoon-controlled breakdown of oxygen-minimum conditions during the past 30,000 years documented in laminated sediments off Pakistan. *Palaeogeography, Palaeoclimatology, Palaeoecology.* 152: 129–161.

Voosen, P. 2017. Record-shattering 2.7-million-year-old ice core reveals start of the ice ages. *Science.* 357: 6351.

Walker, M. J. C. 2005. *Quaternary Dating Methods.* John Wiley & Sons, Chichester, UK.

Wang, P., and Z. An. 2008. Millennial and orbital-scale changes in the East Asian monsoon over the past 224,000 years. *Nature.* 451: 1090–1093.

Wang, P., B. Wang, H. Cheng, J. Fasullo, Z. Guo, T. Kiefer, and Z. Liu. 2014. The global monsoon across time scales: Coherent variability of regional monsoons. *Climate of the Past* 10: 1–46.

Wang, P., B. Wang, H. Cheng, J. Fasullo, Z. Guo, T. Kiefer, and Z. Liu. 2017. The global monsoon across time scales: Mechanisms and outstanding issues. *Earth-Science Reviews.* 174: 84–121.

Wang, P., R. Tada, and S. Clemens. 2018. Global monsoon and ocean drilling. *Scientific Drilling.* 24: 87–91.

Wang, P., S. C. Clemens, R. Tada, and R. W. Murray. 2019. Blowing in the monsoon wind *Oceanography.* 32: 48–59.

White, W. M. (Ed.). 2018. *Encyclopedia of Geochemistry: A Comprehensive Reference Source on the Chemistry of the Earth.* Springer Nature, Cham, Switzerland.

Wyrtki, K. 1971. *Oceanographic Atlas of International Indian Ocean Expedition.* National Science Foundation, Washington, DC.

Yatheesh, V. 2020. Structure and tectonics of the continental margins of India and the adjacent deep ocean basins: Current status of knowledge and some unresolved problems. *Episodes.* 43: 586–608.

Yu, Z., C. Colin, S. Wan, R. Saraswat, L. Song, Z. Xu, P. Clift, H. Lu, M. Lyle, D. Kulhanek, A. Hahn, M. Tiwari, R. Mishra, S. Miska, and A. Kumar. 2019. Sea level controlled sediment transport to the eastern Arabian Sea over the past 600 kyr: Clay minerals and Sr Nd isotopic evidence from IODP site U1457, *Quaternary Science Reviews.* 205: 22–34.

Zachos, J., B. Opdyke, T. Quinn, C. Jones, and A. Halliday. 1999. Early Cenozoic glaciation, Antarctic weathering and seawater 87Sr/86Sr: Is there a link? *Chemical Geology*. 161: 165–180.

Zachos, J., M. Pagani, L. Sloan, E. Thomas, and K. Billups. 2001. Trends, rythms and aberrations in global climate 65 Ma to Present. *Science*. 292: 686–693.

Zhuang, G., M. Pagani, and Y. G. Zhang. 2017. Monsoonal upwelling in the western Arabian Sea since the middle Miocene. *Geology*. 45: 655–658.

8 Pollen Analysis and Paleoenvironmental Studies of Archeological Deposits from the Konkan Coast of India

Satish S. Naik
Deccan College Post-Graduate and Research
Institute, Deemed to be University

CONTENTS

INTRODUCTION: BACKGROUND AND DRIVING FORCES

Pollen analysis is a specialized discipline in the reconstruction of past vegetation, paleoecology, and paleoclimatology through the study of fossil assemblages of palynomorphs that have been isolated from sediments deposited in the recent past or as far back as the Paleozoic Era. Hyde and Williams (1944) originally introduced the term "palynology," which represents the aspects related to the study of pollen and spores. Although there are many branches of palynological research, this chapter focuses on the theme of pollen analysis and paleoenvironmental studies concerned with the reconstruction of past environments through the recovery, identification, and quantitative analysis of microfossils preserved in the archaeological deposits.

The outer wall of pollen grain and spore contains a very durable compound called sporopollenin, which is composed of an oxidative copolymer of carotenoid and carotenoid esters. Because of their durability and resistance to destruction by most types of weathering agents, pollen grains have been recovered from sediments of great antiquity. The variation in exine structure and sculpture together with number and distributional pattern of apertures provides the morphological diversity of the grains. Besides the pollen and spore, the palynological analyses also involve other microfossils such as dermal appendages, cuticles, vascular elements, orbicules, dinoflagellate cysts, acritarchs, chitinozoans, and scolecodonts, together with particulate organic matter and kerogen found in sedimentary rocks and sediments (Brooks and Shaw 1968, Shaw 1971, Brooks 1971).

Fossil pollen has been recovered from sediments as old as Pennsylvanian >300 million years old. Fossil vascular plant spores have been recovered from Silurian deposits over 400 million years old, and fungal and algal spores have been found in sediments over a billion years old. The pre-Quaternary pollen types where the relationship between fossil and living plant is uncertain, the fossil taxa are referred to as form genera and form species (using "ites" as suffix). In the Quaternary, however, the pollen grains and spores can be directly referred to the extant vegetation due to proximity with the present, proving present is key to the past.

The reconstruction of past communities from the microfossil record represents a major step towards the reconstruction of the past ecosystems. The biotic components represented by communities are the most complex part of the ecosystem, and when they have been reconstructed, inferences can be made about the environment of the past ecosystem. Such reconstructions require knowledge about the ecological requirements and tolerances of the species and community involved.

Aided by the palynological data, the riddles of paleoecology, paleoenvironment, paleophytogeography, biostratigraphic correlation, dating of ages, etc. can be established based on the depositional pattern of the fossil pollen grains and spores. Pollen analysis is therefore an extremely powerful tool for the investigation of floristic and climatic changes that took place in the recent past. The minerogenic deposits are usually poor in microfossil preservation, whereas organic-rich acidic deposit from humid zone normally yields the well-preserved pollens (Faegri and Iversen 1964). There is a need to generate enough data on the Holocene sediments from the low-lying coastal regions in order to tackle problems related to climatic variations.

Paleopalynological processing is a highly specialized technique to extract palyno-morphs from the geological as well as archaeological deposits. The extracted palyno-morphs are useful in environmental interpretations, and hence, the technique has got great importance. Pollen grains are tiny objects; their size ranges from 5 to 300 µm, and they cannot be closely studied using only the unaided eye. Ancient cultures such as the Assyrians recognized the importance of pollen in the production of date palms more than 2,000 years ago. Yet the first critical study of pollen had to await the devel-opment of the compound microscope in the 1600s by Hooke. In the late 1800s and early 1900s, there was a great controversy among European scientists concerning the most accurate methods of reconstructing past environments. Some gave emphasis on study of megafossils, and others believed that it could be best accomplished through the study of geology and environmental conditions that led to the formation of the sediment under consideration.

The first step towards the development of the technique of pollen analy-sis took place during the examination and identification of plant remains from Scandinavian peat bogs. A research assistant Gustav Lagerheim represented the first actual pollen statistical count. Von Post (1916) presented the results of his initial pollen analysis at a convention of Scandinavian naturalists. Since then, palynologists in many areas of the world have continued to use this technique to attempt paleoenvironmental reconstruction and to use the resulting data to infer past climatic changes. It was Huntington (1906) who initiated the pollen analysis in India from sediments of Pangong Lake in Ladakh. Later, Wodehouse (1935) and De terra and Paterson (1939) worked out palynology of Indian Quaternary sediments. The investigations on pollen analysis to establish the Quaternary veg-etation history, paleoecology, biostratigraphy, and mapping in India have been adequately done by many scholars.

The chemical factors such as soil pH, oxidation potential (Eh) (Garrels 1960, Kajale and Deotare 1989), degree of microbiological activities, and varying com-position of sporopollenin play an important role in the selective preservation of pollens and spores (Deotare and Kajale 1996). The cumulative effect of these chemical and environmental factors in the preservation or degradation of pollens and spores has been recorded by Faegri and Iversen (1964), highlighting that even slightly acidic nature of the deposit favors microfossil preservation. Any individual factor is not as effective as collective factors for the pollen preservation (Dimbleby 1985, Mandavia 1982).

The investigations on Quaternary pollen analysis in connection with the vegeta-tion history so far done in India with greatly diversified floral and ecological distribu-tion are limited in relation to the vastness of the country. As a result, unlike North America and North European countries, in India we do not have a comprehensive picture of the Quaternary vegetation history and climate in a chronological sequence. The depositional vegetation and environment as already emerged, conveyed a picture of divergence in the characteristics of past vegetation and paleoenvironment from place to place in India ranging from tropical, desert, coastal to temperate conditions. This gets reflected in the present distribution of Indian climatic conditions with a consequential heterogeneity in the floral distribution. Thus, the role and significance of microfossils as a source of environmental reconstruction and their distribution and preservation, which are controlled by changes, have revealed out a comprehensive picture of the Quaternary flora in India (Naik 2014).

After 50 years of research, the tropics and subtropics remain an undersampled set of ecosystems, and are still full of challenges. In some ways, the challenges are growing as the most important lakes have already been cored and their pollen histories are described. These developments have taken place simultaneously with the major efforts to apply computer simulations and pollen dispersal models in order to achieve a better understanding of the spatial representation of the pollen records. Also, it is necessary to discuss the potential of DNA analysis of pollen for investigating the ancestry and past migration pathways of the plants. A better understanding of past changes in the earth's climate system is significant for us to improve our ability to understand future trends of global change under the influence of anthropogenic activity. This is especially important for the prediction of future climate and environmental changes in monsoonal South Asia. Regional and local correlations have also been made in order to have a broad vision of the paleovegetation and paleoclimate in different regions of India. From the point of view of Stone Age occupation, it is important to note that geographical diversity in India is tremendous and that, as at present, this factor led to much regional variability in both climatic regimes and vegetation. It would also seem that periodic changes in climate and vegetation did not disturb continuity of human occupation at any stage and in any area. More intensive and detailed region-based paleobotanical investigations are called for in order to place human–plant relationships in different periods on a sound footing. An issue of crucial importance is the reconstruction of paleodiet, particularly considering the realization now that plant foods played a major role in the prehistoric hunter-gatherer economies of the tropical areas (Naik 2017).

As far as pollen analysis of coastal swamp deposits from Konkan Coast of India is concerned, much of the work in paleopalynology and paleoecology along the western coastal swamp deposits in Maharashtra is contributed by the scholars like Vishnu-Mittre and Guzder (1975), Ghate (1985, 1990, 2007), Sakurkar (1999), Kumaran et al. (2001, 2004a–c, 2005), Limaye (2004), Limaye et al. (2007), and Shindikar et al. (2004). Vishnu-Mittre and Guzder (1975) first tried to extract pollen from swamp area near Mumbai and recovered few pollen types, which were not good enough to draw any statistically significant conclusions. Subsequently, Ghate (1985) while working in Konkan area for her PhD dissertation could recover little pollen from Sopara tidal swamp. Caratini et al. (1980) investigated palynologically a 2.70-m-deep profile near Thane, Bombay (Mumbai). The analysis showed a poor representation of mangrove elements and abundance of grasses. The occurrence of Casuarina pollen from bottom to top of the profile suggests the recent age of sediments. Ratan and Chandra (1984) investigated recent sediments of the continental shelf off Mumbai, of which all the 22 samples had pollen grains of mangrove, tropical evergreen, and mixed deciduous types of forest. Pollen of non-arboreal plants and pteridophytic spores were also recorded in huge quantities. It was observed that the samples collected near the coast have yielded more pollen grains. While working on the Holocene deposits of Dhamapur, Sindhudurg district, Maharashtra, Kumaran et al. (2001) have recovered microfossils like pteridophytic and fungal spores suggestive of swampy and marshy habitat. Recently, Kumaran et al. (2004c) studied floristic composition, palynology, and sedimentary facies of Hadi mangrove swamp (Maharashtra), and yielded a fairly rich and diversified microfossil assemblage. The evolutionary process and phases at the differential rate can be addressed by taking a look at the paleoecology of the mangrove sediments as they form good archives from Quaternary events and particularly for the Holocene.

Because of highly oxidizing environment, pollen analysis of habitational deposits from archaeological sites is not as common in India as in Western countries. Archaeological deposits are considered as a minerogenic deposit and hence poor in microfossil preservation because of highly oxidizing conditions and alkaline nature of the deposits (Deotare and Kajale, 1996). This fact triggered the author to test the pollen potential in the habitational deposits. Accordingly, the concept has been immediately applied on the soil samples collected by Prof. B. C. Deotare from Mr. Raut's square well at Chaul, Raigad district. This resulted in the discovery of many types of palynomorphs (Naik et al. 2009). These palynomorphs are the paleoenvironmental indicators to study and develop new database in order to understand paleoecological changes during Late Holocene. It has been suggested that human activity occurred on periodically drying nonsaline wetland as indicated by the total absence of mangrove pollen in clayey silts (Naik 2012, Naik and Deotare 2008–2009).

KONKAN COAST OF INDIA

Konkan (15°45': 20°0' N and 72°40': 73°0' E) extends from Gujarat in the north to Goa in the south, and its coast line covers 530 km. It is a narrow strip of land, which widens about 50 km towards north and narrower to about 30 km in the south. It includes Thane, Greater Mumbai, Mumbai, Raigad, Ratnagiri, and Sindhudurg districts (Deshpande 1998). All these districts are segmented by a large number of estuaries, creeks, and backwater areas. Geomorphologically, the Konkan region is characterized by beaches and rocky headlands. It is also characterized by hills and plains, and between the foothills and the Arabian Sea lies a narrow strip of rugged land broken by numerous rapid flowing rivers and streams, creeks, and isolated ranges of hills (Kolaba District Gazetteer 1964).

GEOLOGY

The geology of the entire Konkan Coast consists of dark colored volcanic lava flows and laterites. Approximately 80–100 million years ago, the lava flows were poured out of the long and narrow fissures in the earth's crust, at the close of Mesozoic Era. These are spread out in the form of horizontal sheets or beds and constitute the innumerable spur hills and hill ranges; bold, flat-topped ridges; softy peaks and plateaus with impressive cliffs. The hill ranges and the plateau form a part of the famous Western Ghats. In the plains and valleys, the lava flows occur below thin blankets of soil of variable thickness. The basalts are dark grey to grey and bluish-grey in color, and are hard, compact tough, and fine to coarse-grained in texture. Next to basalt comparatively softer amygdular and scoriaceous traps, purple to grayish in color, usually shows rounded, elongated or tubular cavities and geode with infillings of secondary minerals like calcite, zeolites, and variety of secondary quartz like agate, jasper, chalcedony, etc. These generally occupy in the lower portions of the ridges and slopes, and usually in the valleys and plains. A red clayey bed often termed as "redbole" is also noticeable in some places. Beds of "laterites" usually formed by the mechanical and chemical disintegration brought about by the atmospheric agencies on the underlying trap cap, the several peaks, and the lofty ridges are present in the district. They are found at places in the lower regions also (Kolaba District Gazetteer 1964).

SOILS

The following categories of soils are found in the Konkan Coast.

Forest soils: Forest soils are put to cultivation but yield valuable forest produce like teak, myrobalan, baheda, pepper etc.

Varkas soils: Located just before the forest soils are poor in organic matter and nitrogen and are suitable for the growth of the millets.

Rice soils: Occupy the largest area of Konkan region.

Khar soils: Situated on the flat-leveled surfaces and near the creeks and are being brought under reclamations best suited for garden crops, e.g., areca nut, coconut, etc.

Coastal alluvial soils: All along the coast.

Laterite soils: Laterite soils are observed among the Sahyadri hill ranges suitable for valuable species of evergreen and semi-evergreen forests.

DRAINAGE PATTERN

Many rivers and streams originate in the Sahyadris and flow westward to the Arabian Sea. The chief rivers are the Ulhas with its main tributaries – Posheri, Chillar, and Pej in Karjat range; the Patalganga and its tributaries – Balganga and Bhogeshwari in Panvel and Pen ranges; and the Amba in Nagothana range; the most important river concerns with study area are Kundalika in Roha range and also Savitri with its tributary "Kal" in Mangaon and Mahad ranges.

HILLS

The chief hills in the Konkan region are of Sahyadris, the east boundaries of the Konkan; the Sahyadri range is crossed by several passes or ghats, beginning from Malshej Ghat in north to Amboli Ghat in south; Varandha Ghat joins Mahad and Bhor for useful wheeled traffic. Besides hill ranges south of Pen is Sarasgad (407 m) and west of Alibag are more prominent elevations in northern portion of the Raigad district.

FLUVIAL MORPHOLOGY

The geomorphic history and relative chronology of fluvial deposits are generally based on both erosional and depositional features. In the Konkan area, studies on fluvial morphology in the context of archaeology were restricted to the observations made by Todd (1939) and Malik (1959) on the sections of the Kandivli nullahs and Dahisar River.

Attempts have been made to explain climatic change on the basis of evidence for aggradations and erosion. Zeuner (1950) interpreted the gravel deposits in the river valleys of Gujarat as reflecting increased aridity during the Pleistocene. Sankalia (1962), on the other hand, equated erosion and aggradations with the dry and wet phases of Late Pleistocene climate. Rajaguru (1970) initiated further systematic study of alluvial deposits in the western Maharashtra, and he suggests that most probably there was some degree of climatic variation in the Peninsula during the Pleistocene. The most significant depositional feature is the alluvium, which occurs in the form of cut-and-fill terraces. Fluvial deposits are for the most part confined to the middle and lower reaches of the river, composing flood plains, estuarine, and nearshore formation.

CLIMATE AND VEGETATION

The climate of Konkan is similar to that of west coast of India with plentiful and regular seasonal rainfall, oppressive weather in the hot months, and high humidity throughout the year. The summer season from March to May is followed by the southwest monsoon from June to September – October and November form the post-monsoon or the retreating monsoon season, while the period from December to February is the cold season. Being a heavy rainfall area, the precipitation is ranging from 2,500 to 4,500 mm. The Konkan region occupies just 10% of geographical area but receives 46% of total precipitation of the Maharashtra state (National Institute of Disaster Management 2014, Mandale et al. 2017). The rain increases rapidly from coast towards the Western Ghats. The maximum summer temperature rarely rises above 38.5°C, and the minimum is 13.5°C. Humidity is rarely less than 55%, and during monsoon, it rises above 90% (Gaikwad 2013). The climate not only affects vegetation and landscape development but also affects economy and social aspects of the population. The climate is, however, generally pleasant from December to March (Kolaba District Gazetteer 1964).

The wide variations in physiography, geology, soil, and climate of the Konkan region have governed various types of vegetation. Although this is essentially botanical and paleoecological approach, it may also be complementary to ecologists and archaeologists/historians for understanding the vegetational history of the region. The vegetation is mainly distributed in four belts.

Tropical Semi-evergreen to Evergreen Forest

This type of forest is confined to the hilly regions, particularly in the Western Ghats and some of its spurs. There are three types of forest composition. The plateau forest or top layer is composed of the evergreen species, viz., *Actinodaphne angustifolia, Ficus racemosa, Glochidion hohenackeri, Mangifera indica, Memecylon umbellatum, Olea dioica, Syzygium cumini* etc.; the middle layer or terraces below the plateau consists of the species like *Bridelia retusa, Careya arborea, Heterophragma quadriloculare, Lagerstroemia microcarpa,* and *Terminalia bellirica* ; and the ground layer or the undergrowth is composed of *Carissa congesta, Strobilanthes callosa, Gnidia glauca, Leea indica, L. macrophylla, Pavetta indica, Woodfordia fruticosa* etc. A number of lianas or climbers such as *Abrus precatorius*, *Acacia pinnata*, *Acacia. concinna*, *Butea superba*, *Cansjera rheedei*, *Celastrus paniculatus*, *Combretum ovalifolium*, *Dioscorea* spp., *Entada rheedii*, *Gnetum ula*, and *Mucuna pruriens* are also occurred in forests. Various types of grasses are usually located on steep, slopes, edges of the plateau, the cliffs tops and bases etc.

Tropical Moist Deciduous Forest

This forest type is usually found on hill slopes and dominantly scattered throughout the Konkan. Here also, the vegetation is observed in three different storeys. The top storey is composed of species like *Tectona grandis* with the association of *Acacia catechu*, *Albizia lebbeck*, *Bombax ceiba*, *Dalbergia latifolia*, *Pterocarpus marsupium*, *Terminalia paniculata*, *Trewia nudiflora* etc.; the middle storey consists of species like *Careya arborea*, *Erythrina variegata*, *Gmelina arborea*, *Grewia tiliaefolia*, and *Haldina cordifolia*, while the ground layer is composed of the undergrowth species like *Carissa congesta*, *Strobilanthes callosa*, *Helicteres isora*, *Wrightia antidysenterica*, and *Wrightia tinctoria*.

Tropical Dry Deciduous Forests

This type of forest is found in middle and ground layers, as described in the moist deciduous forest. These are also called fuel forests and consist of species like *Terminalia elliptica* associated with *Bombax ceiba*, *Haldina cordifolia*, *Holoptelea integrifolia*, *Mitragyna parviflora*, *Pterocarpus marsupium*, and *Terminalia chebula*.

Littoral and Swamp Forests

This type of forest consists of riparian flora and mangrove forests. Riparian flora is found along the river banks, and river beds on stony, sandy, or alluvial habitat. The species – such as *Capillipedium huegalli*, *Homonoia riparia*, *Indigofera cordifolia*, *Rotula aquatica*, *Tamarix ericoides*, and *Vitex negundo* are found along the sandy soils and stony river banks. On alluvial soil, a number of wetland and marshy plants *Ageratum conyzoides*, *Ammannia baccifera*, *Caesulia axillaris*, *Chenopodium album*, *Typha angustata*, *Chrozophora tinctoria*, *Commelina benghalensis*, *Commelina diffusa* species of *Cyperus* and *Polygonum* are met with. Mangrove forests found along the muddy seashores and sandy saline areas consist of the species like *Avicennia marina var. acutissima* and *A. officinalis* associated with *Acanthus ilicifolius*, *Aegiceras corniculatum*, *Carallia brachiata*, *Rhizophora mucronata*, *Sonneratia* sp. etc. On muddy soil, *Salvadora persica*, *Sesuvium portulacastrum*, and *Suaeda monoica* are frequent, while on sandy seashore *Aeluropus lagopoides*, *Cressa cretica*, *Cyperus bulbosus*, *Hyphaene dichotoma*, *Pandanus tectorius*, *Pedalium murex*, *Spinifex littoreus*, *Tribulus terrestris* etc. are found.

A number of hydrophytes, epiphytes, and parasites are a few insectivorous plants of biological interest, and pteridophytes are also found in the area covered by the above-mentioned forest types (Kothari and Moorthy 1993).

ARCHAEOLOGICAL SITES

Intensive explorations carried out in Konkan Coast led to the discovery of many archaeological sites: some excavated and many unexcavated. The sites from north to south are Chinchani, Nalasopara, Chaul, Mandad, Kelshi, Palshet etc. However, coastal region remained secluded to some extent for proper understanding of the interaction between humans and environment in archaeological perspective. Subsequently, pollen analyses and paleoenvironmental studies were carried at two archaeological sites on the Konkan Coast, namely, Chaul and Mandad (Figure 8.1).

ANCIENT CHAUL

The ancient site of Chaul (18°33'13.06" N, 72°56'15.17" E) is located 17 km south of Alibag, a district headquarter of Raigad district in Maharashtra. The site lies at the western end on the right bank of Kundalika River. The river flows about 0.5 km to the south, and the Arabian Sea is about 1 km to the west of the site. Ancient Buddhist caves are located about 2 km to the northeast of the site. Presently, the coastal lowland is covered by rich vegetation of natural mangrove forests. These forests comprise species including *Avicennia marina var. acutissima* and *A. officinalis* associated together with *Acanthus ilicifolius, Aegiceras corniculatum, Carallia brachiata, Rhizophora mucronata*, species of *Sonneratia* etc. Palms like coconut, betel nut, and other economic plants are cultivated by thickly populated local inhabitants.

Archaeological excavations at Chaul were conducted by Deccan College, Pune, under the direction of Prof. V.D. Gogte, and this revealed substantial evidence of ancient habitation spanning a period of nearly 2,000 years, from 300 BC to 1700 AD (Figure 8.2). The Medieval Period occupation is not affected by flood and/or beach dune activity as it is located on the older mud flat, which is about a meter higher than the present high tide level. However, earlier occupations are within the present intertidal zone on the right bank of Kundalika River. This is within the present tidal range of 2.5 m. Later, Silahara habitation (c. 1000–1200 AD) is often the regression of sea yet in a strong tidal effect as indicated by unoxidized tidal mud flat. Subsequently, the tidal effect seems to be considerably reduced in the Late Silahara Period, indicating more flood environment of Kundalika than tidal environment as reflected in the form of yellowish silt deposit. The first habitation took place on older mudflat, occurring around 3.10 m below the surface. The soil samples were collected on the basis of nature of deposit and cultural material exposed in the section of trench profiles (Gogte 2003, Gogte et al. 2006).

EARLY HISTORIC SITE AT MANDAD

Mandad (18°17'42.67" N, 73°02'58.47" E) is an ancient port site located along the margin of northern branch of Rajapuri Creek in Raigad district, Maharashtra. The place can be identified as Mandragora, which figures in the list of ports on the western coast given in the *Periplus of Erythraean Sea* (Schoff 1912). The ancient habitation site is situated towards western side of the present Mandad village and northwest side of Kuda caves. The group of Kuda caves have been explored by different scholars

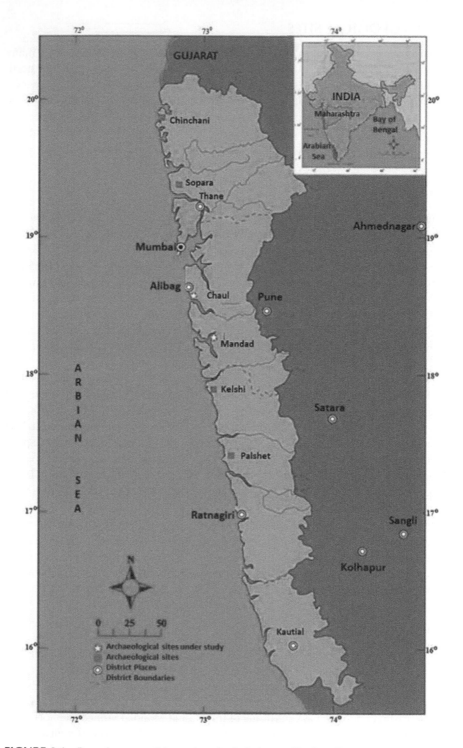

FIGURE 8.1   Location map of the archaeological sites on Konkan Coast.

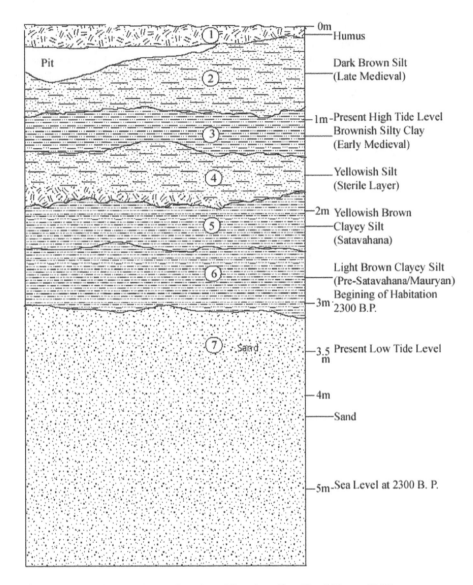

FIGURE 8.2   Cultural lithostratigraphy of Rout's well at Chaul (Gogte 2003).

like Burgess (1881), Nagaraju (1981), Dhavalikar (1984), Gupta (2007), and Gogte
(2004, 2006–2007). Dhavalikar (1984) proposed that the site must be a place of some
importance in the early historic period probably due to its location on the creek and
may have served as port in the Satavahana Period. The site of Mandad has a lot
of potential for the archaeological record but it is not yet excavated because high-
tide sea water covers the site. The vegetation along the muddy creek is dominated
by mangrove forests. The species include *Avicennia marina* var. *acutissima* and *A.
officinalis* associated together with *Acanthus ilicifolius*, *Aegiceras corniculatum*,
Carallia brachiata, *Rhizophora mucronata*, *Sonneratia* spp. among others.

The surface findings as well as cultural material exposed in the section are dat-able to the early centuries of the Christian Era. It involves indigenous black and red ware pottery, red polished ware, mica washed red ware, slipped black ware, glass beads, bricks, broken four-legged saddle quern, and broken double-handled Roman Amphorae. The occurrence of the Roman amphorae at Mandad clearly proves mari-time trade of Mandad with the Roman Empire during 1st century BC. During explo-ration, the exposed pit dug by farmer was noticed on the site. Hence, the soil samples were collected by marking the stratigraphy on the basis of nature of deposit and cultural material exposed in the pit section (Figure 8.3).

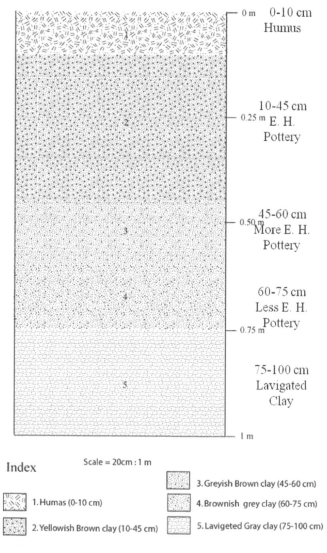

FIGURE 8.3   Cultural lithostratigraphy of pit at Mandad.

MATERIAL AND METHODS

The palynological technique developed by Faegri and Iverson (1964), further adopted by Traverse (1988) and Moore et al. (1991), was modified and specially designed for minerogenic sediments by Deotare (1995). The same technique is modified and employed in processing all the soil samples to get rid of the matrix material as far as possible without any loss of organic matter, and the author has tried to recover undamaged pollen and non-pollen palynomorphs, which are the representatives of different samples collected from archaeological deposits from Konkan Coast of India (Naik 2012).

FIELD METHODS

The soil samples for pollen analysis were collected from archaeological deposits of the exposed stratigraphic sections of the well dug by Mr. Raut and from the trenches excavated by Gogte et al. (2006), *viz.*, Loc. CHL-V (Trench B), Loc. CHL-R (Trench E), and Loc. CHL-C (Trench F) at Chaul, and from the exposed section of pit at Mandad.

LABORATORY METHODS

The method of treating the soil samples varies from sample to sample on the basis of nature, chemical composition, cultural period, lithology, and presence of organic matter, etc. According to Faegri and Iversen (1964), the following steps are involved in maceration technique to extract palynomorphs. Initially, the pH of each and every soil sample was checked and then treated with 10% HCl to dissolve carbonates. Some samples were rich in tenacious clay; then the clay is removed by repetitive washing with 5% sodium hexametaphosphate $(NaPO_3)_6$. Later, the samples were treated with 10% KOH or NaOH (potassium/sodium hydroxide) to remove extra organic substances. The use of hydrofluoric acid (HF) helped to digest silicates and minerals. The residues were then treated with hot 20% HCl to remove hydrofluoric acid. The residues were filtered with distilled water through a microsieve of 5 μm size to remove the debris of less than 5 μm in size with the help of ultrasonicator. The filtrates were then concentrated for the preparation of slides. The organic matter is mounted on glass slides using glycerol or glycerin jelly. These slides were examined under light microscope (Leica DMRBE); the palynomorphs were recorded qualitatively and quantitatively and then photographed satisfactorily. The measurements have also been documented digitally by using Leica Q Win Software. The total numbers of palynomorphs were plotted quantitatively, layerwise, and trenchwise, and the results have been summarized.

RESULTS

RESULTS OF THE CHEMICAL ANALYSIS

The chemical analysis was performed to find out the nature of archaeological deposit. The following samples collected from the site of Chaul were analyzed for pH, organic

TABLE 8.1
Results of Chemical Analysis

Lab. No.	Description	Depth (m)	pH	% of CaCO$_3$	% of O.C.	% of P
2256	Yellowish brown silt	1.70–1.75	7.2	6	0.158	0.34
2255	Brownish clayey silt	2.25–2.30	7.4	5	0.262	0.35
2254	Greyish brown clay – compact	2.65–2.70	7.8	13	0.252	0.26
2253	Brownish silty clay	2.95–3.00	7.9	12	0.288	0.28
2252	Karal	>3.30	8.2	24	0.058	0.13

carbon, calcium carbonate, and total phosphorus as per procedure described by Jackson (1962), Piper (1966), and Black et al. (1965). The analyzed data indicates that the nature of deposit is slightly alkaline. Sample No. 2252 is of relatively loose well-laminated sand deposit without pottery, indicating beach sand deposit. Phosphate content of this sample is very low (0.13%), confirming the absence of human interference. On the contrary, the remaining samples contain relatively high percentage of phosphates indicating habitational nature of deposits with a lot of pottery, brick bats, beads etc. Organic carbon (O.C.) content in sample Nos. 2253–2255 is relatively more than the overlying and underlying samples, and pH values are just above 7, indicating a slightly alkaline nature of the deposits (Table 8.1). The slightly alkaline nature and relatively high organic carbon might have favored the preservation of microfossils in these minerogenic deposits, which is reflected in a significant pollen recovery.

RESULTS OF THE POLLEN ANALYSIS

Pollen analysis results into well-preserved palynomorphs belonging to pteridophytes, angiosperms, fungi, microforaminifera, diatoms, and dinoflagellate cysts gave a good picture of palynoflora at two archaeological sites, *viz.*, Chaul and Mandad.

PALYNOFLORA AT CHAUL

The pollen analysis of 18 soil samples collected from Mr. Raut's square well at Chaul comprises pteridophytic spores, angiospermous pollen, fungal remains, foraminiferal linings, and microforaminifera of marine environment. Diatoms, dinoflagellate cysts, and fungal remains in the form of variety of spores, aseptate and septate mycelium, fungal fruiting bodies, and sporogenous tissues have also been encountered.

The qualitative and quantitative analyses of palynoflora reveal the dominance of angiospermous pollen over pteridophytic spores. The assemblage of palynoflora consists of 3 families of pteridophytic spores, 25 families of angiospermous pollen, and 24 forms of fungal spores belonging to different taxa. The analysis of the present palynofloral assemblage indicates that the fungal remains are dominant constituents, whereas angiospermous pollen and pteridophytic spores are poorly represented.

Quantitative analysis has been done on the basis of frequency of various paly-notaxa in a count of 100 specimens per sample. This revealed that the genera of pteridophytic spores belong to Lygodiaceae, Polypodiaceae, and Selaginellaceae, and constitute 1.11% of the assemblage. The angiospermous pollen is represented by 25 families and constitutes 21.70% of the assemblage. The dominant families are Asteraceae, Bombacaceae, Brassicaceae, Chenopodiaceae/Amaranthaceae, Cyperaceae, Euphorbiaceae, Poaceae, Arecaceae, and Palmae. The fungal assem-blage is common to abundant in nature and is represented by 24 genera, constituting 56.86% of the assemblage. Other microfossils comprising monocotyledonous cells, tracheids of vascular plants, dinoflagellate cysts, and insect egg-like structure – oospore – constitute 2.41% of the assemblage. The microforaminifera are repre-sented by 7 genera, viz., *Ammonia, Bolivina, Elphidium, Globogerina, Lagena, Spiroloculina*, and *Quinqueloculina*, which constitute 16.35%, and foraminiferal linings constitute 1.57% of the assemblage (Figure 8.4).

The percentage frequencies of pteridophytic families (Lygodiaceae, Polypodiaceae, and Selaginellaceae) and angiospermous families (Brassicaceae, Chenopodiaceae/Amaranthaceae, Cyperaceae, Euphorbiaceae, Poaceae, Arecaceae, and Palmae) are dominant throughout all the cultural periods of the deposit. Along with the dominant angiospermous families, the total pollen count of other fami-lies like Amaryllidaceae, Apiaceae, Melastomataceae/Combretaceae, Fabaceae, Mimosaceae, Moraceae/Urticaceae, Myristicaceae, Pandanaceae, *Polyadopollenites* sp., *Retipollenites* sp., Rubiaceae, Sapotaceae, and Sapindaceae (*Schleichera oleosa*) represents their appearance throughout all the cultural periods except the sterile layer of the profile (Figure 8.5).

Microforaminifera represent their appearance restricted to pre-habitational deposit of karal, which dates back to 2,190 ± 95 years BP (Vishnu-Mittre and Guzder 1975), while foraminiferal linings are encountered not only from karal but also from lower levels of early historic deposit.

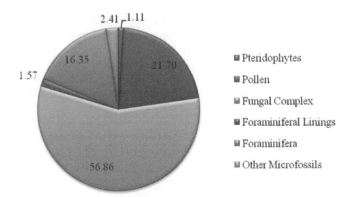

Average Relative % of Abundance

2.41 ┌1.11

16.35   21.70

1.57

56.86

■ Pteridophytes
■ Pollen
■ Fungal Complex
■ Foraminiferal Linings
■ Foraminifera
■ Other Microfossils

FIGURE 8.4 Composite Pie diagram showing the average relative percentage abundance of palynomorphs at Chaul.

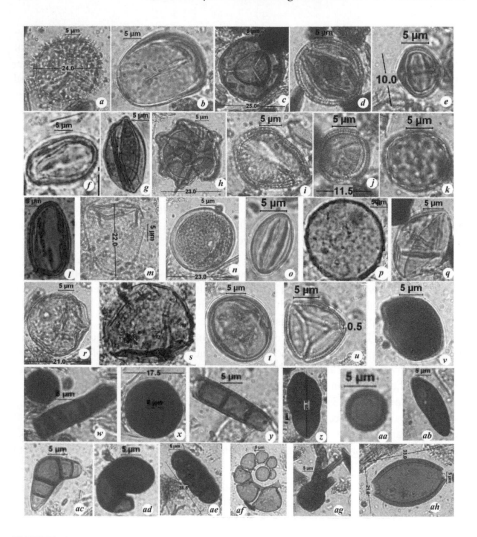

FIGURE 8.5 Representative palynomorphs recovered at Chaul: (a) Lygodiaceae, (b) Polypodiaceae, (c) Selaginellaceae, (d) Avicenniaceae, (e) Rhizophoraceae, (f) Apiaceae, (g) Arecaceae, (h) Asteraceae, (i) Bombacaceae, (j) Brassicaceae, (k) Chenopodiaceae/ Amaranthaceae, (l) Meliaceae, (m) Cyperaceae, (n) Euphorbiaceae, (o) Fabaceae, (p) Moraceae/Urticaceae, (q) Palmae (*Cocos nucifera*), (r) Poaceae, (s) Rubiaceae, (t) Sapotaceae, (u) Sapindaceae (*Schleichera oleosa*), (v) *Retipollenites* sp., (w) fungal septate hyphae, (x) *Nigrospora* sp., (y) *Multicellaesporites* sp., (z) *Dicellaesporites* sp., (aa) *Circinella* sp., (ab) *Inapertisporites* sp., (ac) *Curvularia* sp., (ad) *Trichocladium* sp., (ae) *Meliola* sp., (af) Microforaminiferal lining, (ag) *Tetraploa* sp., (ah) *Trichuris trichiura* egg. (All photomicrographs magnified at 100×, unless otherwise stated.)

PALYNOFLORA AT MANDAD

Pollen analysis of six soil samples collected from the exposed pit section in habitational deposit at Mandad has yielded different types of pteridophytic spores, angiospermous pollen belonging to freshwater-dwelling plants, and mangroves.

Foraminiferal linings and microforaminifera of marine environment have been found. Diatoms, dinoflagellate cysts, and fungal remains in the form of variety of spores, aseptate and septate mycelium, fungal fruiting bodies, and sporogenous tissues have also been encountered.

The qualitative and quantitative analyses of the present palynofloral assemblage indicate that the fungal remains are dominant constituents, whereas angiospermous pollen and pteridophytic spores are poorly represented. The assemblage of palynoflora consists of 5 families of pteridophytic spores, 39 families of angiospermous pollen of mangroves and freshwater plants, and 44 forms of fungal spores belonging to different taxa. The qualitative and quantitative analyses of palynoflora reveal the dominance of angiospermous pollen over pteridophytic spores.

Quantitative analysis has been done on the basis of frequency of various palynotaxa in a count of 100 specimens per samples. This revealed that the fungal remains, represented by 44 genera, constitute 57.83% of the assemblage. The genera of pteridophytic spores belong to Adiantaceae, Lygodiaceae, Polypodiaceae, Selaginellaceae, and Osmundaceae, and constitute 1.85% of the assemblage. The mangrove pollen belongs to the families Acanthaceae, Avicenniaceae, Rhizophoraceae, and Sonneratiaceae, and constitutes 5.48% of the assemblage. The freshwater angiospermous pollen is represented by 30 families and 4 different genera, and constitutes 30.26% of the assemblage. The dominant families are Brassicaceae, Chenopodiaceae/Amaranthaceae, Cyperaceae, Euphorbiaceae, Fabaceae, Palmae, and Poaceae. The other microfossils are represented by monocotyledonous cells, tracheids of vascular plants, stomata of fossil leaves, diatoms, dinoflagellate cysts, insect egg-like structure – oospore – and constitute 3.61% of the assemblage. The microforaminifera are represented by 5 genera, *viz.*, *Ammonia*, *Lagena*, *Bolivina*, *Elphidium*, and *Spiroloculina*, which constitute 0.36%, and foraminiferal linings constitute 0.62% of the assemblage (Figure 8.6).

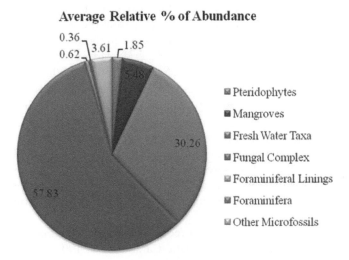

Average Relative % of Abundance

- Pteridophytes
- Mangroves
- Fresh Water Taxa
- Fungal Complex
- Foraminiferal Linings
- Foraminifera
- Other Microfossils

FIGURE 8.6   Composite Pie diagram showing the average relative percentage abundance of palynomorphs at Mandad.

The percentage frequencies of pteridophytic families (Lygodiaceae, Selaginellaceae), angiospermous families (Brassicaceae, Chenopodiaceae/Amaranthaceae, Cyperaceae, Poaceae), and mangrove families (Acanthaceae, Avicenniaceae, Rhizophoraceae) are dominant throughout the early historic period deposit. The percentage frequencies of the families Polypodiaceae, Fabaceae, Palmae, and Sonneratiaceae are prominent in lower levels, while those of Euphorbiaceae are high in upper levels of the early historic period deposit.

Along with dominant families of pteridophytic spores, the other families Adiantaceae and Osmundaceae (*Todisporites flavatus*) are represented in pre-habitational and early historic period deposit, respectively.

Along with the dominant freshwater plant families, the total pollen count of other families like Anacardiaceae, Annonaceae, Araliaceae, Bombacaceae, Boraginaceae, Combretaceae (*Terminalia arjuna*), Gentianaceae, Meliaceae, Menispermaceae, Mimosoideae (*Acacia chundra*), Myrtaceae (*Syzigium cumini*), Palmae (*Cocos nucifera*), Palmae (*Phoenix Sylvestris*), *Palmaepollenites kerlensis*, Polygonaceae, Sapotaceae type, Sapindaceae (*Schleichera oleosa*), and Tiliaceae (*Retitricolporites guinensis*) represents their appearance limited to the early historic period, while the families like Amaryllidaceae, Bignoniaceae, *Dicellaepollenites* sp., Moraceae/ Urticaceae, Myristicaceae, Nymphaeaceae, Plumbaginaceae, *Retipollenites* sp., and *Salix* sp., show their occurrence in pre-habitational deposit, i.e., in lavigated clay.

Among the mangrove families, Sonneratiaceae is represented in lower level of early historic period and pre-habitational deposit, while other families are dominant and are represented throughout the profile (Figure 8.7).

Microforaminifera are restricted to pre-habitational deposit of lavigated clay, while foraminiferal linings are encountered throughout the profile.

DISCUSSION

The pollen, being plant-oriented (i.e., microscopic male reproductive entity of flowering plants), remains intact and recoverable from the sediments, and it is one of the best and most reliable parameters available in the present context to study vegetation and climatic aspects of the past. Several factors are considered to recon-struct the Quaternary paleoenvironment of a particular region through pollen anal-ysis. They include plant habitat; dispersal and pollination mechanisms of plants; meteorological conditions including temperature, wind speed, relative humidity and rainfall; and influence of glaciation. The differential preservation of sediments and the destruction of pollen grains and spores are fairly well known. Among oth-ers, microbial action and oxidation play important roles to understand the nature of depositional environment (Deotare 1995). For the Quaternary Period, palynology has played a key role in reconstructing climatic and vegetational history but some-times the botanical markers selected are not fiddle indicators of climate (Meher-Homji 1994).

Attempts were made in the past to reconstruct the paleoclimate for south and southwestern India based on offshore data (Caratini et al. 1994, Van Campo 1986, Saxena and Misra 1990) and for the Nilgiris using peat deposits (Sukumar et al. 1993, 1995, Geeta et al. 1997, 1999). Nevertheless, these data have relevance to paleoclimatic and phytogeographic studies, while dealing with the Late Holocene

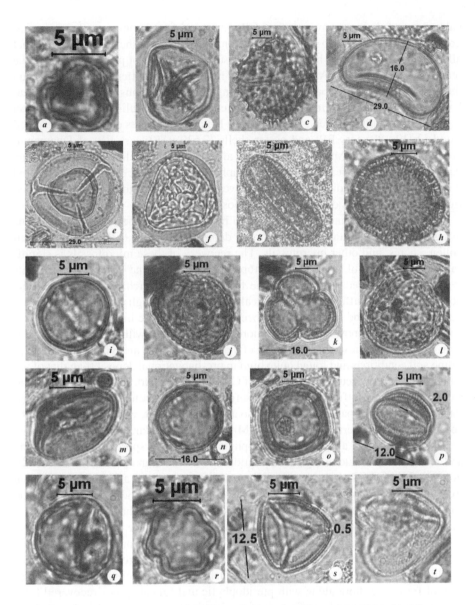

FIGURE 8.7 Representative palynomorphs recovered at Mandad: (a) Myrtaceae (*Syzigium cumini*), (b) Adiantaceae, (c) Lygodiaceae (*Lygodium* sp.), (d) Polypodiaceae, (e) Selaginellaceae (*Selaginella* sp.), (f) Osmundaceae (*Todisporites flavatus*), (g) Acanthaceae, (h) Avicenniaceae, (i) Rhizophoraceae, (j) Sonneratiaceae, (k) Anacardiaceae, (l) Annonaceae, (m) Araliaceae, (n) Bombacaceae, (o) Boraginaceae, (p) Brassicaceae, (q) Chenopodiaceae/Amaranthaceae, (r) Combretaceae (*Terminalia arjuna*), (s) Sapindaceae (*Schleichera oleosa*), (t) Cyperaceae. (All photomicrographs magnified at 100×, unless otherwise stated.)

deposits of Kerala. Tissot (1990) carried out studies on Late Holocene environment in Coondapur area, Karnataka, and on the basis of preliminary palynological results, he produced evidence of continuous existence of the well-developed mangrove forest during the recent past. Vishnu-Mittre and Guzder (1975) studied stratigraphy and palynology of the mangrove swamp of Bombay and found some members of mangrove vegetation and also upland vegetation. Furthermore, they also inferred that alkalinity of the soil may be responsible for the destruction of pollen and hence the reason for poor recovery. Archaeologists and historians have attempted to understand the growth and decay of pastoral and farming cultures in terms of climate and sealevel changes during the Holocene (Deotare et al. 2007, Dhavalikar 2002). Kumaran et al. (2001) recorded pollen grains from the Holocene deposits dated by carbonized wood (2,110 ± 80 years BP) and oyster shells (7,620 ± 110 years BP) of Dhamapur, Sindhudurg district, Maharashtra, show affinity with families of Acanthaceae, Arecaceae, Bombacaceae, Asteraceae, Euphorbiaceae, Fabaceae, Malvaceae, and Poaceae, which are indicative of lowland vegetation similar to evidence recovered from Chaul and Mandad archeological deposits.

In this study, the main objective was to interpret the paleoecological conditions using archaeological and pre-habitational deposits which bear the well-preserved pollen, spores, and other microfossils. The recovery of these microfossils is significant, first because of minerogenic nature of the deposit and second because this evidence is recovered for the first time from the coastal archaeological sites. The pollen recovery is sufficient and statistically significant enough from the habitational deposits of Chaul and Mandad. It has been observed that the interplay between pollen and spores is almost uniform and more or less coherent with the existing vegetation. The siltation caused heavy growth of mangroves, which are typical indicators of saline soil conditions and found quite high in pre-habitational deposit than that of the natural vegetation occurred at the sites of Chaul and Mandad. This pollen analytical study has proved fruitful results and is useful for reconstructing the vegetational history in detail during past 2,500 years or so and related paleoenvironmental studies of coastal region in general and Chaul and Mandad in particular.

The plant species along with pteridophytes and fungal spores indicate swampy and marshy habitat due to high precipitation rate with hot and moist climatic conditions during the early historic time. There is no recovery of any mangrove pollen from the ancient deposits so far analyzed, and this is a clear indication of nonsaline condition that prevailed during the early historic period at Chaul, but not really at Mandad, due to the contamination of mangrove pollens in the deposit. This is also well supported by the work of Kumaran et al. (2005) in the Konkan Coast inferring that the monsoon started weakening during Late Holocene around 4,000 years BP, resulting in a drying up of many marginal marine mangrove ecosystems. The members of freshwater flora along with pteridophytic and fungal spores recovered from archaeological sites are the indicators of swampy and marshy habitat and may imply high precipitation rate with hot and moist climatic conditions.

CONCLUSION

The research reported here is based on paleopalynological data and analysis of palynomorphs. It attempts to investigate the possible state of pollen preservation in various environmental conditions. Thus, pollen analysis has proved to be a reliable tool for environmental studies with reference to archaeological deposits in the Konkan Coast of India.

The pollen composition of Chaul assemblage is largely composed of freshwater flora that supports good vegetational cover during early historic as well as Medieval Period. This assemblage consists of fungal remains (including spore and fruiting bodies), pteridophytic spores, and angiospermous pollen. Further, the pteridophytic composition along with freshwater flora indicates swampy and marshy habitats. The palynoflora suggests a warm humid (tropical–subtropical) climate with heavy rainfall during the early historic period. The environment of deposition before the early historic habitation has been inferred as one of brackish water mangrove swamps, as proved by the presence of mangrove pollen, dinoflagellate cysts, microforaminifera, and foraminiferal linings. The early historic deposit is overlain by sterile deposit, which shows a negligible amount of preservation of palynofossils indicating transportation of the material. The medieval deposit shows the presence of freshwater swampy environment. The rich representation of fungal remains noticed from early historic, medieval, and late Medieval Periods suggests a warm and humid tropical–subtropical climate.

The palynoflora of Mandad is particularly significant as it has been dated on the basis of archaeological material and represents the early historic period. The freshwater elements are represented along with mangrove elements, rich fungal complex; presence of diatoms and dinoflagellate cysts; microforaminifera; and foraminiferal linings. The assemblage attests to strong marine influence for Rajapuri Creek, probably coinciding with the slightly higher sea level. The retrieved palynoflora from Mandad suggests a humid and tropical climate with plenty of rainfall during the deposition. The rich representation of fungal remains noticed from the early historic period suggests warm and humid tropic climatic conditions.

ACKNOWLEDGMENTS

The author is grateful to the authorities of Deccan College Post-Graduate and Research Institute, Pune, for providing necessary facilities to complete this piece of research work. He is indebted to Padma Shri Prof. K. Paddayya, Emeritus Professor and Former Director, Deccan College, Pune, for his continuous encouragement and vital suggestions. The author is grateful to Prof. B. C. Deotare for his kind support and guidance. The author is thankful to Prof. V. D. Gogte for his kind permission to collect soil samples for pollen analysis during the course of archaeological excavations at Chaul. The author is also grateful to Dr. Neloy Khare, Advisor/Scientist – G, Ministry of Earth Sciences, for his kind invitation to contribute this chapter to his proposed edited book titled *Quaternary Climate Change over Indian Subcontinent* to be published by Taylor and Francis.

REFERENCES

Black, C. A., D. D. Evans, J. L. White, L. E. Ensminger, and F. E. Clark (Eds.). 1965. *Methods of Soil Analysis Part, II*. Madison, WI: American Society of Agronomy Inc.

Brooks, J. 1971. Some chemical and geochemical studies on sporopollenin. In *Sporopollenin*, eds. J. Brooks, P. R. Grant, M. Muir, P. van Gijzel, and G. Shaw, 351–407. London: Academic Press.

Brooks, J., and G. Shaw. 1968. Chemical structure of the exine of pollen walls and a new function for carotenoids in nature. *Nature*, 219: 532–533.

Burgess, J. 1881. *Inscriptions from the Cave Temples of Western India.* Bombay: The Government Central Press.

Caratini, C., G. Thanikaimoni, and C. Tissot. 1980. Mangroves of India: Palynological study and recent history of the vegetation. *Proceedings of the 4th International Palynological Conference, Lucknow (1976–1977)*, 3: 49–59. Lucknow: Birbal Sahni Institute of Palaeobotany.

Caratini, C., I. Bentaleb, M. Fontugene, M. T. Morzadee-Kerfourn, J. P. Pascal, and C. Tissot. 1994. A less humid climate since 3500 yr BP, from marine cores off Karwar, western India. *Palaeogeography, Palaeoclimatology, Palaeoecology*, 109: 371–384.

De Terra, N., and T. T. Paterson. 1939. *Studies on the Ice Age in India and Associated Human Cultures*, 1–354. Washington, DC: Carnegie Institution of Washington. Publication no. 693.

Deotare, B. C. 1995. Pollen recovery from minerogenic sediments: A methodological approach. *Man and Environment*, 20(2): 101–105.

Deotare, B. C., and M. D. Kajale. 1996. Pollen analysis. In *Kuntasi – A Harappan Emporium on West Coast*, eds. M. K. Dhavalikar, M. R. Rawal, and Y. M. Chitalwala, 291–296. Pune: Deccan College.

Deotare, B. C., S. G. Deo, P. P. Joglekar, S. Ghate, and S. N. Rajaguru. 2007. Early historic sites in western India: A geo and bio archaeological perspective. In *Archaeology of Early Historic South Asia*, eds. G. Sengupa, and S. Chakraborty, 81–98. New Delhi: Pragati Publications.

Deshpande, G. G. 1998. *Geology of Maharashtra*. Bangalore: Geological Society of India.

Dhavalikar, M. K. 1984. *Late Hinayana Caves of Western India*. Pune: Deccan College Post-Graduate and Research Institute.

Dhavalikar, M. K. 2002. *Environment and Culture: A Historical Perspective*. Pune: Bhandarkar Oriental Research Institute.

Dimbleby, G. W. 1985. *The Palynology of Archaeological Sites*. London: Academic Press Inc.

Faegri, K., and J. Iversen. 1964. *Text Book of Pollen Analysis*. Oxford: Blackwell Scientific Publication.

Gaikwad, M. A. 2013. *Estimation of Crop Water Requirement under Varying Climatic Conditions for Dapoli Tahsil*. M. Tech diss., Dr. Balasaheb Sawant Konkan Krishi Vidyapeeth, Dapoli.

Garrels, R. M. 1960. *Mineral Equilibria at Low Temperature and Pressure*. New York: Harper and Brothers.

Geeta, R., R. Ramesh, and R. Sukumar. 1999. Climatic implications of $\delta^{13}C$ and $\delta^{18}O$ ratios from C_3 and C_4 plants growing in a tropical montane habitat in southern India. *Journal of Bioscience*, 24: 491–498.

Geeta, R., R. Sukumar, R. Ramesh, R. K. Pant, and G. Rajgopalan. 1997. Late Quaternary vegetational and climatic changes from tropical peats in southern India – An extended record up to 40,000 years BP. *Current Science*, 73: 60–63.

Ghate, S. N. 1985. *Late Quaternary Palaeogeography of the littoral zone of North Konkan, Maharashtra*. PhD diss., University of Pune.

Ghate, S. N. 1990. Palaeogeography of Chaul: A coastal town of North Konkan coast. *Bulletin of Deccan College Research Institute*, 49:145–150.

Ghate, S. N. 2007. Geomorphic and environmental changes around Sopara: An early historic port site in North Konkan Maharashtra: A review. *Man and Environment*, 32(2): 74–88.

Gogte, V. D. 2003. Discovery of the ancient port of Chaul. *Man and Environment*, XXVIII(1): 57–74.

Gogte, V. D. 2004. Discovery of the ancient port: Palaepatmai of the Periplus on the West Coast of India. *Journal of Indian Ocean Archaeology*, 1: 124–132.

Gogte, V. D. 2006–2007. Ancient port of Chaul: Semulla of the Periplus of the Erythraean Sea. *Bulletin of Deccan College Research Institute*, 66–67: 161–182.

Gogte, V. D., S. Pradhan, A. Dandekar, S. Joshi, R. Nanji, S. Kadgaonkar, and V. Marathe. 2006. The ancient port at Chaul. *Journal of Indian Ocean Archaeology*, 3: 62–79.

Gupta, S. 2007. Archaeological sites on the Indian Ocean Rim – A growing database. *Journal of Indian Ocean Archaeology*, 4: 102–109.

Huntington, E. 1906. Pangong: A glacial lake in the Tibetan Plateau. *Journal of Geology*, 15: 599–617.

Hyde, H. A., and D. A. Williams. 1944. Use of gravity slide to detect air spora. *New Phytology*, 43: 49–61.

Jackson, M. L. 1962. *Soil Chemical Analysis*. New Delhi: Asia Publishing House.

Kajale, M. D., and B. C. Deotare. 1989. Pollen analysis and chemical examination of a prehistoric lake site at Damdama. *Puratattva*, 19: 27–30.

Kolaba District Gazetteer. 1964. Maharashtra State Gazetteers. Bombay: Government of Maharashtra.

Kothari, M. J., and S. Moorthy. 1993. *Flora of Raigad District, Maharashtra State*. Culcutta: Botanical Survey of India.

Kumaran, K. P. N., K. M. Nair, Mahesh Shindikar, R. B. Limaye, T. R. Mudgal, and D. Padmalal. 2005. Stratigraphical and palynological appraisal of the Late Quaternary mangrove deposits of the west coast of India. *Quaternary Research*, 64: 418–431.

Kumaran, K. P. N., M. R. Shindikar, and R. B. Limaye. 2004a. Mangrove associated lignite beds of Malvan, Konkan: Evidence for higher sea level stand during the Late Tertiary (Neogene) along west cost of India. *Current Science*, 86(2): 335–340.

Kumaran, K. P. N., M. R. Shindikar, and R. B. Limaye. 2004b. Fossil record of marine manglicolous fungi from Konkan, India. *Indian Journal of Marine Sciences*, 33(3): 257–261.

Kumaran, K. P. N., M. R. Shindikar, and T. R. Mudgal. 2004c. Floristic Composition, palynology and sedimentary facies of Hadi mangrove swamp (Maharashtra). *Journal of Indian Geophysics Union*, 8(1): 55–63.

Kumaran, K. P. N., R. B. Limaye, C. Rajshekhar, and G. Rajgopalan. 2001. Palynoflora and Radiocarbon dates of Holocene deposits of Dhamapur, Sindhudurg district, Maharashtra. *Current Science*, 80(10): 1331–1336.

Limaye, R. B. 2004. *Contirbution to palaeopalynology of the coastal deposits of Maharashtra, India*. PhD diss., University of Pune.

Limaye, R. B., K. P. N. Kumaran, K. M. Nair, and D. Padmalal. 2007. Non-pollen palynomorphs as potential palaeoenvironmental indicators on the Late Quaternary sediments of the west coast of India. *Current Science*, 92(10): 1370–1382.

Malik, S. C. 1959. *Stone Age industries of the Bombay and Satara Districts*. Baroda: M. S. University. Archaeology Series No. 4.

Mandale, V. P., D. M. Mahale, S. B. Nandgude, K. D. Gharde, and R. T. Thokal. 2017. Spatio-temporal rainfall trends in Konkan region of Maharashtra state. *Advanced Agricultural Research and Technology Journal*, I(1): 61–69.

Mandavia, C. 1982. Pollen preservation and extraction from Loessic samples. *Man and Environment*, 7: 92–93.

Meher–Homji, V. M. 1994. Climate changes over space and time: Their representation repercussions on the flora and vegetation. *The Palaeobotanist*, 42(2): 225–240.

Moore, P. D., J. A. Webb, and M. E. Collinson. 1991. *Pollen Analysis*. Oxford: Blackwell Scientific Publications.

Nagaraju, S. 1981. *Buddhist Rock Cut Architecture of Western India (C. 250 B.C.-c.AD)*. Delhi: Agam Kala Prakashan.

Naik, S. S. 2012. *Pollen Analysis and Palaeoenvironmental studies of Archaeological deposits from Konkan and Malabar Coast of India*. PhD diss., Pune: Deccan College Post-Graduate and Research Institute (Deemed University).

Naik, S. S. 2014. An overview of quaternary palynology in India. *Gondwana Geological Magazine*, 29(1 and 2): 95–110.

Naik, S. S. 2017. *India: Vegetational History and Environment of the Pleistocene*. Bengaluru: The Mythic Society.

Naik, S. S., and B. C. Deotare. 2008–2009. Value of non-pollen palynomorphs as palaeoenvironmental indicators from archaeological site of Chaul, Maharashtra. *Bulletin of the Deccan College Research Institute*, 68/69: 125–136.

Naik, S. S., V. Kathale, and B. C. Deotare. 2009. Pollen potential in the habitational deposits of ancient Chaul (Maharashtra). *Man and Environment*, XXXIV(1): 101–108.

National Institute of Disaster Management. 2015. *Annual Report of NIDM 2014–15 Third edition*. New Delhi: National Institute of Disaster Management, Ministry of Home Affairs, Government of India. http://www.nidm.gov.in (accessed July 10, 2020).

Piper, C. S. 1966. *Soil and Plant Analysis*. Bombay: Hans Publishers.

Rajaguru, S. N. 1970. *Studies in the Late Pleistocene of the Mula-Mutha Valley*. PhD diss., University of Poona.

Ratan, R., and A. Chandra. 1984. Palynological investigations of the Arabian Sea: Pollen and spores from the recent sediments of the continental shelf off, Bombay, India. *The Palaeobotanist*, 31: 218–233.

Sakurkar, C. V. 1999. *Contribution to Stratigraphy and Palynology of the Quaternary sediments of Coasts of Maharashtra and Goa, India*. PhD diss., University of Pune.

Sankalia, H. D. 1962. Stone age industries of Bombay: A re-appraisal. *Journal of Asiatic Society of Bombay (New Series)*, 34–35: 120–131.

Saxena, R. K., and N. K. Misra. 1990. Palynological investigation of the Ratnagiri beds of Sindhu Durg District, Maharashtra. In *Proc. Symp. Vistas in Indian Palaeobotany, The Palaeobotanist*, eds. K. P. Jain, and R. S. Tiwari, 38: 263–276.

Schoff, W. H. 1912. *The Periplus of Erythraean Sea*. New York: Longmans, Green and Co.

Shaw, G. 1971. The chemistry of sporopollenin. In *Sporopollenin*, eds. J. Brooks, P. R. Grant, M. Muir, P. van Gijzel, and G. Shaw, 305–348. London and New York: Academic Press.

Shindikar, M. R., R. B. Limaye, K. P. N. Kumaran, and V. R. Gunale. 2004. Past and present mangrove flora along Konkan, West Coast, India. *Proceedings of National seminar on New frontiers in Plant Taxonomy and Biodiversity Conservation*, Thiruvananthapuram, India, 46.

Sukumar, R., R. Ramesh, R. K. Pant, and G. Rajgopalan. 1993. A δ 13C record of the Late Quaternary climatic change from tropical peats in south India. Nature, 364: 703–706.

Sukumar, R., H. Suresh, and R. Ramesh. 1995. Climate change and its impact on tropical montane ecosystem in southern India. *Journal of Biogeography*, 22: 533–537.

Tissot, C. 1990. Late Holocene environment in Coondapur area, Karnataka: Preliminary palynological results. In *Proc. Symp. Vistas in Indian Palaeobotany, The Palaeobotanist*, eds. K. P. Jain, and R. S. Tiwari, 38: 348–358.

Todd, K. R. U. 1939. Paleolithic industries of Bombay. *Journal of the Royal Anthropological Institute of Great Britain and Ireland*, 69(2): 257–72.

Traverse, A. 1988. *Paleopalynology*. Boston, London, Sydney, Wellington: Unwin Hyman.

Van Campo, E. 1986. Monsoon fluctuations in two 20,000 Yr B.P. Oxygen-isotope/pollen records off southwest of India. *Quaternary Research*, 26: 376–388.

Vishnu-Mittre, and S. J. Guzder. 1975. Stratigraphy and palynology of the mangrove swamps of Bombay. *The Palaeobotanist*, 22(2): 111–117.

Von Post, L. 1916. Einige südschwedischen Quellmore. *Bull Geol Inst Univ Uppsala*, 15: 219–278.

Wodehouse, R. P. 1935. *Pollen Grains; Their Structure, Identification and Significance in Science and Medicine*. New York and London: McGraw-Hill Book Company, Inc.

Zeuner, F. E. 1950. *Stone Age Pleistocene Chronology in Gujarat*. Poona: Deccan College.

9 Revisiting Late Quaternary Sea Levels along the Indian Sub-Continent with a Novel Approach in the Face of Climate Change

Rajani Panchang and Abhilash Sen
Savitribai Phule Pune University

Sea level rise is an invisible Tsunami, building force while we do almost nothing.

—*Benjamin H. Strauss*

CONTENTS

INTRODUCTION

For the better part of human history, man has been ignorant to the subtle changes of nature that influence, if not control, the direction in which the civilization progresses. It is only since the past few centuries have we begun to observe, record, and attempt to decipher the cause and the impact of these forces of nature. Of them, the greatest enigma is probably our oceans! Humanity from its inception has had a very intimate relationship with the oceans. Be it the existence of land-bridges due to low sea levels that facilitated migration of early hominids across the globe or be it the bustling ports of today that use the sea as the chief mode of cheap transport, the sea has been with man at every step of the way. Sometimes, it has been our ally, and many a times, it has hindered our progress; archaeological findings of entire cities engulfed by the sea can testify to this (Guzder, 1980; Rao, 1983; Stanley et al., 2004; Giosan et al., 2006, Gaur et al., 2009; Stanley and Toscano, 2009).

India being a peninsular country is blessed with a 7,516.6 km long coastline. Along with this, India is situated in the tropical zone, having a variety of climatic conditions and physiographic settings supporting diverse flora and fauna. India ranks third in fishery production and second in aquaculture. Fisheries alone have employed 145 million people and contributed to 1.07% of the GDP and generated export earnings of Rs. 334.41 billion as per a recent estimate of National Fisheries Development Board (Dey, 2020). The Government of India foresees tremendous potential in maritime tourism in India and the need for its promotion. So it has set up a committee of senior officials to explore the avenues for coastal tourism in India and work out ways to promote the same (News Bharati, 2019). In April 2020, the World Bank approved a $400 million multi-year financing envelope to help India enhance its coastal resources, protect coastal populations from pollution, erosion, and sea level rise, and improve livelihood opportunities for coastal communities.

However, in the light of climate change and predictions of unprecedented sea level rise, this very resourceful coastline is highly vulnerable. Developmental and mitigative programmes requiring huge investments, such as the one being supported by the World Bank, need a thorough understanding of the coastal processes and sea-level history that would continue to shape its landscape. In order to maximize the benefits of the rich coastal resources, asset building is the key. Past case studies, observations, records, and lessons learned need to be put to practical use, while planning sites of economic investments along coastlines. India lacks precise sea level records. With the advent of advanced techniques of observation, measurement, modelling, and prediction, meaningful knowledge needs to be extracted from sea-level records prior to instrumental era. This chapter is aimed at establishing the status of sea level data along the Indian subcontinent. It discusses the need for a novel approach to revisit archives and generate data that can be practically applied for problem solving and sustainable development.

FACTORS INFLUENCING SEA LEVELS AROUND THE INDIAN SUBCONTINENT

'Sea level' is a dynamic datum. It does not only change temporally, but spatially too. In order to improvise upon the employability of the sea level data, a thorough understanding of the factors controlling them and their interrelationships is primary. Processes with higher frequency tend to make a larger coastal manifestation, while deep sea processes may not always be recorded at the coast. However, 'coastal variability' should not always be considered as 'short spatial scale variability' but can be the result of signals transmitted along the coast from 1,000 km away (Woodworth et al., 2019).

Certain phenomena controlling sea levels or coastal configurations across the world occur over millions of years are a product of tectonic plate movements. Subduction and sea floor spreading at Mid Oceanic ridges are the principal tectonic processes that change the shape and size of ocean basins, thereby influencing sea levels. These operate over millions of years and are so slow that short-term global satellite records do not record them. Glacial isostatic adjustment (GIA) also occurs over thousands of years but is comparatively a much shorter-term process. Currently, the effect has been estimated to be −0.3 mm/yr of equivalent sea level rise due to increasing ocean basin size. This effect is corrected in the satellite altimeter global mean sea level time series and contributes 0.3 mm/yr to the estimated global mean sea level. This is considered a small effect since it is less than our estimated error of 0.4 mm/yr (Peltier, 2009). The correction for glacial isostatic adjustment (GIA) accounts for the fact that the ocean basins are getting slightly larger since the end of the last glacial cycle, due to the rebound effect on the lithosphere due to the melting of the Pleistocene ice sheets. Thus, the effective relative sea level rise on coasts will be very small (Tamisiea et al., 2010). Understanding the varied forms of vertical land motions (VLM) and where they are occurring today are both critical and make the prediction of local and regional changes in sea level difficult. Seismicity drives uplift or sinking (subsidence) of the earth either suddenly during abrupt events or gradually over thousands to millions of years due to long-term plate motion. Sediment compaction driven by extraction of groundwater or oil and gas leads to two to three times the sea level rise than the current global mean sea level rise.

There are several causes apart from tectonic and glacio-isostatic adjustments of sea-level change. Hydrostatic deformation functions very similar to GIA, where the weight of the sea water over continental landmass may cause subsidence, further accelerating/facilitating transgression and vice versa. River runoff has been neglected till recent, as a contributor for sea-level variability other than seasonal time scales. They influence sea levels across 10–100 km or wider and can cause change in several centimeters in sea level, especially contributing to the MSL seasonal cycle in large deltas (Meade and Emery, 1971; Laiz et al., 2014; Wijeratne et al., 2008; Piecuch et al., 2018). Construction of huge dams almost stalled the rising seas in the 1970s because of the amount of water they prevented from entering the oceans. Without them, the annual rate of rise would have been around 12% higher (Frederikse et al., 2020). The supply of sediments to ocean causes a unidimensional positive shift to sea levels. This factor is extremely crucial in enclosed and semi-enclosed basins. Large rivers such as the Indus (~250 MT/yr), Narmada (~35 MT/yr), Ganges (413 MT/yr), Brahmaputra (648 MT/yr), Mahanadi (11.5 MT/yr), Krishna (14.4 MT/yr), and Godavari (187 MT/yr) (Milliman and Syvitski, 1992; Subramanian, 1996; Gupta and Chakrapani, 2005) supplying sediments to the Northern Indian Ocean enclosed at the northern extremity play an extremely important factor influencing sea levels along the coast of the Indian subcontinent.

Long-term climate change that can cause global sea level changes to the scale of several tens of meters with regional variations (due to local variability) has been driven by thermal expansion of the water column or change in air temperature induced by orbital forcing. Such changes in the geological past have occurred over millions of years. Currently, the major factors influencing global sea-level changes are thermal expansion caused by warming of the oceans (as oceans absorb more than 90% of the excess heat generated due to anthropogenic emissions) and increase in melting of ice trapped on land in the form of glaciers or ice sheets (NOAA, 2019). With current trends in global warming, 1–3 m rise in global sea level is expected by the year 2100. The instrumental record of such sea level change mainly comprises of tide gauge measurements over the past two to three centuries and, since the early 1990s, of satellite-based radar altimeter measurements (Church et al., 2013). They indicate that sea level shows substantial regional variability at decadal to multi-decadal time scales (e.g., Carson et al., 2017; Hamlington et al., 2018). These regional changes are essentially due to changing winds, air-sea heat and freshwater fluxes, atmospheric pressure loading and the addition of melting ice into the ocean, which alters the ocean circulation (Stammer et al., 2013; Forget and Ponte, 2015; Meyssignac et al., 2017b). The addition of water into the ocean also changes the geoid, alters the rotation of the Earth, and deforms the ocean floor, which in turn changes sea level (e.g., Tamisiea, 2011; Stammer et al., 2013). In India, the extraction of groundwater has generated a local decrease in geoid that has slowed down the local SLR in the 20th century (Meyssignac et al., 2017a.). Part of this regional sea level rise is due to global sea level rise of which a majority is attributable to anthropogenic greenhouse gas emissions (Slangen et al., 2016). The remaining part of the regional sea-level rise in ocean basins is a combination of the response to anthropogenic GHG emissions and internal variability (e.g., Stammer et al., 2013). The Southern Annular Mode (SAM) operative in the Southern Ocean particularly influences the sea levels in the Indian sector, which may have contributed to the regional asymmetry of decadal sea-level variations during most of the 20th century (Thompson and Mitchum, 2014). In the Indian Ocean, the surface wind anomalies responsible for the sea level spatio-temporal variability are associated with the ENSO and Indian Ocean Dipole (IOD) modes (Nidheesh et al., 2013; Han et al., 2014; Thompson et al., 2016; Han et al., 2017). Singh et al. (2001) report increase in the frequency of tropical cyclones in the Northern Indian Ocean. A threefold increase in widespread extreme weather events over India has been reported over the past century (Roxy et al., 2017). Climate change is expected to worsen the frequency, intensity, and impacts of some types of extreme weather events such as storm surges, tropical cyclones, droughts, excessive precipitation resulting in increased riverine discharge, increased wind speed, wave heights, and coastal inundation. All of these have implications for sea-level rise along the Indian Coast.

While there is consensus among scientists that there is sea-level rise along both, Arabian Sea and Bay of Bengal coasts on the basis of global averages, India lacks robust baseline data for future studies and predictions. India currently depends on limited tide gauge stations, which have been providing data for about the past 100 years and some satellite altimetry data which have their own inconsistencies. Projections of mean significant wave height lack consensus over the Indian Ocean.

Poor knowledge of vertical land motion can bias estimates of global mean sea level rise and sea level projections, which in turn hinders the assessment of the impacts on coastal population and assets (Ballu et al., 2011; Hanson et al., 2011; Hallegatte et al., 2013; Han et al., 2015; Le Cozannet et al., 2015). Therefore, increasing our knowledge of vertical land motion for both the detection and attribution of climate change signals in sea level records as well as the assessment of coastal impacts of future sea levels has become a pressing issue (Wöppelmann and Marcos, 2016).

GEOMORPHOLOGY OF PENINSULAR INDIA AS A PRODUCT OF PAST SEA LEVELS AND AN INFLUENCE ON FUTURE SEA LEVELS

The Indian Peninsula that we see today is a result of various stages of tectonic development which started from the Late Jurassic (14–120 Mya) with the breaking up of India and Madagascar from Antarctica and Australia. This resulted in the formation of the present East coast of India. The Western coastline had formed (~70 Mya) following the rifting between India and Madagascar. The western continental margin is a passive rifted margin formed in two stages, the Southern part formed during the rifting between India and Madagascar and then during the rifting between India and Seychelles-Laxmi ridge in the Late Cretaceous (Bhattacharya et al., 1994; Chaubey et al., 2002; Dubey et al., 2019; Minshull et al., 2008; Yatheesh et al., 2009; Bhattacharya and Yatheesh, 2015). The then present Northern coastline of India also disappeared nearly 50 Mya when the Indian and Asian Plate collided (Curray and Moore, 1974). Currently, India has a 7,500 km long coastline and has been undergoing morphological changes throughout the geological past. It has mainly evolved in the background of the post-glacial transgression over the pre-existing topography of the coast and offshore (Baba and Thomas, 1999). The continental shelf of India is wide (about 340 km) toward the North and tapering toward the South (60 km). The tidal range is about 11 m in the Gulf of Cambay (Nayak and Shetye, 2003), 4–6 m in the Ganges-Bramhaputra Delta (Haque and Nicholls, 2018) and about 1 m toward the south of the peninsula (Unnikrishnan, 2010). About 100 rivers along with their tributaries drain the Indian peninsula. The major rivers flowing into the Arabian Sea are Narmada, Tapi, Sabarmati, Purna and those flowing into the Bay of Bengal are Brahmaputra, Yamuna, Ganga, Meghna, Mahanadi, Godavari, Krishna, and Cauvery. They supply large volumes of fresh water and sediment to the Northern Indian Ocean contributing as important factors affecting local sea levels along the Indian coastline.

These two present coastlines after subsequent changes through erosion and sedimentation over time have developed their individual distinctive features (Figure 9.1). Almost the entire western coast, barring the regions of Gujarat and Kerala show dominantly a rocky coastline with barriers and embayments. The northern and central western coast of India is dominated by low dissected plateaus. Many of the rivers flowing west over the Sahayadri's follow a trellis drainage pattern suggesting structural control over their drainage. The rivers and their tributaries follow numerous fractures & faults, which are trending NNW-SSE to NNE-SSW. Neotectonic movements have been reported along these weak planes (Powar and Patil, 1980; Widdowson and Cox, 1996; Widdowson, 1997; Gunnell, 2001). The southern coast

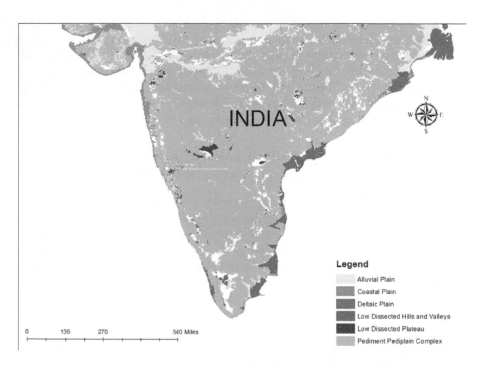

FIGURE 9.1 The coastal geomorphology map of peninsular India has been prepared with the help of ArcGIS using geospatial data from 'Bhukosh' developed by the Geological Survey of India. Bhukosh is a web application acting as client for Spatial Data Infrastructure (SDI) services so that multiple geospatial datasets can be viewed and integrated into a single map. For the given map, geomorphology data collected by the Geological Survey of India on a 1:250,000 scale have been used to identify the major coastal geomorphological features along peninsular India.

of Maharashtra and Karnataka is dominated by pediment peneplain complex. This is attributable to the higher rates of erosion and weathering prevalent in that area as it is fed by a large number of rivers compared to the northwestern coast. The coastal region of Kerala is distinctly different from the rest of the Western coast as it is mostly dominated by coastal plains and in some areas, deltaic plains. In Central Kerala, at the mouth of the Periyar River, satellite imagery shows evidence of building up of new delta over paleo-delta showing a gradual change to a regressive coastal environment along the Central coast of Kerala (Narayana et al., 2001). In comparison to the Western coast, the Eastern coast is a significantly regressive coast comprising of deltaic plains.

All the west-flowing rivers have funnel-shaped estuarine bays with extensive mud flats, mangroves, salt marshes, and estuarine islands. The absence of deltas on the west coast is the main geomorphic distinction between the east and west coasts of India. It is instead characterized by beaches, coastal dunes, promontories (with sea caves, wave cut platforms and cliffs), bays, and lagoons at irregular intervals. Some sections bear distinct evidence of neotectonism (Vaidyanathan, 1967). On the contrary, the Ganga-Brahmaputra Delta along coast of West Bengal, Mahanadi Delta along

Odisha Coast, and the Krishna Godavari Delta along the coast of Andhra Pradesh dominate the East Coast of India. The large East-flowing rivers form well-developed deltaic and estuarine systems. The deltas are covered with mangrove swamps, intra-deltaic lakes, alluvial plains and lagoons. Ancient river channels, beach ridges, former confluences and strandlines are evidence along the coast of Tamil Nadu and Andhra Pradesh. Longshore drift of deltaic sediments along the east Coast have formed conspicuous spits along the coast of Odisha and Tamil Nadu – Andhra Pradesh giving rise to India's two largest brackish water lagoons, Chilika and *Pulicat,* respectively. The cause for such a diversity in coastal characters may be attributed to the fact that there are a greater number of large sediments carrying rivers dominating the Eastern coast of India and also the fact that the Western coastline of India has been subjected to a greater amount of tectonic activity and subsequent sea level changes which can be inferred from the multiple marine terraces that have been observed near the coast of Ratnagiri (Nair, 1975). Further evidence of uplift along the Western coastline can be inferred by elevated oyster reefs, beach ridges, and also tidal gauge data (Subrahmanya, 1996). In contrast, the Eastern coastline shows evidence of subsidence. This subsidence is a result of isostatic adjustment, dehydration of sediments, and to an extent, anthropogenic activity (extensive agriculture on deltaic land, construction of large dams, unchecked exploitation of groundwater, and upscaling of oil and gas exploration) which help facilitate influx of a large amount of unconsolidated sediments into the basin. This subsidence is a risk to inhabitants living in the coastal regions and there is need for close monitoring of the subsidence rates along the East coast.

The coastline of Gujarat is quite unique in its respect to the rest of India. The geomorphology of the coastline is actively controlled by tectonic activity occurring in this region since post-Mesozoic (Biswas, 1987; Biswas and Khattri, 2002). Active faulting has divided the coastline into segments each showing a different geomorphological history of evolution and presently also causes a lot of seismic activity. The coastal alluvial plain of Kutch was formed during the Late Pleistocene by fluvial sediments deposited on peneplained Tertiary rocks in two separately created tectonic basins. Tectonic uplift along E-W trending faults during Early Holocene creating deeply incised fluvial valleys. The present coastal features comprising raised beaches, raised mudflats, abandoned coastal cliffs, uplifted estuarine tidal to fluvial terraces, and stabilized coastal dunes developed because of uplift during the last 2 kyr (Merh, 1993; Maurya et al., 2008). These features become extremely valuable for the study of past sea levels. The Saurashtra region based on its geomorphology and sedimentological evidence has preserved a record of sea level stands in the Late Quaternary Period (Prizomwala et al., 2018). The prominent beaches and well-developed sand dunes along the coasts of India are rich placer deposits of monazite, zircon, rutile, and ilmenite (Nayak, 2005).

LATE PLEISTOCENE-HOLOCENE SEA LEVELS ALONG THE INDIAN SUB-CONTINENT: A COMPARATIVE APPROACH

Several attempts have been made to understand the late Quaternary sea level fluctuations along the Indian subcontinent. The transgression and regression history for major part of the Pleistocene is poorly known worldwide. Whatever little is known to the scientific fraternity about the Early Pleistocene sea-level changes has been derived directly from the few coastal deposits preserved. Signatures of high stand have been destroyed by subsequent regressions. Preservation has been poor in areas that have remained unglaciated or subject to minimal fluvial erosion. Deposits of presumed early Pleistocene age can rarely be dated with confidence. The time interval from 2.6 Ma to 780 ka is beyond the range of most geochronological methods, and finding materials suitable for dating by methods that are applicable to this window of Earth history represents a further difficulty. Commonly, many of the physicochemical requirements or assumptions of relevant geochronological methods are not upheld in their field contexts, making it extremely difficult, to derive absolute ages for deposits. As a result, many attempts to assign ages to sequences of presumed early Pleistocene age have relied on palaeontological evidence of palaeontologically significant fossil taxa and their biostratigraphical correlation with the marine record (Murray-Wallace and Woodroffe, 2014). Such approaches, however, are of low temporal resolution and may span several marine isotope stages. More recently, $^{87}Sr/^{86}Sr$ analyses have been applied to date early Pleistocene successions (James et al., 2006; Wehmiller et al., 2012). Indirect evidence for early Pleistocene sea levels has been derived from oxygen isotope analyses on benthic foraminifera (Shackleton et al., 1990; Raymo, 1992).

Outcrops older than 18,000 kya are rarely found, either because they are either submerged under water or its exposures are eroded/degraded on land. After the last interglacial period (1.4 Mya), there has been significant lowering of the sea level (Pandey and Guha, 1982). Pandey and Guha (1982) reported that offshore sediments from Tarapur which have been dated to be of the Early Pleistocene represent inner shelf facies. This points to the fact that during the Early Pleistocene, the sea levels were more or less congruent to the present levels. Kale and Rajaguru (1985) studied microfauna from a Tarapur oil well and concluded that the sea level 250 kya was similar to the present sea level. Available data for 220 kya suggests that sea level was 55–60 m higher than present (Verma and Mathur, 1979).

Late Pleistocene (125,000–26,500 Years BP)

Also known as the last interglacial, or the penultimate interglacial, proxy evidence suggests that the global mean annual surface temperatures during Late Pleistocene was 1° to 2° warmer than the preindustrial era and the higher latitudes were 2° warmer than present. While these changes can be attributed to the eccentricity of the earth's orbit, the impact was not uniform across the globe, much similar to climatic changes during the Holocene. The exact nature of the transgressional and regression patterns during the Pleistocene is poorly known.

The Pleistocene deposits of Goa and Konkan bear signatures of high sea levels and are found 3–5 km inland and 100 m APSL (Chatterjee, 1961; Poscoe, 1964; De Souza, 1965; Ahmed, 1972; Dikshit, 1976; Guzder, 1980; Karlekar, 1981).

In the western coastal margin, evidence of Quaternary sea level occur only in patches, e.g., Warkalli beds in Kerala, Ratnagiri beds in Maharashtra, and marine deposits in Kathiawar. At the same time, the magnitude of sea level rise was much larger along the West coast of India. Dating of relict corals on the Saurashtra coast it was established the sea level was 2–6 m higher than present 120 kya (Gupta, 1972;

Gupta and Amin, 1974). Kale and Rajaguru, 1985 studying foraminiferal limestone along the Kathiawar coast found the sea level to be 25 km inland 116 kya.

Brückner (1989) found the sea level at Cape Comorin, Gulf of Mannar to be 1.2 m above present 124 kya by dating beach conglomerate. Dating *Acropora* at Rameshwaram Island too yielded similar inferences; 1.1 m above present level 125.6 kya (Banerjee, 2000). Porites at Narikulam and Rameshwaram island terrace indicated the sea level to be above 2.4 m APSL 108 kya and 2.9 m APSL between 92 and 104 kyr BP.

The Warkalli beds indicate high sea level during the Pleistocene which is also supported by the Kathiwar beds in Saurashtra, indicating a major transgressive phase between 80,000 and 65,000 years. These deposits are recorded 25 km inland (Kale and Rajaguru, 1985). Nigam et al. (1992) dated ooids at 92–101 m BPSL along the west coast marking the depth of sea level 50 kya. A carbon date from Panambur harbor reveals that the sea level was few meters below present level at 37–38 kya. C^{14} dates from Mahim, Uran, and Saurashtra suggest that the sea level was close to the present level 36 kya (Guzder, 1980; Kale et al., 1984). Radiometric dates from Cape Comorin indicate the sea level was a few meters higher than present 32.5 kya. Banerjee (2000) studying foraminiferal aggregate found the sea level at Cape Comorin at the Gulf of Mannar to be 4.9 m above present sea 29.3 kya. These pieces of evidence collectively point to the transgressive phase along the northern west coast and southern tip of India.

However, the period between 36 and 25 kyr was recorded as a regressive phase along the southwest coast of India. Singh et al. (2001) created a palaeobathymetric curve by studying *Limacina inflata/Creseis Spp.*, Pteropods/Planktonic foraminiferal ratios and Benthic/Planktic foraminiferal ratios in 13, marine sediment cores collected along the coast of Kerala. The study demonstrated that the sea level was ~25 m BPSL around 36 kya BP and 30 kya the sea level stabilized at 40 m BPSL. In between 36 and 25 kya the rate of sea level fall was relatively slow (Singh et al., 2001) drawing concurrence with sea level curve proposed for Huon Peninsula, Papua New Guinea by Chappell and Shackleton (1986). By dating oolitic sands at −100 to −145 m depth Subba Rao (1964) by dating of oolitic sands at −100 to −145 m depth indicated a low sea stand during the Late Pleistocene off the East coast of India.

It is believed that most of the palaeo-wetland present along the coast was developed due to the land-sea interaction during the last interglacial prior to the Late Pleistocene. Some Late Pleistocene deposits have also been reported from the coastal lands of Rameshwaram. This area has also been subjected to severe coastal erosion and cliff retreat during Late Pleistocene, which is evident from extensive coastal plain sand development (Maya et al., 2017).

Comprehensive dating of sea level indicators from various parts of the eastern coast of India has revealed the evidence of marine terraces which indicate the sea level was 2.9 m above present during the Last Interglacial. After this, the sea level fall was neither rapid nor unidirectional.

LGM (26,500–18,000 Years BP)

This coincides with the fact that Global sea levels during this time was >110 m below MSL.

Southwest India: Singh et al. (2001) studied sediment cores along the Kerala coast. They suggested that the rate of sea level fall increased after 22 kya, and the sea level attained ~100 m below present sea level during the LGM. According to Maya et al. (2017) the formation of Komallur wetland in the Southwest of India 20 kya coincides with the regressive phase of LGM. Palynological and micropalaeontological data show that these palaeo-wetlands were formed as a consequence of land-sea interactions during the last interglacial prior to the Late Pleistocene. During the Last Glacial Maxima (~18 kyr BP), evidence along the Kerala coast shows that sea level was approximately 120 meters than what is at the present. Later, there was a rise in the sea level of about 20 m in between 17.1–12.5 ka which indicated the first phase of deglaciation (Maya et al., 2017).

East and South Indian Coast: Banerjee (1993) also has reported vertical tectonic movements along the northern part of Pulicat Lake, uplifted marine terraces along the Coromondal coast as well as the rocky sea front at Visakhapatnam and Bimlipatnam. His studies on the basis of coastal ironstone, multimineral placer sand on the inner shelf, differential vertical movements, and littoral sediment budget of the East and South Indian coast of India has suggested a delayed sea level fall in response to the LGM along the East coast of India. There is an occurrence of multimineral sand, rich in rutile, zircon, garnet, ilmenite and monazite along the East and south Indian coastal margin. Development of ironstone in the coastal region indicates a further, sharp fall in sea level (15.5 kya) which can be also correlated to the sharp fall in terrigenous input. This is a very unique feature, and a detailed analysis of the sand bodies to understand their genesis indicates that there was a slow fall in sea level over an extended period of time (18–16 kya) which facilitated erosion and release of altered heavy minerals (Banerjee, 1993).

This indicates a much-delayed effect of the LGM on the Eastern coastline, which is in contrast to the west coast where the sea level fell coinciding with the LGM (Banerjee, 1993). Khandelwal et al. (2008) from palynological evidence studies at Chilka Lake have observed that after the LGM with the subsequent rise in sea level the sea level did not reach Chilka immediately which was at that time a river or a river delta with freshwater vegetation.

Presence of anhydrite of gypsum ~7 m below present sea level at Kolleru indicates that during 18.4 kya, the sea level was much below the present sea level at that time. At Kolleru, the shoreline is believed to be at least 65 km outwards from its present position during this time (Rao et al., 2020).

Maurya et al. (2008) have indicated a low sea level (~15 ka) during the upper part of Late Pleistocene in the coastal region of Kachchh. The Gulf was submerged at 15 ka despite the sea level being low. The 14.3 kyr aged carbonate dolomite crust occurring at 35 m within the Gulf of Kutch is an evidence of transgression at 15 kya (Rao et al., 2003).

HOLOCENE (11,600–PRESENT)

Though the Holocene is formally considered the geological epoch that began 11,600 years BP that marks the end of the last glacial period where major climatic changes took place, it is preceded by the Early Holocene Sea Level Rise (EHSLR) that changed the face of the earth as known to early humans. The rapid melting of the Pleistocene glacial ice sheets manifested in a 60–100 m sea level rise across the globe, submerging land bridges that had enabled cross-continent migrations. The present sea level was attained in several episodes, with its trends as well as rates of rise and fall varying throughout the Holocene, primarily driven by climate.

The signatures of the Holocene marine fluctuations occurring along continental margins all across the globe, constitute one of the best high-resolution evidences of sea level history on Earth (Murray-Wallace and Woodroffe, 2014: 484). Several submerged terraces, detected by echo-sounding records have been on the Western continental shelf of India at depths of 92, 84, 71, 65, 55 and 31 m below present mean sea level indicating past sea-stands and sea levels lower than present. (Nair, 1975; Wagle et al., 1994; Rao et al., 1996). Dredged samples of these sites have recovered coral specimen confirming existence ridge like/barrier coral reefs all along the coastline of India, parallel to it at shelf break (Nair and Qasim, 1977). Coral banks like the Gaveshani and the Angria originating from the modern-day shelf edge are testimony of keep-up corals that have survived the sea level rise. Relict sand zones consisting of coated grains, ooids, benthic foraminifera, larger foraminifera with intertidal barnacles, soft coral sclerites, bryozoa representing coastal or shallow marine facies have been found at a depth range of 60–100 m below PSL along Indian Continental shelf (Hashimi and Nair, 1976; Nigam et al., 1993; Banerjee, 2000; Loveson and Nigam, 2019). The evidence of sea levels being higher than present is evidenced by the presence of marine/marginal marine deposits like oyster shell beds, sand dunes, beach-rocks, shelly middens bearing marine gastropods and shells, peat, and coal beds several kilometers inland and several meters above present MSL (Hashimi et al., 1995; Banerjee, 2000; Loveson and Nigam, 2019; Mathur et al., 2004). The sea levels higher than present have also been manifested as erosional signatures in the form of nick points, wave-cut platforms, sea caves, etc. (Kale and Rajaguru, 1985; Rao et al., 1996; Bhatt and Bhonde, 2006; Subramanya, 1996; Maurya et al., 2008). The sea level signatures of the Holocene lie exposed in the seabed or on land, that can be easily observed/accessed through geophysical or physical observations. This is one of the reasons that the Holocene sea level history is widely studied and well documented in the literature.

CONUNDRUM OF RADIOCARBON DATES

In an era of conventional and bulk radiocarbon dating, which had large range of errors, it is credible that early workers attempted to date signatures of past sea levels from outcrops either submerged on the continental shelves or exposed on the present-day coast. Such dates were far and few and were reported from isolated segments of the extensive coastline of India. Table 9.1 is a ready reconnaire of all the radiocarbon dates produced from samples collected along the coast of India. The dates available along both the coasts compiled against a single timeline. These dates create a conundrum of applicability, as different samples collected from the same depth of the same region yielded different radiocarbon ages, while samples dated at different depths yielded same ages. Similarly, same sea level events seemed to have different age in a different sector of the coastline. Kale and Rajaguru, 1985 inferred the sea level to

TABLE 9.1

A Ready Reconnaire of All the Radiocarbon Dates Produced from Samples Collected along the Coast of India

	West Coast				East Coast			
Age (Cal Years BP)	Sampling Location	Evidence/ Material Used for Radiocarbon Dating	Reference	Elevation (m) w.r.t. Present Sea Level	Sampling Location	Evidence/ Material Used for Radiocarbon Dating	Reference	Elevation (m) w.r.t. Present Sea Level
92					Rameshwaram island terrace	Porites	Banerjee (2000)	2.9
100					Karikovil TN	Shells	Banerjee (2000)	2.9
104					Rameshwaram island terrace	Porites	Banerjee (2000)	2.9
104					Rameshwaram island terrace	Pyrites	Banerjee (2000)	2.9
108					Narikulam beach, Southern tip of India	Porites	Banerjee (2000)	2.4
112					Rameshwaram island terrace	Porites	Banerjee (2000)	2.4
116					Rameshwaram island terrace	Porites	Banerjee (2000)	2.9
125.6					Rameshwaram island	Acropora	Banerjee (2000)	1.1
140					Ramkrishnapuram, South TN	Coral	Loveson (1994)	0.2
430					Surangadu, South TN	Shell	IAR 75–76 p. 38	3
750					Karikovil, South TN	Shells	Banerjee (2000)	0.9
955					Mamallapuram, Spit S of Chennai	Charcoal	Rajendran et al. (2006)	−1
1,050					Paniya Tivu	Shell	IAR 75–76 p. 38	3

(Continued)

TABLE 9.1 (Continued)
A Ready Reconnaire of All the Radiocarbon Dates Produced from Samples Collected along the Coast of India

Age (Cal Years BP)	West Coast				East Coast			
	Sampling Location	Evidence/Material Used for Radiocarbon Dating	Reference	Elevation (m) w.r.t. Present Sea Level	Sampling Location	Evidence/Material Used for Radiocarbon Dating	Reference	Elevation (m) w.r.t. Present Sea Level
1,085					Mamallapuram, Spit S of Chennai	Organic matter	Achyuthan and Baker (2002)	0.3
1,115					Marakkanam, N of Puducherry	Shell	Hameed et al. (2006)	−0.4
1,470	Uran, Mumbai Maharashtra	Sand	Brückner (1989)	0.5				
1,563					Marakkanam	Shell	Hameed et al. (2006)	−0.86
1,581					Mamallapuram	Charcoal	Rajendran et al. (2006)	−1
1,620					Mamallapuram	Organic matter	Achyuthan and Baker (2002)	0.5
1,674					Mamallapuram	Charcoal	Rajendran et al. (2006)	−1
1,765	Erangal-Bhatti (Mumbai)	Littoral concrete	Aggarwal and Guzder (1972)	3.5				
1,800	Uran, Mumbai Maharashtra	Shells	Brückner (1989)	2				
1,900					Mamallapuram	Organic matter	Achyuthan and Baker (2002)	0.9

(Continued)

TABLE 9.1 (Continued)

A Ready Reconnaire of All the Radiocarbon Dates Produced from Samples Collected along the Coast of India

Age (Cal Years BP)	West Coast				East Coast			
	Sampling Location	Evidence/Material Used for Radiocarbon Dating	Reference	Elevation (m) w.r.t. Present Sea Level	Sampling Location	Evidence/Material Used for Radiocarbon Dating	Reference	Elevation (m) w.r.t. Present Sea Level
1,972					Chervu kandriga, Pulicat, Andhra	Sandy clay	Farooqui and Vaz (2000)	1.5
2,000					Pichavaram, Pondicherry	Mangrove root tips		0
2,030	North West Ratnagiri	Shells	Brückner (1989)	3.5				
2,050	Mandwa (Colaba)	Littoral concrete	Aggarwal and Guzder (1972)	0.05				
2,115	Madh Point Bhatti	Littoral concrete	Aggarwal and Guzder (1972)	0.55				
2,130					Pandiyan Tivu, Muttayyapuram, S of Rameshwaram	Coral	IAR 75–76 p. 83	0.5
2,180	Chaul (Colaba)	Littoral concrete	Aggarwal and Guzder (1972)	−3				
2,200	Konkan	Radiometric dates	Kale and Rajaguru (1985)	−1.5				
2,260	Valvati Creek	Shell	Brückner (1989)	−1	Attakanitippa, Pulicat Sanctuary, TN	Pelecypods	Farooqui andf Vaz (2000)	−0.9
2,305	Mirya Bhatti (Ratnagiri)	Littoral concrete	Aggarwal and Guzder (1972)	6				

(Continued)

TABLE 9.1 (Continued)

A Ready Reconnaire of All the Radiocarbon Dates Produced from Samples Collected along the Coast of India

Age (Cal Years BP)	West Coast				East Coast			
	Sampling Location	Evidence/ Material Used for Radiocarbon Dating	Reference	Elevation (m) w.r.t. Present Sea Level	Sampling Location	Evidence/ Material Used for Radiocarbon Dating	Reference	Elevation (m) w.r.t. Present Sea Level
2,316								−1
2,410	Valvati Khatan Creek	Shell	Brückner (1989)	2				
2,410	Korlai-Borlai	Littoral concrete	Aggarwal and Guzder (1972)	0.5				
2,600	Mandve Estuary Maharashtra, near Alibaug; 18°41′49″, 73°1′50″	Shell	Brückner (1989)	−3				
2,615					Sankrail, Howrah, WB	Peat	Mittre and Gupta (1970)	−0.275
2,630					Mandapam	Coral	Loveson (1994)	0.62
2,730	Erangal-Bhatti (Mumbai)	Littoral concrete	Aggarwal and Guzder (1972)	3				
2,740					Mandapam, Rameshwaram, TN	Cardium	Brückner (1989)	1
2,740					Mandapam, Rameshwaram, TN	Shell above HTL	Brückner (1989)	1
2,750					Odinur, TN	Sediment	Hameed et al. (2006)	−1.9
2,755					Korkai, S of Rameshwaram, TN	Wood	IAR 69–70, p. 68	−3.17

(Continued)

TABLE 9.1 (Continued)
A Ready Reconnaire of All the Radiocarbon Dates Produced from Samples Collected along the Coast of India

Age (Cal Years BP)	West Coast				East Coast			
	Sampling Location	Evidence/ Material Used for Radiocarbon Dating	Reference	Elevation (m) w.r.t. Present Sea Level	Sampling Location	Evidence/ Material Used for Radiocarbon Dating	Reference	Elevation (m) w.r.t. Present Sea Level
2,790					kamarajapuram ramnathapuram TN	Shell	IAR 75–76, p. 84	6
2,800	Mirya Bhatti (Ratnagiri)	Littoral concrete	Aggarwal and Guzder (1972)	5.9				
2,800	Konkan	Radiocarbon dates	Kale and Rajaguru (1985)	–1.5				
2,800	St. Xavier Bhatt	Littoral concrete	Aggarwal and Guzder (1972)	1.55				
2,900					Gangasagar, Hoogly, WB	Silty clay	Chakrabarti (1992)	–1.7
3,100					Sriharikota, Pulicat Lake, Andra Pradesh	Shell	Farooqui and Vaz (2000)	–2
3,145					Muthukadu, N of Puducherry, TN	Tidal clay	Achyuthan (1997)	–0.5
3,170					Namkhana, WB	Silty lay	Gupta (1981)	–1.75
3,410	Honnebail, offhore, Near northern coast of Karnataka; 14°35′59.033″, 74°13′49.744″	Shells	Nambiar et al. (1991)	–19				
3,475					Muthukadu N of Puducherry, TN	Tidal clay	Achyuthan (1997)	–0.7

(Continued)

TABLE 9.1 (Continued)

A Ready Reconnaire of All the Radiocarbon Dates Produced from Samples Collected along the Coast of India

Age (Cal Years BP)	West Coast				East Coast			
	Sampling Location	Evidence/Material Used for Radiocarbon Dating	Reference	Elevation (m) w.r.t. Present Sea Level	Sampling Location	Evidence/Material Used for Radiocarbon Dating	Reference	Elevation (m) w.r.t. Present Sea Level
3,660					Mandapam, Rameshwaram, TN	Coral above HTL	Brückner (1989)	0.5
3,670					Mandapam, Rameshwaram, TN	Coral	Loveson (1994)	0.42
3,800					Pamban coast, Rameshwaram, TN	Shells	Banerjee (2000)	2.9
3,820					Korkai, TN Near deltic river	Shell	IAR 75–76, p. 84	6.5
3,840					Vilundi theertham, Rameshwaram	Shells	Banerjee (2000)	2.4
3,920					Pamban, Rameshwaram, TN	Coral	Loveson (1994)	0.48
4,020					Pamban, Rameshwaram, TN	Coral	Stoddart and Pillai (1972)	0.47
4,080	Porbandar & Mithapur	Cerethium & Turbo shells	Mathur et al. (2004)	2.5				
4,162					Munaikkadu, Rameshwaram, TN	Shell scaphara	Banerjee (2000)	1.7
4,245	Manori (Mumbai)	Littoral concrete	Aggarwal and Guzder (1972)	−1				

(Continued)

TABLE 9.1 (*Continued*)
A Ready Reconnaire of All the Radiocarbon Dates Produced from Samples Collected along the Coast of India

Age (Cal Years BP)	West Coast				East Coast			
	Sampling Location	Evidence/ Material Used for Radiocarbon Dating	Reference	Elevation (m) w.r.t. Present Sea Level	Sampling Location	Evidence/ Material Used for Radiocarbon Dating	Reference	Elevation (m) w.r.t. Present Sea Level
4,320					Illankalanvadi, Kanyakumari	Shell	IAR75–76, p. 83	6
4,325	Near Majali, offshore, North Karnataka coast; 14°49.43', 73°59.31'	Sediment carbonate	Nigam and Khare (1992)	−26				
4,385	Manori (Mumbai)	Littoral concrete	Aggarwal and Guzder (1972)	3				
4,350					Rameshwaram islands	Shell arca	Banerjee (2000)	2.9
4,400					Kovakulam, Cape Camorin TN	Shell scaphara	Banerjee (2000)	2.4
4,500	Saurashtra	Strandline		~2.5				
4,540	Manori (Mumbai)	Littoral concrete	Aggarwal and Guzder (1972)	1				
4,608					Kasdreddinilem, Pulicat Andhra	Peat	Farooqui and Vaz (2000)	1
4,860					Villundi theertam, Rameshwaram TN	Cardita	Banerjee (2000)	2.4
4,900					Chilka, Odisha	Pollen	Rao et al. (2020)	−2.2

(Continued)

TABLE 9.1 (Continued)

A Ready Reconnaire of All the Radiocarbon Dates Produced from Samples Collected along the Coast of India

Age (Cal Years BP)	West Coast				East Coast			
	Sampling Location	Evidence/Material Used for Radiocarbon Dating	Reference	Elevation (m) w.r.t. Present Sea Level	Sampling Location	Evidence/Material Used for Radiocarbon Dating	Reference	Elevation (m) w.r.t. Present Sea Level
5,000	Madhuvanty Creek	Sand	Brückner (1989)	2	Panighata WB	Peat	Shahidul (1995)	−0.57
5,000					Muthukadu, Chennai	Oyster shell	Achyuthan (1997)	−1.85
5,070	Manori (Mumbai)	Littoral concrete		0.8				
5,140					Kakinad lighthouse, Andhra, N of Godavari Delta	Shell arca	Banerjee (2000)	3.9
5,310					North of Pamban	Porites	Banerjee (2000)	1.4
5,410					Munaikkadu, Rameshwaram TN	Porites	Banerjee (2000)	1.2
5,440					Ariyan Kundu	Coral	Loveson (1994)	.
5,470					Dubash Chetti	Shell	IAR 75–76, p. 84	6
5,540					Odinur TN	Sediment	Hameed et al. (2006)	−2.1
5,670					Kaikallur, Kolleru lake, Andhra Inland between K-G rivers	Shell	Banerjee (2000)	3.9
5,700					Villundi theertam	Porites	Banerjee (2000)	1.4
5,710					Dubash Chetti	Shell	IAR 75–76, p. 84	3

(Continued)

TABLE 9.1 (Continued)
A Ready Reconnaire of All the Radiocarbon Dates Produced from Samples Collected along the Coast of India

Age (Cal Years BP)	West Coast				East Coast			
	Sampling Location	Evidence/ Material Used for Radiocarbon Dating	Reference	Elevation (m) w.r.t. Present Sea Level	Sampling Location	Evidence/ Material Used for Radiocarbon Dating	Reference	Elevation (m) w.r.t. Present Sea Level
6,210					Villundi theertam, Rameshwaram TN	Porites	Banerjee (2000)	1.4
6,240					Tiruchendur, S. of Rameshwaram	Shells in ridge	Brückner (1989)	7
6,500	Saurashtra	Strandline		~2.5				
6,650					Sulurpet, W of Pulicat, Andhra	Peat	Farooqui and Vaz (2000)	4.98
6,450					Pisasu munai, Rameshwaram, TN	Porites	Banerjee (2000)	1
6,900					Kolaghat WB	Peat	Das et al. (1996)	−0.69
7,000					Chilka	Pollen	Rao et al. (2020)	−3.2
7,030					Calcutta	Peat	Barui et al. (1986)	−1.26
7,130					Kadiapatnam, W of Kanyakumari, TN	Fossil wood	Brückner (1989)	−8
7,600	Madhuvanty Creek Saurashtra	Marshy soil	Brückner (1989)	1.5				
7,685					Chilka	Peat	Rao et al. (2020)	−4.41
7,845	Offshore, from Devgad Maharashtra; 16°18′, 73°2′	Colaba	Rajagopalan et al. (1980)	−60				

(Continued)

TABLE 9.1 (Continued)
A Ready Reconnaire of All the Radiocarbon Dates Produced from Samples Collected along the Coast of India

Age (Cal Years BP)	West Coast				East Coast			
	Sampling Location	Evidence/Material Used for Radiocarbon Dating	Reference	Elevation (m) w.r.t. Present Sea Level	Sampling Location	Evidence/Material Used for Radiocarbon Dating	Reference	Elevation (m) w.r.t. Present Sea Level
7,900					Chilka	Pollen	Rao et al. (2020)	−7.96
8,200					Marakkanam, N of Puducherry	Shell	Hameed et al. (2006)	−1.5
8,300	Offshore, Malvan Maharashtra	Colaba	Rajagopalan et al. (1980)	−68				
8,315	Kochi, Kerala; Willington Island	Wood	Aggarwal and Gosh (1973)	−16.75				
8,340	Offshore Mumbai	Pelletal Sand	Rao et al. (1994)	−80				
8,380	Inland, Near Malvan; 16°0′20″, 73°51′	Colaba	Rajagopalan et al. (1980)	−97				
8,395	Offshore, Kasheli, Maharashtra coast; 16°40′, 72°48′	Colaba	Rajagopalan et al. (1980)	−68				
8,395	Offshore Ratnagiri	Algal nodules	Vora and Almeida (1990)	−80				
8,465	Shelf-break Gulf of Cambay	Halimeda limestone	Rao et al. (1994)	−95				
8,620	Offshore, near Ankola Karnataka coast; 14°40′8.631″, 73°54′9.569″	Carbonized wood	Nambiar et al. (1991)	−26.8				

(Continued)

TABLE 9.1 (Continued)
A Ready Reconnaire of All the Radiocarbon Dates Produced from Samples Collected along the Coast of India

	West Coast				East Coast			
Age (Cal Years BP)	Sampling Location	Evidence/ Material Used for Radiocarbon Dating	Reference	Elevation (m) w.r.t. Present Sea Level	Sampling Location	Evidence/ Material Used for Radiocarbon Dating	Reference	Elevation (m) w.r.t. Present Sea Level
8,910	Offshore, Karwar, Karnataka coast; 14°43.806', 74°2.649'	Peat	Nigam (1991)	−29				
8,960	Offshore, Velan, coast of Gujrat; 19°30.5', 70°34.0'	Oolitic limestone	Nair and Hashimi (1980)	−82				
9,116	Offshore, Malvan Maharashtra; 15°59.35', 72°57.56'	Algal limestone	Hashimi and Nair, unpublished	−70				
9,135	Western continental shelf	Composite shell	Nair and Hashimi (1980)	−58				
9,170					Marakkanam, N of Puducherry	Sediment	Hameed et al. (2006)	−1.52
9,200	Near shelf break coast of Mumbai; 18°36', 70°39.3'	Oolitic limestone	Nair and Hashimi (1980)	−98				
9,285	Offshore Gulf of Cambay	Halimeda	Rao et al. (1994)	−85				
9,435	offshore, Malvan Maharashtra; 16°9'39", 72°50'	Colaba	Rajagopalan et al. (1980)	−90				
9,560	Offshore, Karwar, Karnataka coast; 14°43.806', 74°2.649'	Peat	Nigam, unpublished	−29				

(Continued)

TABLE 9.1 (Continued)
A Ready Reconnaire of All the Radiocarbon Dates Produced from Samples Collected along the Coast of India

Age (Cal Years BP)	West Coast				East Coast			
	Sampling Location	Evidence/ Material Used for Radiocarbon Dating	Reference	Elevation (m) w.r.t. Present Sea Level	Sampling Location	Evidence/ Material Used for Radiocarbon Dating	Reference	Elevation (m) w.r.t. Present Sea Level
9,630	Offshore, Alvekodi Karnataka coast; 14°25'37.091", 74°12'35.501"	Carbonized wood	Nambiar et al. (1991)	−32				
9,830	Offshore, Mul Dwarka, coast of Gujrat; 20°10', 70°26'59"	Sediment	Rajagopalan et al. (1978)	−73				
9,960	Offshore Gulf of Cambay	Oolites	Nair (1974); Nair et al. (1979); Nair and Hashimi (1980)	−80				
10,000	Offshore Kerala	Microfossils	Singh et al. (2001)	−20				
10,075	Offshore Alibaug	Pelletal sand	Rao et al. (1994)	−100				
10,400	Offshore Veraval, coast of Gujrat; 20°24', 69°41"	Ooid concentrate	Nair and Hashimi (1980)	−85	East coast	Placer sand	Banerjee (1993)	−120
10,790					Vishagapathnam shelf, Andhra	Carbonate rock	Rao and Rao (1994)	−85
10,415	Near shelf break, Ratnagiri coast; 17°0', 71°59"	Algal Bryozoan Limestone	Nair and Hashimi (1980)	−180				
11,040	Offshore Benaulim, Goa; 15°15', 73°36"	Limestone	Hashimi and Nair (1986)	−95				

(Continued)

TABLE 9.1 (Continued)
A Ready Reconnaire of All the Radiocarbon Dates Produced from Samples Collected along the Coast of India

	West Coast				East Coast			
Age (Cal Years BP)	Sampling Location	Evidence/Material Used for Radiocarbon Dating	Reference	Elevation (m) w.r.t. Present Sea Level	Sampling Location	Evidence/Material Used for Radiocarbon Dating	Reference	Elevation (m) w.r.t. Present Sea Level
11,040	Shelf-break Gulf of Cambay	Algal nodules	Nair (1975)	−110				
11,150	At shelf break from Mumbai; Coast; 19°5′, 69°45′	Algal pellet limestone	Nair and Hashimi (1980)	−150				
11,500	Offshore, coast of Mumbai; 19°2′10.44″, 71°30′36.36″	Ooids	Nigam et al. (1992)	−81				
11,850	Shelf break, Gulf of Cambay		Rao et al. (1996)	−120				
11,980	Shelf break, Gulf of Cambay	Beachrock	Rao et al. (1996)	−130				
12,000	Offshore date west coast	Sea level curve plot	Kale and Rajaguru (1985)	<−138				
12,010					Rameshwaram north	Coral	Loveson (1994)	0.8
12,510	WCSI	Oolitic limestone	Nair, unpublished	−78				
12,539					Vishakhapatnam shelf	Carbonate rock	Mohan Rao and Rao (1994)	1994
13,000	Offshore, coast of Mumbai; 19°2′10.44″, 71°30′36.36″	Ooids	Nigam et al. (1992)	−92				
13,170					Olaikuda marine terrace, Rameshwaram	Porites	Banerjee (2000)	1.9
14,500	Offshore, coast of Mumbai; 19°2′10.44″, 71°30′36.36″	Ooids	Nigam et al. (1992)	−101				

(Continued)

TABLE 9.1 (*Continued*)
A Ready Reconnaire of All the Radiocarbon Dates Produced from Samples Collected along the Coast of India

	West Coast				East Coast			
Age (Cal Years BP)	Sampling Location	Evidence/ Material Used for Radiocarbon Dating	Reference	Elevation (m) w.r.t. Present Sea Level	Sampling Location	Evidence/ Material Used for Radiocarbon Dating	Reference	Elevation (m) w.r.t. Present Sea Level
14,740	WCSI	Coralline algae	Nair, unpublished	−173				
15,000	Off Shore Kerala	Microfossils	Singh et al. (2001)	−80				
18,000					Kolleru Lake, Andhra; shoreline was at least 65 km away from present	Organic sediment	Rao et al. (2020)	−7
22,000	Off shore Kerala	Microfossils	Singh et al. (2001)	−110				
29,300					Cape Camorin, Gulf of Mannar	Foraminifera aggregate	Banerjee (2000)	4.9
80,000–65,000	Kathiawar coast	Foraminiferal limestone	Kale and Rajaguru (1985)	25 km inland				
92,000					Rameshwaram island terrace	Porites	Banerjee (2000)	2.9+5
124,000					Cape comorin, Gulf of Mannar	Beach conglomer-ate	Brückner (1989)	1.2
~250,000	Tarapur oil well	Microfauna	Kale and Rajaguru (1985)	0				

be less than 138 m below present sea level at 12 kya, from offshore sea level data. However, from micropaleontological evidence from sediment cores off the Kerala, it was deciphered that the sea level was already much higher, at 110 m depth at 22 kya already.

Inconsistent sea levels have been reported for the same time from same regions by different workers and shallower sea levels have been reported from older times. Opiods dated from offshore of Mumbai suggest the sea level 81 m below present 11.5 kya (Nigam et al., 1992), Algal-pellet limestone from the shelf break off Mumbai suggests MSL to be 150 m below present 11.15 kya (Nair and Hashimi, 1980) and limestone dated from off shore Benaulim Goa points to sea level 95 m below present 11.04 kya. Limestone from inland Malvan indicates a sea level 97 m below present while that from offshore Malvan suggests it was at 68 m below present MSL, 8.3 kya. (Rajagopalan et al., 1980). Nair and Hashimi (1980) reported different sea levels along Ratnagiri at shelf break (algal bryozan limestone, 180 m below present) and offshore Veraval, Gujarat (ooid concentrate, 85 m below present) at 10.415 kya, pointing to almost a 100 m synchronous difference in sea level at the beginning of the Holocene between the northwest and central west coast of India. Proxy records suggest that the sea rose faster in the north west of the Indian Coast than in the central coast. While sediments dated at 9.83 kya. from offshore Mul Dwarka Gujarat Coast indicate sea level to be 73 m below present (Rajagopalan et al., 1978), limestone from offshore Malvan, Maharashtra suggests the sea level to be 90 m below present 9.435 kya. (Rajagopalan et al., 1980) and 98 m below present 9.2 kya near shelf break off the coast of Mumbai (Nair and Hashimi, 1980). Similarly, though oolitic limestone from Velan Coast, Gujarat indicates sea level to be 82 m below present 8.96 kya. (Nair and Hashimi, 1980), limestones from offshore of Kasheli, Thane, Maharashtra indicate sea levels at 68 m below present 8.395 kya and 60 m below present 7.845 kya Devgad offshore (Rajagopalan et al., 1980). Interestingly, all dates derived south of Goa indicate much higher sea levels as compared to the coast of Maharashtra and Gujarat. The sea level had already attained a height of ~20 m below present MSL 10 kya along Southwestern coast (Singh et al., 2001).

ERA OF COMPREHENSIVE SEA LEVEL CURVES

It was only in the 1980s that workers attempted to collectively look at these dates and to compile a single sea level curve for the entire east (Banerjee, 2000; Loveson and Nigam, 2019) or west coast of India (Kale and Rajaguru, 1985; Hashimi et al., 1995; Rao et al., 1996) capable of reflecting regional trends in comparison with published eustatic sea level curves (Fairbanks, 1989; Chappell and Shackleton, 1986). Proposing a single sea level curve was a practical approach to simplify the many palaeoclimatic studies gaining trend and settling archaeological questions along the Indian coastline (Rajguru and Guzder, 1973; Guzder, 1980; Rao, 1983).

The following section summarizes the five sea level curves (Figure 9.2) proposed by earlier workers for the East and West Coast of India:

1. Responding to the need to compare regional trends of sea level rise along the Indian coast with the eustatic sea level rise, Kale and Rajaguru (1985) developed a sea level curve for the western continental margin of India,

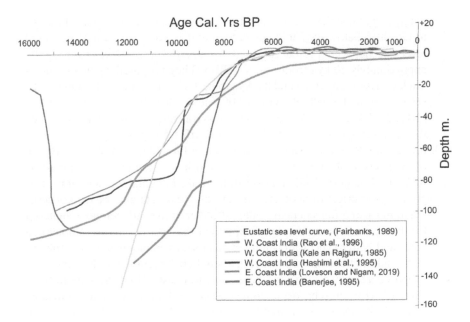

FIGURE 9.2 A comparative composite of the different sea level curves proposed by previous workers with an aim to reflect comprehensive sea level trends over the Holocene, along the entire east or west coast of India, plotted against a eustatic sea level curve.

accounting for the duration of 12 kya to present. While doing so they envisaged three hinderances: (1) The erratic scatter obtained by plotting age to depth relationship of over 50 dates previously reported dates; (2) Reduced or exaggerated sea-level fluctuations inferred on the basis of contaminated C-14 dates and (3) Tectonically active coastline of Saurashtra and Kutch. Thus, they preferentially used certain radiometric ages reported from Mumbai and further south for the formulation of their sea level curve. They transformed the sea levels and their corresponding ages logarithmically to develop an equation which exhibited that the sea level along the west coast of India, during the Holocene, had the tendency to rise with decrease in time, in relation to the current sea-level. They suggested that the Holocene was marked by a meso-phase of transgression along the West coast, as observed in other parts of the world, fixing the sea level at 138 m below present approximately 12 kya. Their curve corresponded relatively well with the eustatic curves proposed by Fairbridge (1961), Shepard (1963), Mörner (1969) and Bloom (1971) Hence, they attributed the Holocene sea level rise trends along the Indian Coast to the eustatic sea level rise associated with end of Pleistocene glaciation. The biggest drawback of this curve was that it could not explain the many other signatures of sea level along the entire stretch of the coast from Saurashtra to Kerala reported by previous workers, which they had not considered for the construction of their curve.

2. Hashimi et al. (1995) compiled a sea level curve for the western continental margin of India representing the trends from 14.5 to 1.5 kya by carefully selecting 58 published radiocarbon dates generated on specimen collected

below and above the PMSL along the western coast. They reported the sea to be 100 m below PSL at about 14.5 kya followed by a gradual rise in sea level rise between 14 and 10 kyr (0.5 cm/yr). The sea level rose rapidly approximately 10 kya from −80 to −30 m. They proposed a general rate of sea level rise between 12 and 8.5 kya to be 1.42 cm/yr. The data suggested a sea level stand for a duration of ~1,500 years at 30 m below present between ~10 and 8.5 kya. The present sea level was transgressed at 7 kya after which the sea level remained 1–3 m above PMSL till 1,500 years ago.

3. Rao et al. (1996) proposed a sea level curve for 12,000–9,000 years BP for the west coast of India. He then compared it with the curves proposed by Kale and Rajaguru (1985) and Hashimi et al., 1995 and the need for incorporating the effect of neotectonism on the curve. He also raised concerns regarding using diagenetically altered material and allochtonous material for dating past sea levels. They expressed the need for more radiocarbon dates and better synchronization between sedimentologists, geomorphologists, and radiochemists in bringing about better accuracy in the curve.

4. Proposed a sea level curve for the entire east coast of India and South India since LGM to present, using conventional geological sea level indicators and with the support of available absolute age determinations. The indicators he used were coastal ironstones, multimineral placer sand on the inner shelf facies studies of channel fill deposits on the inner shelf, differential vertical movements, and littoral sediment budgets as records of weathering, erosion, and sedimentation rather than using a model-driven approach. He emphasized on using regionally extensive features in order to eliminate local noise. As per his curve, the sea level was similar to present sea levels at the onset of LGM, about 22 kya and demonstrated a delayed sea level fall (w.r.t. global timing of LGM signatures) between 18 and 14 kyr BP. The sea level dropped to −120 m and stayed low till about 9 kyr BP. It rose steeply to a depth of about −10 m 8 kya. The present sea level was attained 7 kya after which it has fluctuated continuously by a few meters above and below PSL. Banerjee emphasized that fluctuations in freshwater discharge along different sectors of the east coast in response to global atmospheric temperatures during the Holocene controlled shoreline processes in the Bay of Bengal.

5. Loveson and Nigam (2019) reconstructed a comprehensive sea level curve to represent the trends of the sea level along the east coast of India from ~15 kya to present, using 75 published C-14 dates generated from on coastal plains and offshore regions of the East Coast. They reported a that the sea level was at −100 m at about 14.8 kya after which it rose to a depth of about 28 m BPSL at kyr BP at the rate of about 2.17 cm/yr till 9.2 kya. They suggest a sluggish sea level rise for a period of about 1,200 years 9.2 and 8 kyr BP. The rate of sea level once again rose to 2.2 cm/yr between 8 and 7 kyr. The sea transgressed the PSL about 7 kya after which it has fluctuated between +4 and −1 m w.r.t. PSL. They show 5 episodes of transgression and regressions since 7 kyr which show the sea level is on a receding trend.

To begin with, it is interesting to note that none of the workers propose a unified single sea level curve for the entire coastline of India, recognizing that the two coastlines are intrinsically different. However, a close comparison of these sea level curves undoubtedly shows different trends not only between the two Indian coasts, but also within the different curves proposed for the same coastline. What seems to lead the inferences of sea level trends is the bias of each author toward the choice of data selected to plot the curve. It is evident from Table 9.1 that radiocarbon dates generated on wood and peat from inner shelf regions of Konkan, Maharashtra and Karnataka yield much older ages at shallower depths (Agrawal and Ghosh, 1973; Nigam, 1991; Nambiar et al., 1991) in comparison to carbonates such as limestones, microfossils and pellet sands from deeper depths yield younger ages (Rao et al., 1994; Rajagopalan et al., 1980; Vora and Almeida, 1990). Ellison and Stoddart (1991) report the appearance of mangroves only after 7 kyr BP on the tropical shelves. So, these older ages of plant material could be debated. Similarly excluding the role of neotectonism in the construction of sea level curves does not offer scope for realistic projections for predictions in the light of climate change (Rao et al., 1996). Mathur et al. (2004) express concern over the inability of the proposed curves to give a clear picture for the entire coast and that different local sea level curves for different segments of the Indian coast may be able to deliver more realistic scenarios. Figure 9.1 and the related section on the geomorphology of the peninsular coastline supports this logical line of thinking. Banerjee (1993) too suggested that the controls on sea levels along the East coast of India would be dependent on more complex dynamics that vary across its length.

UNDERSTANDING SECTOR-WISE TRENDS IN LIGHT OF CLIMATE CHANGE

In order to tease out sector-wise local trends of sea level over the Holocene for the entire coastline, the same set of published chronological data in Table 9.1 was used to plot sea level curves for different segments of the Indian coastline (Figure 9.3a and b). The segments identified were as follows: (1) Gujarat Coast (tectonically significantly active) constituting data from Kutch and Saurashtra (2) Mumbai (inner shelf to coastal zone prone to subsidence for entire Holocene, only reclaimed for urbanization 200 years ago) (3) Konkan coast comprising Maharashtra coast south of Mumbai, Goa, Karnataka (4) Kerala Coast characterized by deltaic and coastal plains prone to submergence (5) Tamil Nadu Coast (6) Andhra Pradesh Coast (7) Odisha Coast (8) West Bengal Coast. The four segments along the East coast have been identified on the basis that they all belong to different river basins.

Figure 9.3a shows a comparison of the four sectors along the west coast of India. Kerala shows a complete offset from the trends of rest of the coast, which could be attributed to his extremely low slopes and coastal plains hardly above sea level. Saurashtra has an evident offset throughout the length, which can be attributed to tectonic uplift or downthrow. Otherwise all the three curves, namely for Konkan, Mumbai and Saurashtra follow a similar trend, but are separated by a systematic vertical shift, attributable to geomorphology of neotectonics. The sparse data points along the Kerala coast suggest that sea levels have been higher than all the other

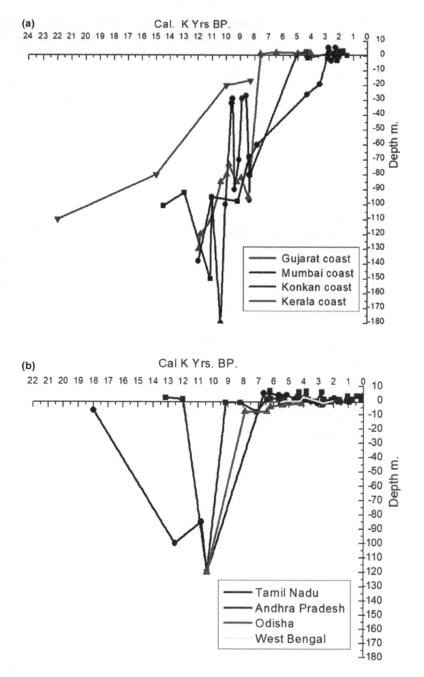

FIGURE 9.3 Sector-wise sea level trends plotted age vs. depth by using previously published data on dating of sea level indicators along the coasts of India (Table 9.1). (a) A composite for the West coast of India, which identifies the coast of Gujrat, Mumbai, Central West Coast (segment between Alibaug and Kerala), and Kerala as different sectors. (b) A composite for the East Coast of India which identifies the coast of Tamil Nadu, Andhra Pradesh, Odisha, and West Bengal as separate sectors.

sectors throughout the Holocene. Its trends are a unidirectional rise, seemingly unrelated to factors influencing the rest of the coastline. At 12 kyr BP the sea level all along the coast barring Kerala coast was at ~130 m below MSL. However, at 10 kyr BP, the sea was at −100 m for the Central West coast of India while at Saurashtra it was at −80 m BPSL, suggesting a tectonically influenced 20 m uplift. The validity of the four data points for Konkan within a dept range of 28–32 m and aged between 8.5 and 9.5 kyr have been derived on peat/carbonized samples and thus doubtful. Excluding these data points, all the curves for Mumbai, Konkan, and Gujarat coast show fluctuations of different magnitudes between 9 and 7 kyr with sea levels dropping by ~2–3 m alongside the coast of Mumbai and Konkan, while the sea level drops by 30 m indicating a tectonic subsidence along the Gujarat Coast. The sea level along all the sectors transgressed the PSL at different times, Kutch being the earliest to do so at about 7,500 years BP and has stayed about 1–2 m higher. It attained PSL at 4 kyr BP. The sea rose at a rate of almost 8.5 cm/yr between 8.5 and 7.5 kyr. BP. The sea level transgressed PSL at Mumbai ~5 kyr. BP and the Central west coast of India at around 2.7 kyr. BP.

Figure 9.3b is less complex than 3a and maybe could be attributed to the more homogenous geomorphology along the East coast of India. Few observations are very clear; the sea levels were same as or higher than present sea levels prior to LGM. The drop in sea levels as a response to LGM was delayed in the Bay of Bengal, namely 18 kya along the Andhra Pradesh coast and 14 kya along Tamil Nadu coast. All the sectors along East Coast show a maximum sea level fall stabilized at −120 m at ~10.5 kya after which the rise is sea level is steep. The sea levels transgress the onshore at Tamil Nadu at 9 kyr after which it again drops to a few meters below PSL and finally rises above PSL at 7 kyr BP and has been fluctuating close to the modern day SL. Interestingly, the sea levels rose rapidly to about −5 m along the Odisha coast about 8 kya, after which it slowed down to attain present sea level at about 4 kyr BP. The sea levels along Andhra Pradesh Coast rose rapidly to about +5 m APSL at 7 kyr BP after which it has been on a falling trend. It seems to be fluctuating close to PSL between 4 and 2 kyr BP. Unfortunately, not enough data are available for the coast of West Bengal. Limited data available documents the fact that sea levels have been very close to PSL between 6 and 2 kyr BP. This figure points to the differential progradation of deltaic coasts along these three sectors. The delayed arrival of the sea at the Odisha coast is indicative of increased freshwater influx and sediment build-up in response to the Holocene warming.

IMPLICATIONS OF PREDICTING FUTURE SEA LEVEL RISE ALONG THE INDIAN SUBCONTINENT

Sea level fluctuations that prevailed on the Earth's surface though-out the geological history have a changed connotation in the human history. The advent of the intelligent species *homo-sapiens* and their dominance converted the earth to a resource. Being sites of natural resource, namely food gathering, hunting, and fishing, fertile soils, trade routes and easy and cheap modes of transport, coastlands have been ideal sites for settlement historically and for urbanization and development in the modern times. Currently 40% of the world's population lives within 100 km of the coastlines.

More than 600 million people (around 10% of the world's population) live in coastal areas with low elevation i.e. less than 10 m above sea level. GMSL from tide gauges and altimetry observations increased from 1.4 mm/yr from 1901 to 1990, 2.1 mm/yr from 1970 to 2015, to 3.2 mm/yr from 1993 to 2015 and 3.6 mm/yr from 2006 to 2015 (IPCC, 2019). Global mean sea level has risen about 21–24 cm since 1880, with about a third of that coming in just the last two and a half decades (Lindsey, 2009). Studies confirm that even with sharp, immediate cuts in GHG emissions, the sea levels will continue to rise (Church et al., 2013) by another 0.5 m by end of this century (Javrejeva et al., 2012; Stocker et al., 2013; Kopp et al., 2014; Jackson and Jevrejeva, 2016; Nauels et al., 2017). It is estimated that if the emissions remain uncontrolled, it could trigger early-onset of Antarctic icesheet melting, leading to an extreme case scenario of sea level rise of about 2 m. (Kopp et al., 2017; Wong et al., 2017).

Sea level rise leads to coastal erosion, inundations, storm floods, tidal waters encroachment into estuaries and river systems, contamination of freshwater reserves due to sea water intrusion, detrimental to food crops due to salinization of soil, loss of fish nurseries and nesting grounds, as well as displacement of coastal lowlands and wetlands. In particular, sea level rise poses a significant risk to coastal regions and communities. Ocean warming has been linked to extreme weather events such as high waves, storm surges, and more intense tropical cyclones globally. In contrast to pre-historical implications of sea level rise, the economic value of the losses due to sea level rise is a now considered a direct measure of vulnerability of coastlines. While changes in temperatures and precipitation qualify as potential factors capable of influencing vulnerability of ecosystems with respect to sea level rise on the global scale (Garner et al., 2015; Osland et al., 2017), community practices, conversion of coastal land, non-sustainable practices of exploiting ecosystem services, etc. affect vulnerability on local scales (Foster et al., 2017). Multi-hazard risk assessments have been undertaken to estimate flooding and inundation of coastal lands in India (Kunte et al., 2014).

A policy research document prepared by the World Bank identified 84 countries that will be affected by the predicted sea level rise and stated that the most seriously impacted will be the countries along South East Asia. It estimates that a total land area of 74,000–178,000 km^2 will be completely inundated and a population ranging from 37 to 90 million people may have to be displaced (Dasgupta et al., 2007). Recent studies using high resolution data states that the global vulnerability to sea level rise and coastal inundation was triple the previous estimates. Approximately 190 million people currently occupy global land below projected high tide lines for 2100 under scenarios of low carbon emissions. As indicated by CoastalDEM, Bangladesh, India, and Vietnam rank only second to China in the average number of people living on land implicated by 2100, threatening to displace 21–30 million even under the low emissions scenario (K14/RCP 2.6), as compared to 9–19 million today, and with another 7–20 million on land threatened by annual storm surge (Kulp and Strauss, 2019).

Figures 9.4 and 9.5 are the predicted coastal inundation maps of India prepared using the Climate Central, Coastal Risk Screening Tool (Climate Central Inc., 2020). The software utilizes CoastalDEM data which is a new DEM developed using a neural network to perform nonlinear, nonparametric regression analysis

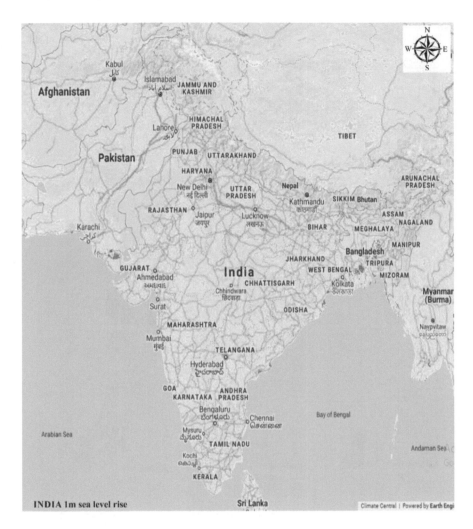

FIGURE 9.4 The map of India which shows the areas susceptible to the projected 1 m rise in sea level.

of SRTM (Shuttle Radar Topographic Model) error. This model incorporates 23 variables, including population and vegetation indices, and was trained using lidar-derived elevation data in the US as ground truth. Previously models, based on satellite data, would mistake treetops or building rooftops as the actual elevation of land at a particular point, resulting in a flawed estimation of the height of land in various parts of the world, especially areas that are densely populated (Kulp and Strauss, 2019). Apart from showing an overall effect of sea-level rise along the coastline of India, the figure also shows the extent of inundation that will take place in some of the most urbanized and inhabited regions India. This clearly puts into perspective the massive economic and ecological crisis that we are headed toward at the turn of this century.

(a)

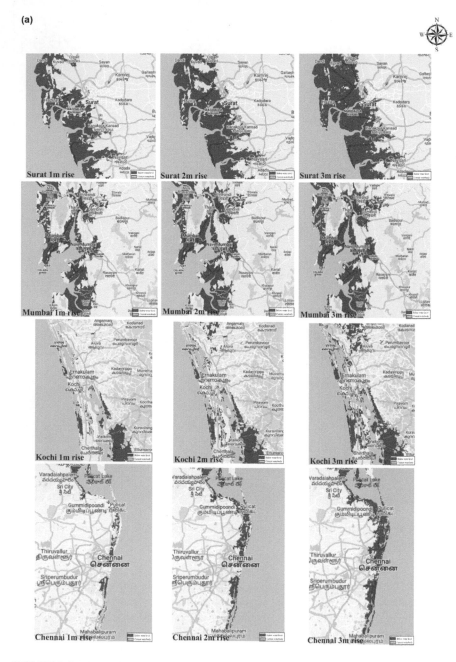

FIGURE 9.5 (a and b) A composite of enlarged portions of this map, depicting scenarios of 1, 2, and 3 m sea level rise of the seven most vulnerable regions along the coastline of India, namely Surat, Mumbai, and Kochi along the west coast of India and Chennai, Pondicherry, Mahanadi Delta, and West Bengal/Sunderban Delta along the east coast of India.

(Continued)

(b)

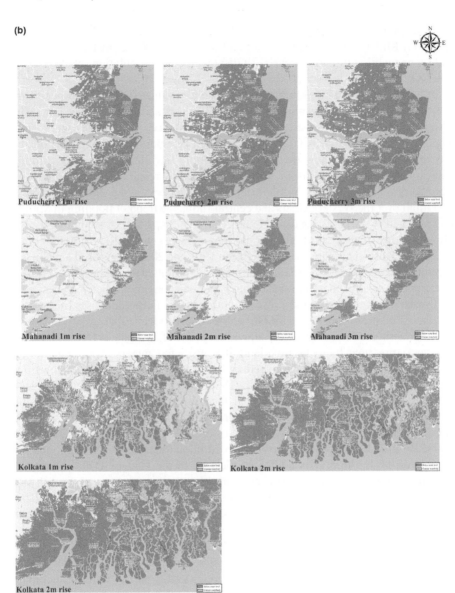

FIGURE 9.5 (*CONTINUED*) (a and b) A composite of enlarged portions of this map, depicting scenarios of 1, 2, and 3 m sea level rise of the seven most vulnerable regions along the coastline of India, namely Surat, Mumbai, and Kochi along the west coast of India and Chennai, Pondicherry, Mahanadi Delta, and West Bengal/Sunderban Delta along the east coast of India.

Large metropolitan cities like Kolkata, Mumbai, Yangon (Rangoon), Shanghai, Dhaka, and Bangkok are counted among the top ten most vulnerable cities due to sea level rise by 2070. Two Indian cities Kolkata (~2 trillion USD) and Mumbai (1.6 trillion USD) feature among the top ten cities that are at risk as deduced in terms of estimated loss of assets and rank fourth and sixth in the list, respectively. However, Kolkata (14 million people) and Mumbai rank (12 million people) first and second while estimating the risk of displacement of population (Nicholls et al., 2007, OECD).

India's financial capital, Mumbai is home to an estimated 20 million people and was reclaimed from the Arabian Sea almost 200 years ago. The city was born when seven neighboring islands were merged through engineering exercises into a single landmass. In places such as Grant Road, even in summer during high tide, one can see water just a foot below the drains that empty into the sea. In 2005 Mumbai witnessed the worst deluge in its history, triggered by relentless rain, coupled with storm surges and high tides. While recent studies have shown a three-fold increase in rainfall over the Western Ghats and central India, the trend of flooding in Mumbai was repeated in the monsoon of 2017 and 2020 (Anima, 2020). A study by NEERI's Mumbai office states that Mumbai could face damages worth Rs. 35,00,000 crores by 2050 because of climate change. Between 1901 and 200, it registered a mean temperature rise of 1.62°C. The sea level around the island city is rising by 2.4 mm/yr. The consequences of these climate change phenomenon are predicted to be a chain of disasters such as flash floods, disease outbreaks, building collapses, dislocation, and increased prevalence of vector-borne diseases (Pandve, 2010).

Kolkata situated right within the alluvial deposits of the Ganges Delta, has grown in several directions by reclaiming land. It is just a few meters above sea level and was originally a marshy wetland, remains of which can still be seen toward its eastern parts. The Ganges-Brahmaputra Delta also known as the Sunderban Delta faces issues of conversion of traditional land use; removal of natural vegetation buffers and degradation of mangroves contribute to high exposure rates to coastal flooding, erosion and salinization (Hossain et al., 2018; Ghosh et al., 2018). The World Wildlife Fund (2009) reports that in addition to sea-level rise, a ground subsidence of 0.6–1.9 mm/yr is adding to the risk in the Ganges Delta. Due to the combined effects of sea-level rise and subsidence, the report predicts saltwater intrusion 100 km inland from the coast, greatly impacting ground water supplies in the Ganges Delta. Compounding the effects of saltwater intrusion, over-exploitation of groundwater in and around Kolkata has led to a drop in its level, leading to further intrusion of sea water and thus making the subsurface ground water and soil resources saline. Currently, riverine flooding dominates the delta. However, the predicted increase in the high tides and cyclones to this part of India is anticipated to generate large coastal flooding events (Rahman and Rahman, 2015). To add to the factors that make Kolkata, Bangladesh, and the Ganges Delta vulnerable is the fact that it is located on a seismic zone, making it prone to earthquakes. Prevention and protection measures on the delta are limited to restoring natural buffers like mangroves (Quan et al., 2018; Rahman et al., 2018) and hardly any efforts are being taken with respect to address the issue of subsidence (Schmidt, 2015; Schmitt et al., 2017). Oppenheimer et al. (2019) opine that because of its massive size, the entire delta will be at risk of flooding, erosion, and salinization only in worst case scenarios of emissions.

Recent assessments of coastal erosion along the Coast of Chennai indicate that land losses currently dominate over land gains and that human interventions are a major driver of shoreline changes. From the analysis it clearly shows that the northern portion of the Chennai port is eroding at rates as high as 4–5 m/yr and the southern portion of the port is accreting at a rate of 2–4 m/yr along the Mariana Beach (Kankara et al., 2018).

For India, the government reports reveal sea levels have risen by 8.5 cm in the past 50 years with an average increase of 1.7 mm/yr (TOI, 2019a). The long-term data, collected at ten Indian ports, indicates sea level at Diamond Harbor in West Bengal with the highest average sea level increase (5.6 mm/yr) followed by Kandla (3.16 mm/yr) in Gujrat, Haldia (2.89 mm/yr) in West Bengal, Port Blair (2.2 mm /yr) in Andaman and Nicobar Islands and Okha (1.5 mm/yr) in Gujrat (TOI, 2019a). The sea level rise in the low-lying areas of Mumbai are expected to be a gradual change and not a sudden event with an increase in intensity and duration of prolonged flooding (TOI, 2019b). Another direct effect of global mean temperature rise is the increase in the amount of sub-daily rainfall across India (Ali and Mishra, 2018) This may have a direct implication in the increase in local sea levels in areas like Mumbai, Kolkata and Odisha which have seen a marked increase in extreme rainfall and devastating cyclones over the past decades. Though some coastal ecosystems show strong resilience to sea level rise (McKee and Vervaeke, 2018), mangrove areas of the India-Pacific region cannot outpace current SLR rates, and are at risk of disappearing (Lovelock et al., 2015). The World Bank has pledged through its $400 million worth programme, ENCORE (Enhancing Coastal and Ocean Resource Efficiency) to safeguard the Indian coastal and marine assets over the next decade. It in its first phase cover eight coastal states (Andhra Pradesh, Gujarat, Goa, Karnataka, Kerala, Odisha, Tamil Nadu, and West Bengal) and three coastal Union Territories (Daman and Diu, Lakshadweep, and Puducherry), where coastal resources are under significant pressure (World Bank, 2020).

The environmental and economic ramification of such an event is nearly incalculable. The Intergovernmental Panel on Climate Change (IPCC, 2014) in its fifth assessment report has called this warming of climate 'unequivocal' and 'unprecedented over decades to millennia'. Without mitigation efforts through an analytical approach which evaluate the expected risks, reap the benefits and to some extent adapt to the situation, human society is facing imminent disaster. Therefore, it is of utmost importance for us to be able to update our centennial scale sea level projection models into at least decadal scale models. For this, a precise understanding of the cause and effects of sea level change is necessary. This requires inclusion of many variable factors into the simulation model which have previously not been considered or have been considered insignificant before now. Only by scrutinizing the subtlest of nuances in sea-level change can we hope to prepare ourselves for the future to come.

CONCLUSIONS

Scanty early to mid-Pleistocene sea level data available around the Indian subcontinent reveal that the sea level was 55–60 m higher than the present level about 220 kya. The last interglacial from 125 to 36 kya was manifested as a transgressive phase, where sea levels 1–4 m above present have been recorded, sometime as far as 25 km inland as in Saurashtra. There do exist contradictory reports of sea level being lower than present

50–25 kyr BP along the southwestern margin of India. The LGM recorded sea levels as low as −120 m at about 20 and 16 kyr BP along the West and East coasts, respectively. The onset of LGM was reportedly much delayed along the East coast of India. The Holocene sea level rise was manifested differently at different locations along the coastline and controlled by tectonics and morphology along the West Coast and sedimentation and fluvial processes in response to Holocene warming along the East Coast. Though a few appreciable attempts have been made to compile comprehensive sea level curves to represent trends along the entire East and West Coasts, there is a need to look at local sea level signatures in order to improve their applicability.

Going by available data, the sea level actually transgressed the current coastline of India approximately 7 kya and since then has, quite a few times, at many locations, been 2–5 m APSL (Table 9.1). Loveson and Nigam (2019) report 5 such episodes since 7 kyr suggesting a sea level rise every 1,400 years. Each of these episodes if decoded well, could serve as a reference helpful in dealing with the professed sea level rise of similar magnitude, expected to hit our coastline at a much faster rate, by the next century. The knowledge regarding past sea levels is crucial in identifying vulnerable areas, processes, and intensity of inundation, that can help mitigate effects of predicted sea level rise in response to climate change. However, there is extreme dearth of well dated, geological, geomorphological and neotectonics data along the Indian coast. Much of the old-dated evidence used conventional bulk radiocarbon dating techniques which had severe errors. The advent of modern techniques of geochronology like AMS radiocarbon dating, thermoluminescence and optically stimulated luminescence need to be exploited to refine our understanding of past sea level signatures, which in turn can enhance our predictability. Site specific study of sea level fluctuations calibrated using better geochronological tools and knowledge of processes influencing sea levels in an area of interest, can help improve high resolution, sea level modelling. Prediction and vulnerability assessments need reliable DEM data, verified using ground truthing on local scales.

ACKNOWLEDGEMENTS

The authors are thankful to the Vice Chancellor of Savitribai Phule Pune University Prof. Nitin Karmalkar and Prof. Suresh Gosavi, Head, Department of Environmental Science, SPPU for extending their cooperation and logistic support for carrying out marine research on the University campus. One of the authors, A.S. acknowledges the financial support provided by the Ministry of Earth Sciences, Government of India under its Palaeoclimate Change Programme in the form of Research Fellowship.

REFERENCES

Achyuthan, H. (1997). Age and formation of oyster beds of Muthukadu tidal flat zone, Chennai, Tamil Nadu. *Current Science*, 73(5), 450–453.

Achyuthan, H., and Baker, V. (2002). Coastal response to changes in sea level since the last 4500 BP on the east coast of Tamil Nadu, India. *Radiocarbon*, 44(1), 137–144.

Agrawal, D. P., and Ghosh, A. (1973). Radiocarbon and Indian archaeology. *International Symposium on "Radiocarbon and Indian Archaeology"*, Bombay, 1972, Tata Institute of Fundamental Research, Bombay, xxiv + 526 pp. 18 leaves of plates: ill., maps; 25 cm.

Ahmed, E. (1972). *Coastal Geomorphology of India*, Orient Longman. New Delhi.

Ali, H., and Mishra, V. (2018). Increase in subdaily precipitation in India under 1.5°C and 2.0°C warming worlds. *Geophysical Research Letters*, 45, 6972–6982. DOI: 10.1029/2018GL078689.

Anima, P. (2020). Sea level rise: Is India ready for the challenge?. https://www.thehindu-businessline.com/blink/cover/sea-level-rise-is-india-ready-for-the-challenge/article30818917.ece (Published 14/02/2020)

Baba, M., and Thomas, K. V. (1999). Geomorphology of the Indian Coast, Strategy for Sustainable Development in the Coastal Area. New Delhi: Ministry of Environment and Forests.

Ballu, V., Bouin, M. N., Siméoni, P., Crawford, W. C., Calmant, S., Boré, J. M., and Pelletier, B. (2011). Comparing the role of absolute sea-level rise and vertical tectonic motions in coastal flooding, Torres Islands (Vanuatu). *Proceedings of the National Academy of Sciences*, 108(32), 13019–13022.

Banerjee, P. K. (1993). Imprints of late quaternary climatic and sea level changes on East and South Indian coast. *Geo-Marine Letters*, 13(1), 56–60.

Banerjee, P. K. (2000). Holocene and late Pleistocene relative sea level fluctuations along the east coast of India. *Marine Geology*, 167(3–4), 243–260.

Barui, N. C., Chanda, S., and Bhattacharya, K. (1986). Late Quaternary vegetational history of the Bengal basin. In: Samanta, B. K. (Ed.), Proceedings of the XI Colloquium. *Micropalaeontological Stratigraphy*, Geological, *Mining, and Metallurgical* Society *of India*, 54, 197–201.

Bhatt, N., and Bhonde, U. (2006). Geomorphic expression of late Quaternary sea level changes along the southern Saurashtra coast, western India. *Journal of Earth System Science*, 115(4), 395–402.

Bhattacharya, G. C., Chauhey, A. K., Murthy, G. P. S., Srinivasan, K., Sarma, K. V. L. N. S., Suhrahmanyam, V., and Krishna, K. S. (1994). Evidence for sea-floor spreading in the Laxmi basin, north-eastern Arabian Sea, Earth and Planet. *Science Letters*, 125, 211–220.

Bhattacharya, G. C., and Yatheesh, V. (2015). Plate-tectonic evolution of the deep ocean basins adjoining the Western continental margin of India-A proposed model for the early openingscenario. In: Mukherjee, S. (Ed.), *Petroleum Geosciences: Indian Contexts*. Cham: Springer International Publishing, pp. 1–61.

Biswas, S. K. (1987). Regional tectonic framework, structure and evolution of the western marginal basins of India. *Tectonophysics*, 135(4), 307–327.

Biswas, S. K., and Khattri, K. N. (2002). A geological study of earthquakes in Kutch, Gujarat, India. *Journal of the Geological Society of India*, 60(2), 131–142.

Bloom, A. L. (1971). Glacial-eustatic and isostatic controls of sea level since the last glaciation. In: Turekian, K. K. (ed.), *The Late Cenozoic Glacial Stages*. New Haven: Yale University Press, 355–379.

Brückner, H. (1989). Late Quaternary shorelines in India. In: Scott, D. B., Pirazzoli, P. A., and Honig, C. A. (Eds.), *Late Quaternary sea-level correlation and applications*. Dordrecht, Netherlands: Springer, 169–194.

Carson, M., Köhl, A., Stammer, D., Meyssignac, B., Church, J., Schröter, J., and Hamlington, B. (2017). Regional sea level variability and trends, 1960–2007: A comparison of sea level reconstructions and ocean syntheses. *Journal of Geophysical Research: Oceans*, 122(11), 9068–9091.

Chakrabarti, P. (1992). Geomorphology and Quaternary Geology of Hoogly Estuary, West Bengal plain. India. Unpublished Ph.D Thesis, University of Calcutta, 105 pp.

Chappel, J., and Shackleton, J. J. (1986). Pleistocene sea levels and the oxygen isotope record: A reconciliation. *Nature (London)*, 324, 137–140.

Chatterjee, S. P. (1961). Fluctuation of sea level around the coasts of India during the Quaternary period. *Z. Geomorph. n.F., Suppl.*, 3, 48–56.

Chaubey, A. K., Dyment, J., Bhattacharya, G. C., Royer, J. Y., Srinivas, K., and Yatheesh, V. (2002). Paleogene magnetic isochrons and palaeo-propagators in the Arabian and Eastern Somali basins, NW Indian Ocean. In: Clift, P. D., Croon, D., Gaedicke, C., & Craig, J. (Eds.), *The Tectonic and Climatic Evolution of the Arabian Sea Region. Geological Society of London, Special Publication*, 195, 71–85.

Church, J. A., Clark, P. U., Cazenave, A., Gregory, J. M., Jevrejeva, S., Levermann, A., Merrifield, M. A., Milne, G. A., Nerem, R. S., Nunn, P. D., Payne, A. J., Pfeffer, W. T., Stammer D., and Unnikrishnan, A. S. (2013). Sea level change. In: Stocker, T. F., Qin, D., Plattner, G.-K., Tignor, M., Allen, S. K., Boschung, J., Nauels, A., Xia, Y., Bex, V., and Midgley, P. M. (Eds.), *Climate Change 2013: The Physical Science Basis. Contribution of Working Group I to the Fifth Assessment Report of the Intergovernmental Panel on Climate Change*. Cambridge, UK: Cambridge University Press, 1137–1216. http://www.climatechange2013.org/report/full-report.

Climate Central. (2021). Coastal risk screening tool. https://coastal.climatecentral.org/map/11/72.8977/19.0032/?theme=sea_level_rise&map_type=coastal_dem_comparison&basemap=roadmap&contiguous=true&elevation_model=best_available&forecast_year=2050&pathway=rcp45&percentile=p50&refresh=true&return_level=return_level_1&slr_model=kopp_2014

Curray, L. R. and Moore, D. G. (1974). Sedimentary and tectonic processes in the Bengal deep-sea fan and geosyncline. In: Burke, C. A. and Drake, C. L. (Eds.), *The Geology of Continental Margins*. Berlin Heidelberg: Springer-Verlag, 617–627.

Das, D., Samanta, G., Mandal, B. K., Roy Chowdhury, T., Chanda, C. R., Chowdhury, P. P., Basu, K. G., and Chakraborti, D. (1996). Arsenic in groundwater in six districts of West Bengal, India. *Environmental Geochemistry and Health*, 18(1), 5–15. DOI: 10.1007/BF01757214.

Dasgupta, S., Laplante, B., Meisner, C., Wheeler, D., and Yan, J. (2007). The Impact of Sea Level Rise on Developing Countries: A Comparative Analysis, World Bank Policy Research Working Paper, 4136, p. 51.

De Souza, N. G. (1965). Some aspects of the geomorphology of Goa. Vasundhara: Journal of Geological Society. University of Saugor, I, 13–20.

Dey, K. (2020). Financial Express India's Blue Economy net getting bigger! Country ranks third in fisheries and second in aquaculture. https://www.financialexpress.com/opinion/indias-blue-economy-net-getting-bigger-country-ranks-third-in-fisheries-and-second-in-aquaculture/1867607/).

Dikshit, K. R. (1976). Geomorphic features of the West coast of India between Bombay and Goa. *Geological Review of India*, 30(3), 260–281.

Dubey, K. M., Chaubey, A. K., Mahale, V. P., and Karisiddaiah, S. M. (2019). Buried channels provide keys to infer Quaternary stratigraphic and paleo-environmental changes: A case study from the west coast of India. *Geoscience Frontiers*, 10(4), 1577–1595.

Ellison, J. C., and Stoddart, D. R. (1991). Mangrove ecosystem collapse during predicted sea-level rise: Holocene analogues and implications. *Journal of Coastal Research*, 7, 151–165.

Fairbanks, R. G. (1989). A 17,000-year glacio-eustatic sea level record: Influence of glacial melting rates on the Younger Dryas event and deep-ocean circulation. *Nature*, 342(6250), 637–642.

Fairbridge, R. W. (1961). Eustatic changes in sea level. *Physics and Chemistry of the Earth*, 4, 99–185.

Farooqui, A., and Vaz, G. G. (2000). Holocene sea-level and climatic fluctuations: Pulicat lagoon-a case study. *Current Science*, 79(10), 1484–1488.

Forget, G., and Ponte, R. M. (2015). The partition of regional sea level variability. *Progress in Oceanography*, 137, 173–195.

Foster, T. E., Stolen, E. D., Hall, C. R., Schaub, R., Duncan, B. W., Hunt, D. K., and Drese, J. H. (2017). Modeling vegetation community responses to sea-level rise on Barrier Island systems: A case study on the Cape Canaveral Barrier Island complex, Florida, USA. *PloS One*, 12(8), e0182605.

Frederikse, T., Landerer, F., Caron, L., Adhikari, S., Parkes, D., Humphrey, V. W., and Wu, Y. H. (2020). The causes of sea-level rise since 1900. *Nature*, 584(7821), 393–397. DOI: 10.1038/s41586-020-2591-3.

Garner, K. L., Chang, M. Y., Fulda, M. T., Berlin, J. A., Freed, R. E., Soo-Hoo, M. M., and Kendall, B. E. (2015). Impacts of sea level rise and climate change on coastal plant species in the central California coast. *PeerJ*, 3, e958.

Gaur, A. S., Sundaresh, and Tripati, S. (2009). New evidence on Maritime Archaeology around Mul Dwarka (Kodinar), Gujarat Coast, India. *Man and Environment*, 34(2), 72–76

Giosan, L., Donnelly, J. P., Constantinescu, S., Filip, F., Ovejanu, I., Vespremeanu-Stroe, A., and Duller, G. A. (2006). Young Danube delta documents stable Black Sea level since the middle Holocene: Morphodynamic, paleogeographic, and archaeological implications. *Geology*, 34(9), 757–760.

Ghosh, R., Gustafson, A., and Schunnesson, H. (2018). Development of a geological model for chargeability assessment of borehole using drill monitoring technique. *International Journal of Rock Mechanics and Mining Sciences*, 109, 9–18. DOI: 10.1016/j.ijrmms.2018.06.015

Gunnell, Y. (2001). Dynamics and kinematics of rifting and uplift at the western continental margin of India: Insights from geophysical and numerical models. *Memoir - Geological Society of India*, 47, 475–496.

Gupta, H., and Chakrapani, G. J. (2005). Temporal and spatial variations in water flow and sediment load in Narmada River Basin, India: natural and man-made factors. *Environmental Geology*, 48(4–5), 579–589.

Gupta, H. P. (1981). Palaeoenvironments during Holocene time in Bengal Basin, India as reflected by palynostratigraphy. *Palaeobotanist*, 27(2), 138–160.

Gupta, S. K. (1972). Chronology of the raised beaches and inland coral reefs of the Saurashtra coast. *The Journal of Geology*, 80(3), 357–361.

Gupta, S. K., and Amin, B. S. (1974). Io/U ages of corals from Saurashtra coast. *Marine Geology*, 16(5), M79–M83.

Guzder, S. (1980). Quaternary environment and Stone Age cultures of the Konkan coastal Maharashtra India; Deccan College, Pune.

Hallegatte, S., Green, C., Nicholls, R. J., and Corfee-Morlot, J. (2013). Future flood losses in major coastal cities. *Nature Climate Change*, 3, 802–806.

Hameed, A., Achyuthan, H., and Sekhar, B. (2006). Radiocarbon dates and Holocene sea-level change along the Cuddalore and Odinur Coast, Tamil Nadu. *Current Science*, 91(3), 362–367.

Hamlington, B. D., Burgos, A., Thompson, P. R., Landerer, F. W., Piecuch, C. G., Adhikari, S., and Ivins, E. R. (2018). Observation-driven estimation of the spatial variability of 20th century sea level rise. *Journal of Geophysical Research: Oceans*, 123(3), 2129–2140.

Han, G., Ma, Z., Chen, N., Yand, J., and Chen, N. (2015). Coastal sea level projections with improved accounting for vertical land motion. *Scientific Reports*, 5, 160085. DOI: 10.1038/srep16085.

Han, W., Meehl, G. A., Hu, A., Zheng, J., Kenigson, J., Vialard, J., and Rajagopalan, B. (2017). Decadal variability of the Indian and Pacific Walker cells since the 1960s: Do they covary on decadal time scales? *Journal of Climate*, 30(21), 8447–8468.

Han, W., Vialard, J., McPhaden, M. J., Lee, T., Masumoto, Y., Feng, M., and De Ruijter, W. P. (2014). Indian Ocean decadal variability: A review. *Bulletin of the American Meteorological Society*, 95(11), 1679–1703.

Hanson, S., Nicholls, R. J., Ranger, N., Hallegatte, S., Corfee-Morlot, J., Herweijer, C., and Chateau, J. (2011). A global ranking of port cities with high exposure to climate extremes. *Climate Change*, 104, 89–111. DOI:10.1007/s10584-010-9977-4.

Haque, A., and Nicholls, R. J. (2018). Floods and the Ganges-Brahmaputra-Meghna Delta. In: Nicholls, R. J., Hutton, C. W., Adger, W. N., Hanson, S. E., Rahman, M. M., and Salehin, M. (Eds.), Ecosystem Services for Well-Being in Deltas: Integrated assessment for policy analysis. Cham: Palgrave Macmillan, 147–159. DOI: 10.1007/978-3-319-71093-8_8.

Hashimi, N. H., and Nair, R. R. (1976). Carbonate components in the coarse fraction of western continental shelf (northern part) of India. *Indian Journal of Marine Sciences*, 5(1), 51–57. http://nopr.niscair.res.in/handle/123456789/39440

Hashimi, N. H., and Nair, R. R. (1986). Climatic aridity over India 11,000 years ago: evidence from feldspar distribution in shelf sediments. *Palaeogeography, Palaeoclimatology, Palaeoecology*, 53(2–4), 309–319.

Hashimi, N. H., Nigam, R., Nair, R. R., and Rajagopalan, G. (1995). Holocene sea level fluctuations on western Indian continental margin: An update. *Journal of the Geological Society of India*, 46, 157–162. http://drs.nio.org/drs/handle/2264/2364.

Hossain, M. A. R., Ahmed, M., Ojea, E., and Fernandes, J. A. (2018). Impacts and responses to environmental change in coastal livelihoods of south-west Bangladesh. *Science of the Total Environment*, 637–638, 954–970.

IPCC. (2014). *Climate Change 2014. Synthesis Report. Contribution of Working Groups I, II and III to the Fifth Assessment Report of the Intergovernmental Panel on Climate Change* [Core Writing Team, Pachauri, R. K., and Meyer, L. A. (Eds.)]. Geneva, Switzerland: IPCC, 151 p.

Jackson, L. P., and Jevrejeva, S. (2016). A probabilistic approach to 21st century regional sea level projections using RCP and high-end scenarios. *Global and Planetary Change*, 146, 179–189.

James, N. P., Bone, Y., Carter, R. M., and Murray-Wallace, C. V. (2006). Origin of the Late Neogene Roe plains and their calcarenite veneer: Implications for sedimentology and tectonics in the Great Australian Bight. *Australian Journal of Earth Sciences*, 53(3), 407–419.

Javrejeva, S., Moore, J., and Grinsted, A. (2012). Sea level projections to AD2500 with a new generation of climate change scenarios. *Global and Planetary Change*, 80–81, 14–20.

Kale, V. S., Kshirsagar, A. A., and Rajaguru, S. N. (1984). Late Pleistocene Beach rocks from Uran, Maharashtra, India. *Current Science*, 53(6), 317–319.

Kale, V. S., and Rajaguru, S. N. (1985). Neogene and Quaternary transgressional and regressional history of the west coast of India: An overview. *Bulletin of the Deccan College Research Institute*, 44, 153–167.

Kankara, R. S., Ramana Murthy, M. V., & Rajeevan, M. (2018). National assessment of shoreline changes along Indian Coast – a status report for 26 years 1990–2016, National Centre for Coastal Research, Chennai, India, pp. 81. https://www.nccr.gov.in/sites/default/files/schangenew.pdf

Karlekar, S. N. (1981). A geomorphic study of South Konkan. PhD. thesis submitted, University of Poona (unpublished).

Khandelwal, A., Mohanti, M., García-Rodríguez, F., and Scharf, B. W. (2008). Vegetation history and sea level variations during the last 13,500 years inferred from a pollen record at Chilika Lake, Orissa, India. *Vegetation History and Archaeobotany*, 17(4), 335–344.

Kopp, R. E., DeConto, R. M., Bader, D. A., Hay, C. C., Horton, R. M., Kulp, S., and Strauss, B. H. (2017). Evolving understanding of Antarctic ice-sheet physics and ambiguity in probabilistic sea-level projections. *Earth's Future*, 5(12), 1217–1233.

Kopp, R. E., Hay C. C., Little C. M., and Mitrovica, J. X. (2015). Geographic variability of sea-level change. *Current Climate Change Reports*, 1, 192–204.

Kopp, R. E., Horton, R. M., Little, C. M., Mitrovica, J. X., Oppenheimer, M., Rasmussen, D. J., and Tebaldi, C. (2014). Probabilistic 21st and 22nd century sea-level projections at a global network of tide-gauge sites. *Earth's Future*, 2(8), 383–406.

Kulp, S. A., and Strauss, B. H. (2019). New elevation data triple estimates of global vulnerability to sea-level rise and coastal flooding. *Nature Communications*, 10(1), 1–12.

Kunte, P. D., Jauhari, N., Mehrotra, U., Kotha, M., Hursthouse, A. S., and Gagnon, A. S. (2014). Multi-hazards coastal vulnerability assessment of Goa, India, using geospatial techniques. *Ocean & Coastal Management*, 95, 264–281.

Laiz, I., Ferrer, L., Plomaritis, T. A., and Charria, G. (2014). Effect of river runoff on sea level from in-situ measurements and numerical models in the Bay of Biscay. Deep sea research Part II. *Topical Studies in Oceanography*, 106, 49–67.

Le Cozannet, G., Rohmer, J., Cazenave, A., Idier, D., van de Wal, R., de Winter, R., Pedreros, R., Balouin, Y., Vinchon, C., and Oliveros, C. (2015). Evaluating uncertainties of future marine flooding occurrence as sea-level rises. *Environmental Modelling & Software*, 73, 44–56.

Lindsey, R. (2009). *Climate and Earth's Energy Budget*. Greenbelt, MD: NASA Earth Observatory.

Lovelock, C. E., Cahoon, D. R., Friess, D. A., Guntenspergen, G. R., Krauss, K. W., Reef, R., and Saintilan, N. (2015). The vulnerability of Indo-Pacific mangrove forests to sea-level rise. *Nature*, 526(7574), 559–563.

Loveson, V. J. (1994). Geological and geomorphological investigations related to sea level variation and heavy mineral accumulation along the southern Tamil Nadu beaches, India. *Unpublished Ph. D. Thesis*, Madurai Kamaraj University, Madurai.

Loveson, V. J., and Nigam, R. (2019). Reconstruction of Late Pleistocene and Holocene sea level curve for the east coast of India. *Journal of the Geological Society of India*, 93(5), 507–514.

Mathur, U. B., Pandey, D. K., and Bahadur, T. (2004). Falling late holocene sea-level along the Indian coast. *Current Science*, 87(4), 439–440.

Maurya, D. M., Thakkar, M. G., Patidar, A. K., Bhandari, S., Goyal, B., and Chamyal, L. S. (2008). Late Quaternary geomorphic evolution of the coastal zone of Kachchh, western India. *Journal of Coastal Research*, 24(3), 746–758.

Kumaran, P., Maya, K., Vishnu Mohan, S., Ruta, B., and Limaye, D. P. (2017). Geomorphic response to sea level and climate changes during Late Quaternary in a humid tropical coastline: Terrain evolution model from Southwest India. *PLoS ONE*, 12(5), 1–32.

McKee, K. L., and Vervaeke, W. C. (2018). Will fluctuations in salt marsh–mangrove dominance alter vulnerability of a subtropical wetland to sea-level rise? *Global Change Biology*, 24(3), 1224–1238.

Meade, R. H., and Emery, K. O. (1971). Sea level as affected by river runoff, eastern United States. *Science*, 173(3995), 425–428.

Merh, S. S. (1993). Neogene-Quaternary Sequence in Gujarat – A review. *Journal of the Geological Society of India*, 41(3), 259–276.

Meyssignac, B., Piecuch, C. G., Merchant, C. J., Racault, M. F., Palanisamy, H., MacIntosh, C., and Brewin, R. (2017a). Causes of the regional variability in observed sea level, sea surface temperature and ocean colour over the period 1993–2011. *Surveys in Geophysics*, 38(1), 187–215.

Meyssignac, B., Slangen, A. A., Melet, A., Church, J. A., Fettweis, X., Marzeion, B., and Palmer, M. D. (2017b). Evaluating model simulations of twentieth-century sea-level rise. Part II: regional sea-level changes. *Journal of Climate*, 30(21), 8565–8593.

Milliman, J. D., and Syvitski, J. P. (1992). Geomorphic/tectonic control of sediment discharge to the ocean: the importance of small mountainous rivers. *The Journal of Geology*, 100(5), 525–544.

Minshull, T. A., Lane, C. I., Collier, J. S., Whitmarsh, R. B. (2008). The relationship between rifting and magmatism in the northeastern Arabian Sea. *Nature Geoscience*, 1, 463–467.

Mittre, V., and Gupta, H.P. (1970). Pollen analytical study of Quaternary deposits in the Bengal Basin. *Palaeobotanist*, 19 (3), 297–306.

Mörner, N.-A. (1969). Estimating future sea level changes. *Global and Planetary Change*, 40, 49–54; (2004). *Sveriges Geoliska Undersökning*, C-640, 1–487.

Murray-Wallace, C. V., and Woodroffe, C. D. (2014). *Quaternary Sea-Level Changes: A Global Perspective*. Cambridge, UK: Cambridge University Press.

Nair. (unpublished/personal communication) in Hashimi et al., 1995.

Nair, R. R. (1975). Nature and origin of small scale topographic prominences on the western continental shelf of India. *Indian Journal of Marine Sciences*, 4, 25–29.

Nair, R. R., and Hashimi, N. H. (1980). Holocene climatic inferences from the sediments of the western Indian continental shelf. P Indian As-Earth. *Proceedings of the Indian Academy of Sciences (Earth and Planetary Sciences)*, 89, 299–315. http://drs.nio.org/drs/handle/2264/6834

Nair, R. R., and Qasim, S. Z. (1977). Occurrence of a bank with living corals off the South-West coast of India. *Indian Journal of Marine Sciences*, 7(1), 55–58.

Nambiar, A. R., Rajagopalan, G., and Rao, B. R. J. (1991). Radiocarbon dates of sediment cores from inner continental shelf off Karwar, west coast of India. *Current Science*, Vol. (61), pp. 35–354.

Narayana, A. C., Priiu, C. P., and Chakrabarti, A. (2001). Identification of paleodelta near the mouth of Perryar river in central Kerala. *Journal of the Geological Society of India*, 57, 545–547.

Nauels, A., Meinshausen, M., Mengel, M., Lorbacher, K., and Wigley, T. M. (2017). Synthesizing long-term sea level rise projections-the MAGICC sea level model v2. 0. *Geoscientific Model Development*, 10(6), 2495–2524. DOI: 10.5194/gmd-10-2495-2017.

Nayak, G. N. (2005). Indian Ocean coasts - Coastal geomorphology. In: Schwartz, M. L., (Ed.), *Encyclopedia of Coastal Science*. Springer. DOI: 10.1007/1-4020-3880-1_179.

Nayak, R. K., and Shetye, S. R. (2003). Tides in the Gulf of Khambhat, west coast of India. *Estuarine, Coastal and Shelf Science*, 57(1–2), 249–254.

News Bharati. (2019). Sea to be the next gateway for tourism in the country', Centre steps up to promote maritime tourism in India, https://www.newsbharati.com/Encyc/2019/8/2/Discourse-over-promoting-marine-tourism-.html.

Nicholls, R. J., Hanson, S., Herweijer, C., Patmore, N., Hallegatte, S., Corfee-Morlot, J., and Muir-Wood, R. (2007). Ranking of the world's cities most exposed to coastal flooding today and in the future. Organisation for Economic Co-operation and Development (OECD), Paris.

Nidheesh, A. G., Lengaigne, M., Vialard, J., Unnikrishnan, A. S., and Dayan, H. (2013). Decadal and long-term sea level variability in the tropical Indo-Pacific Ocean. *Climate Dynamics*, 41(2), 381–402.

Nigam, R. (1991). Foraminiferal variations in cores from inner shelf of Karwar: A key to paleomonsoonal variations during Holocene over Konkan coast, India, Department of Science and Technology Report, New Delhi, 78p.

Nigam, R., Hashimi, N. H., Meneze, E. T., and Wagh, A. B. (1992). Fluctuating sea levels off Bombay (India) between 14,500 and 10,000 years before present. *Current Science*, 62, 309–311.

Nigam, R., Henriques, P. J., and Wagh, A. B. (1993). Barnacle fouling on relict foraminiferal specimens from the western continental margin of India: An indicator of paleosea-level. *Continental Shelf Research*, 13(2–3), 279–286.

Nigam, R., and Khare, N. (1992). The reciprocity between coiling direction and reproduction in benthic foraminifera. *Journal of Micropalaentology*, 11(2), 221–228.

NOAA. (2019). Is sea level rising?, https://oceanservice.noaa.gov/facts/sealevel.
html#:~:text=The%20two%20major%20causes%20of,as%20glaciers%20and%20
ice%20sheets.

Oppenheimer, M., Glavovic, B. C., Hinkel, J., van de Wal, R., Magnan, A. K., Abd-Elgawad,
A., Cai, R., Cifuentes-Jara, M., DeConto, R. M., Ghosh, T., Hay, J., Isla, F., Marzeion, B.,
Meyssignac, B., and Sebesvari, Z. (2019). Sea level rise and implications for low-lying
islands, coasts and communities. In: Pörtner, H.-O., Roberts, D. C., Masson-Delmotte,
V., Zhai, P., Tignor, M., Poloczanska, E., Mintenbeck, K., Alegría, A., Nicolai, M.,
Okem, A., Petzold, J., Rama, B., and Weyer, N. M. (Eds.), *IPCC Special Report on the
Ocean and Cryosphere in a Changing Climate*. In press.

Osland, M. J., Griffith, K. T., Larriviere, J. C., Feher, L. C., Cahoon, D. R., Enwright, N. M.,
and Baustian, J. J. (2017). Assessing coastal wetland vulnerability to sea-level rise along
the northern Gulf of Mexico coast: Gaps and opportunities for developing a coordinated
regional sampling network. *PloS One*, 12(9), e0183431.

Pandey, J., and Guha, D. K. (1982). Termination of last Neogene transgression and
some Quaternary shoreline changes in Gujarat. *National Seminar on Quaternary
Environments*, 1, 396–411.

Pandve, H. T. (2010). Climate change and coastal mega cities of India. *Indian Journal of
Occupational and Environmental Medicine*, 14(1), 22.

Peltier, W. R. (2009). Closure of the budget of global sea level rise over the GRACE era:
the importance and magnitudes of the required corrections for global glacial isostatic
adjustment. *Quaternary Science Reviews*, 28(17–18), 1658–1674.

Piecuch, C. G., Bittermann, K., Kemp, A. C., Ponte, R. M., Little, C. M., Engelhart, S. E., and
Lentz, S. J. (2018). River-discharge effects on United States Atlantic and Gulf coast sea-
level changes. *Proceedings of the National Academy of Sciences*, 115(30), 7729–7734.

Poscoe, E. H. (1964). *A Manual of the Geology of India & Burma*. Vol. III. New Delhi: Govt
of India.

Powar, K. B., and Patil, D. N. (1980). Structural evolution of the Deccan Volcanic Province:
A study based on LANDSAT-1 imageries. *Proceedings 3rd Indian Geology Congress*,
Poona, 235–253.

Prizomwala, S. P., Yadav, G., Bhatt, N., and Sharma, K. (2018). Late Pleistocene relative
sea-level changes from Saurashtra, west coast of India. *Current Science*, 115(12), 2297.

Quan, N. H., Toan, T. Q., Dang, P. D., Phuong, N. L., Anh, T. T. H., Quang, N. X., and Sea, W.
B. (2018). Conservation of the Mekong Delta wetlands through hydrological management. *Ecological Research*, 33(1), 87–103.

Rahman, M. M., Jiang, Y., and Irvine, K. (2018). Assessing wetland services for improved
development decision-making: a case study of mangroves in coastal Bangladesh.
Wetlands Ecology and Management, 26(4), 563–580.

Rahman, S., and Rahman, M. A. (2015). Climate extremes and challenges to infrastructure
development in coastal cities in Bangladesh. *Weather and Climate Extremes*, 7, 96–108.

Rajagopalan, G., Mittre, V., and Sekar, B. (1978). Birbal Sahni Institute radiocarbon measurements I. *Radiocarbon*, 20: 398–404.

Rajagopalan, G., Mittre, V., and Sekar, B. (1980). Birbal Sahni Institute radiocarbon measurements II. *Radiocarbon*, 22(1), 54–60.

Rajguru, S. N., and Guzder, S. J. (1973). A review of research on Quaternary sea level changes
and archaeological sites in India. *Bulletin of the Deccan College Research Institute*,
33(1/4), 193–207.

Rajendran, C. P., Rajendran, K., Machado, T., Satyamurthy, T., Aravazhi, P., and Jaiswal, M.
(2006). Evidence of ancient sea surges at the Mamallapuram coast of India and implications for previous Indian Ocean tsunami events. *Current Science*, 91, 1242–1247.

Rao, M. S. (1964). Some aspects of continental shelf sediments off the east coast of India.
Marine Geology, 1(1), 59–87.

Rao, K. M., and Rao, T. C. S. (1994). Holocene sea levels of Visakhapatnam shelf, east coast of India. *Journal of the Geological Society of India*, 44(6), 685–689.

Rao, P. C., Rajgopalan, G., Vora, K. H., and Almeida, F. (2003). Late Quaternary sea level and environmental changes from relic carbonate deposits of the western margin of India. *Proceedings of Indian Academy of Sciences*, 112, 1–25.

Rao, S. R. (1983). Sunken Ship and Submerged ports. *Science Today*, 17(7), 18–23.

Rao, N., Pandey, K., Kubo, S., Saito, S., Naga Kumar, Y., Demudu, K. C. V., Hema Malini, G., Nagumo, B., Nakashima Rei, N., and Sadakata, N. (2020). Paleoclimate and Holocene relative sea-level history of the east coast of India. *Journal of Paleolimnology*, 64, 71–89. DOI: 10.1007/s10933-020-00124-2.

Rao, V. P., Veerayya, M., Thamban, M., and Wagle, B. G. (1996). Evidences of Late Quaternary neotectonic activity and sea-level changes along the western continental margin of India. *Current Science*, 71(3), 213–219.

Rao, V. P., Veerayya, M., Nair, R. R., Dupeuble, P. A., and Lamboy, M. (1994). Late Quaternary Halimeda bioherms and aragonitic faecal pellet-dominated sediments on the carbonate platform 'of the western continental shelf of India. *Marine Geology*, 121(3–4), 293–315.

Raymo, M. E. (1992). Global climate change: A three million year perspective. In: Kukla, G. J. and Went, E. (Eds.), *Start of a Glacial*. NATO ASI Series, Vol. 1. Berlin, Heidelberg: Springer-Verlag, 207–223.

Roxy, M. K., Ghosh, S., Pathak, A., Athulya, R., Mujumdar, M., Murtugudde, R., and Rajeevan, M. (2017). A threefold rise in widespread extreme rain events over central India. *Nature Communications*, 8(1), 1–11.

Schmidt, C. W. (2015) Delta subsidence: An imminent threat to coastal populations. *Environmental Health Perspectives*, 123(8), A204–A209.

Schmitt, R., Rubin, Z., and Kondolf, G. (2017). Losing ground-scenarios of land loss as consequence of shifting sediment budgets in the Mekong Delta. *Geomorphology*, 294, 58–69.

Shackleton, N. J., Berger, A., and Peltier, W. R. (1990). An alternative astronomical calibration of the lower Pleistocene timescale based on ODP Site 677. *Transactions of the Royal Society of Edinburgh: Earth Sciences*, 81(4), 251–261.

Shahidul, M. (1995). Mid-Holocene environmental and vegetational changes in the coastal region of Bangladesh. In: McLelland, S. J., Skellern, A. R., and Porter, P. R. (Eds.), *Postgraduate Research in Geomorphology, Selected Paper from the 17th BGRG Postgraduate Symposium*. Leeds, UK: School of Geography, University of Leeds, pp. 14–20.

Shepard, F. P. (1963). Thirty-five thousand years of sea level. In Clements, T. (Ed.), *Essays in Marine Geology in Honor of K. O. Emery*. Los Angeles: University of Southern California Press, 1–10

Singh, A., Ramachandran, K., Samsuddin, M., Nisha, N., and Haneeshkumar, V. (2001). Significance of pteropods in deciphering the late Quaternary sea-level history along the southwestern Indian shelf. *Geo-Marine Letters*, 20(4), 243–252.

Slangen, A. B., Church, J. A., Agosta, C., Fettweis, X., Marzeion, B., and Richter, K. (2016). Anthropogenic forcing dominates global mean sea-level rise since 1970. *Nature Climate Change*, 6(7), 701–705.

Stammer, D., Cazenave, A., Ponte, R. M., and Tamisiea, M. E. (2013). Causes for contemporary regional sea level changes. *Annual Review of Marine Science*, 5, 21–46.

Stanley, J. D., Goddio, F., Jorstad, T. F., and Schnepp, G. (2004). Submergence of ancient Greek cities off Egypt's Nile Delta-A cautionary tale. *GSA Today*, 14(1), 4–10.

Stanley, J. D., and Toscano, M. A. (2009). Ancient archaeological sites buried and submerged along Egypt's Nile delta coast: Gauges of Holocene delta margin subsidence. *Journal of Coastal Research*, 25(1), 158–170.

Stocker, T. F., Qin, D., Plattner, G.-K., Tignor, M., Allen, S. K., Boschung, J., Nauels, A., Xia, Y., Bex, V., and Midgley, P. M. (Eds.). (2013). *Contribution of Working Group I to the Fifth Assessment Report of the Intergovernmental Panel on Climate Change.* Cambridge, UK: Cambridge University Press.

Stoddart, D. R., and Pillai, C. S. G. (1972). Raised reefs of Ramanathapuram, south India. *Transactions of the Institute of British Geographers*, 56, 111–125. DOI: 10.2307/621544.

Subramanian, V. (1996). The sediment load of Indian rivers - an update. Erosion and Sediment Yield: Global and Regional Perspectives, *Proceedings of the Exeter Symposium.* IAHS Publ. no. 236, 183–189.

Subramanya, K. R. (1996). Tectonic, Eustatic and Isostatic Changes along the Indian Coast. In Milliman, J. D. and Haq, B.U. (eds.), *Sea-Level Rise and Coastal Subsidence: Causes, Consequences and Strategies.* New York: Kluwer Academic Publishers, pp. 193–203.

Tamisiea, M. E. (2011). Ongoing glacial isostatic contributions to observations of sea level change. *Geophysical Journal International*, 186(3), 1036–1044.

Tamisiea, M. E., Hill, E. M., Ponte, R. M., Davis, J. L., Velicogna, I., and Vinogradova, N. T. (2010). Impact of self-attraction and loading on the annual cycle in sea level. *Journal of Geophysical Research: Oceans*, 115(C7). DOI: 10.1029/2009JC005687. https://agu-pubs.onlinelibrary.wiley.com/doi/full/10.1029/2009JC005687

Tamisiea, M. E., and Mitrovica, J. X. (2011). The moving boundaries of sea level change: Understanding the origins of geographic variability. *Oceanography*, 24, 24–39.

The Times of India. (2019a). India witnessing average sea level rise of 1.7mm/year, https://timesofindia.indiatimes.com/india/india-witnessing-average-sea-level-rise-of-1-7mm/year/articleshow/72134279.cmsm (Published on 20/11/2019).

The Times of India. (2019b). Sea may not submerge all of Mumbai, but there will be long, intense flooding, https://timesofindia.indiatimes.com/home/sunday-times/sea-may-not-submerge-all-of-mumbai-but-there-will-be-long-intense-flooding/articleshow/71986886.cms#:~:text=November%2010%2C%202019-,Sea%20may%20not%20submerge%20all%20of%20Mumbai%2C%20but,will%20be%20long%2C%20intense%20flooding&text=A%20high%2Dprofile%20study%20last,cities%20like%20Mumbai%20and%20Kolkata (Published on 10/11/2019). (Accessed 24th January 2021, 22.37 Hrs. IST).

The World Wildlife Fund. (2009). Mega-stress for mega-cities a climate vulnerability ranking of major coastal cities in Asia. PreventionWeb, p. 40. https://www.preventionweb.net/files/11822_megastresscitiesreport.pdf

Thompson, P. R., and Mitchum, G. T. (2014). Coherent sea level variability on the North Atlantic western boundary. *Journal of Geophysical Research: Oceans*, 119(9), 5676–5689.

Thompson, P. R., Piecuch, C. G., Merrifield, M. A., McCreary, J. P., and Firing, E. (2016). Forcing of recent decadal variability in the equatorial and North Indian Ocean. *Journal of Geophysical Research: Oceans*, 121(9), 6762–6778.

Unnikrishnan, A. S. (2010). Tidal propagation off the central west coast of India. *Indian Journal of Geo-Marine Sciences*, 39(4), 485–488.

Vaidyanathan, R. (1967). An outline of the geomorphic history of India, South of North latitude 18°, *Proceedings of the Seminar on Geomorphological Studies in India, Centre for Advanced Study in Geology*, Sagar, 121–130.

Verma, K. K., and Mathur, U. B. (1979). Quaternary sea-level changes along west coast of India. *Geological Survey of India Miscellaneous Publication*, 45, 263–272.

Vora, K. H., and Almeida, F. (1990). Submerged reef systems on the central western continental shelf of India. *Marine Geology*, 91(3), 255–262.

Wagle, B. G., Vora, K. H., Karisiddaiah, S. M., Veerayya, M., and Almeida, F. (1994). Holocene submarine terraces on the western continental shelf of India; implications for sea-level changes. *Marine Geology*, 117(1–4), 207–225.

Wehmiller, J. F., Harris, W. B., Boutin, B. S., and Farrell, K. M. (2012). Calibration of amino acid racemization (AAR) kinetics in United States Mid-Atlantic Coastal Plain Quaternary mollusks using 87Sr/86Sr analyses: Evaluation of kinetic models and estimation of regional late Pleistocene temperature history. *Quaternary Geochronology*, 7, 21–36.

Widdowson, M. (1997). Tertiary palaeosurfaces of the SW Deccan, Western India: implications for passive margin uplift. *Geological Society*, London, Special Publications, 120(1), 221–248.

Widdowson, M., and Cox, K. G. (1996). Uplift and erosional history of the Deccan Traps, India: Evidence from laterites and drainage patterns of the Western Ghats and Konkan Coast. *Earth and Planetary Science Letters*, 137(1–4), 57–69.

Wijeratne, E. M. S., Woodworth, P. L., and Stepanov, V. N. (2008). The seasonal cycle of sea level in Sri Lanka and southern India. *Western Indian Ocean Journal of Marine Science*, 7(1).

Wong, T. E., Bakker, A. M., and Keller, K. (2017). Impacts of Antarctic fast dynamics on sea level projections and coastal flood defense. *Climatic Change*, 144(2),347–364.

Woodworth, P. L., Melet, A., Marcos, M., Ray, R. D., Wöppelmann, G., Sasaki, Y. N., and Merrifield, M. A. (2019). Forcing factors affecting sea level changes at the coast. *Surveys in Geophysics*, 40(6), 1351–1397.

Wöppelmann, G., and Marcos, M. (2016). Vertical land motion as a key to understanding sea level change and variability. *Reviews of Geophysics*, 54(1), 64–92.

World Bank. (2020). *World Development Report 2020: Trading for Development in the Age of Global Value Chains*. Washington, DC: World Bank. DOI: 10.1596/978-1-4648-1457-0. License: Creative Commons Attribution CC BY 3.0 IGO.

Yatheesh, V., Bhattacharya, G. C., and Dyment, J. (2009). Early oceanic opening off western India–Pakistan margin: The Gop Basin revisited. *Earth and Planetary Science Letters*, 284(3–4), 399–408.

10 Evaluation of Recent and Paleo-Tsunami Deposits over Indian Subcontinents through Multiproxy Clues

Subodh Kumar Chaturvedi
Arba Minch University

Neloy Khare
Ministry of Earth Sciences

CONTENTS

INTRODUCTION

Tsunamis and storm surges are some of the natural events in coastal regions that leave a huge impact on the coast and coastal waters. The severe damage of corals and mangroves, negative impact on diversity and population of marine lives, disfiguring of coastal geomorphology, marine flooding of low-lying coastal areas, and agricultural land and irreparable loss of human life and property are some of the impacts in coastal regions due to such natural events.

The historical records and geological investigations have revealed that many flourished coastal towns got destroyed due to the impact of tsunami events (Rajendran et al. 2011, Wikipedia, n.d.-b). Jaiswal et al. (2008) prepared a catalogue of 90 tsunamis of the Indian Ocean regions that occurred from 326 BC to 2005 AD. They observed that the Sunda Arc in eastern side of Indian Ocean is the most active region for the tsunamis in the Indian Ocean. The source zones of remaining tsunamis are Andaman and Nicobar islands, Burma–Bangladesh region in the eastern side, while Makran accretion zone and Kutch–Saurashtra region are in the western side of the Indian Ocean. The historical observations of Tsunami indicate that eastern side of the Indian Ocean is most vulnerable to tsunami than the west.

Tsunami waves and storm surges on their path of propagation erode and transport huge amount of sediments from offshore and onshore regions, and deposit the same in these regions under conducive environment when energy of waves diminishes. Such records of tsunamis, at least the major ones, are preserved in the marine sediments and fauna contained within the onshore and offshore regions.

A number of researchers utilized the marine sediments and fauna contained within the onshore and offshore regions to infer the tsunami and storm surges of recent and past events (Khare and Chaturvedi 2005, 2006a, b, Hussain et al. 2006, Nigam and Chaturvedi 2006, Rajamanickam et al. 2007, Loveson et al. 2007, Srinivasalu et al. 2009, Gandhi et al. 2014, Chaturvedi and Rajamanickam 2016, Chaturvedi 2018, Salama et al. 2018). On the one hand, tsunamis leave their trails on human history; on the other hand, they often deposit economically valuable heavy minerals on the places of their impact in the coastal plains. Studies have revealed the enrichment of heavy minerals on the Central Tamil Nadu Coast, India, and other coastal regions just after the infamous Indian Ocean tsunami of December 26, 2004 (Angusamy et al. 2005, Rajamanickam et al. 2007).

Under such circumstances, the challenges before the geoscientists are to distinguish such events through sedimentological and faunal signatures. Further, the coastal regions have also been experiencing sea-level changes, shifting of river channels, coastal subsidence, neotectonic activities, etc. Hence, while inferring the paleo-storm and paleo-tsunami deposits, the influence of these factors needs to be taken into consideration.

In view of the above, the present paper evaluates the various proxies, including sediment textures, mineral compositions, and fossil assemblages utilized by the researchers to infer the recent, paleo-tsunami and paleo-storm sediment deposits over Indian subcontinents, and revisit the conceptual models to explain such natural phenomena.

STUDY AREA

Indian Ocean is the third-largest ocean of the world. It is bounded by India, Pakistan, and Iran to the north; the Arabian Peninsula and Africa to the west; and Australia, Sunda Island, and Malaysia to the east. Unlike other oceans, the Indian Ocean is landlocked to its north, which has important consequences on the ocean circulation. The Indian Ocean in its southwest is connected with the Atlantic Ocean, in its east with the equatorial Pacific Ocean through the deep passages of the Indonesian Sea, and in its south with the Southern Ocean.

The northern part of the Indian Ocean is made up of the large basins on either side of the Indian Peninsula, the Arabian Sea in the west, and the Bay of Bengal in the east which receive most of the river runoff from the Himalayas and form the largest deep-sea fans of the world ocean – the Bengal and Indus fans. The prominent rivers of Himalaya such as Indus, Ganges, Brahmaputra, and Irrawaddy Rivers have carved large canyons that extend to the steep continental slope below the shelf break.

The Indian Ocean is primarily characterized by passive continental margins, except for the Makran and Indonesian margins, which are active. The presence of a mid-ocean ridge and numerous aseismic ridges and microcontinents make the Indian Ocean unique (Figure 10.1). The important meridional aseismic ridges such as the Ninety East Ridge, the Chagos–Laccadive, Madagascar, and Mozambique plateaus in this region are not part of the global oceanic ridge system. The continental shelf of the Indian Ocean countries extends to an average width of about 120 km in the Indian Ocean, with its widest points (300 km) off Mumbai on the western coast of India. The island shelves are only about 300 m wide. The shelf break is at a depth of about 140 m (www.britannica.com). In general, the continental shelf of the east coast of India is narrow compared to that of the west coast of India, while the continental

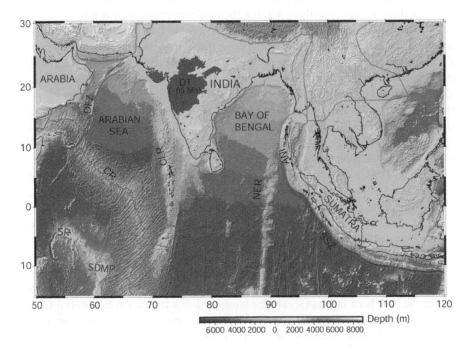

FIGURE 10.1 Map of Northern Indian Ocean. CR: Carlsberg Ridge; SDMP: Says De Malha Plateau; SP: Seychelles Plateau; NER: Ninety East Ridge; CLR: Chagos Laccadive Ridge; DT: Deccan Trap; OFZ: Owen Fracture Zone; SSZ: Sunda/Sumatra Subduction Zone; and ANI: Andaman and Nicobar Islands.

slope is wider in the east coast as compared to the west coast of India. The general surficial sediment distribution pattern of the continental shelf from coast towards the offshore consists of a sandy nearshore zone up to a water depth of ~20 m, followed by a narrow band of clayey silt in the inner shelf in <50 m of water depth. The outer shelf sediments are calcareous sand distributed between the depth of 50 and 100 m water depth, whereas the silty sand is dominant on the upper continental slope (Nair et al. 1978, LaFond 1957). A comparison of sediments of the continent shelf between east and west coast of India indicates that the exposure of the outer shelf relict sand on the east coast of India is confined to smaller areas, while the outer shelf relict sand on the western shelf of India is exposed all along the shelf. Moreover, the nearshore sand zone is wider and extends up to 35 m depth on the east coast, whereas on the west coast, it is narrow and extends only up to less than 10–15 m water depth (Rao and Kessarkar 2001). Various researchers such as Hadley (1964), Ingle and Ames (1966), Draper (1967), Silvester and Mogridge (1970), Swift et al. (1972), many papers in Swift et al. (1972), Stahl et al. (1974), and Schofield (1975) have argued that both the mutually dependent "relatively-undisturbed-relict sediment" and "no-transport-at-depth" hypotheses are largely, if not completely, erroneous. Schofield (1976) while reviewing the sedimentological data and hydrological information of the continental shelf off the east coast of Otago and Southland published by various researchers inferred that a belt of fine sediment between belts of nearshore and offshore coarser sediment does not necessarily mean that the offshore coarse belt is relict from some previous period of low sea level. The offshore coarse belt is related to an offshore current, and the intermediate fine belt is most likely the zone in which nearshore and offshore currents partly nullify each other. Despite the presence of an intermediate fine belt, there must be times when there is a transfer of sediment across it from the offshore coarse belt onto the coast. During stormy periods (most often in winter), the strong inshore-generated currents would tend to force the offshore current seaward similar to that observed with the Southland Current (Jillett 1969). This concept also explains the nearer-inshore position of the offshore current outside Little Omaha Bay, and its part penetration onto land during progradation of the Mangatawhiri Spit.

Schofield (1967) has observed the coastal penetration of the offshore current by incoming of offshore feldspar-enriched sand and explained a single pattern of sedimentation during coastal progradation. During this period of part penetration onto the coast, there were still some inshore-generated currents, and hence the intermediate fine belt, although narrower, persisted.

The circulation in the Northern Indian Ocean is unique because it experiences strong seasonal winds called the monsoons. Hence, unlike in much of the Pacific and Atlantic Oceans, the currents in the Northern Indian Ocean change with season. In the Northern Indian Ocean, the westward-flowing North Equatorial Current (NEC) is prominent in January and March when the northeast monsoon is fully established. It runs as a narrow current from Malacca Strait to Southern Sri Lanka, where it bends southward between 2°S and 5°N in the region between 60°E and 75°E. The South Equatorial Current (SEC) occupies the region south of 8°S. Between the two westward flows, the Equatorial Counter Current (ECC) is occupied. The northeast monsoons drive the water along the coast of Bay of Bengal to circulate in an anticlockwise direction. Similarly, the water along the coast of Arabian Sea also circulates in an anticlockwise circulation. In summer, due to the effects of the strong southwest monsoon and the absence of the northeast trades, a strong current flows from west to east, which completely obliterates the NEC. Hence, there is no counter-equatorial current as well. Thus, the circulation of water in the northern part of the ocean is clockwise during summer (www.pmfias.com).

Source and impact of tsunami waves on coastal oceanography and coastal geomorphic features are described in the following sections.

Source of the Tsunami Waves

The sources for the generation of tsunami waves may be the earthquake-related tectonic activities, volcanic eruptions, submarine landslides, meteorological impact, explosion, or combined impact of these sources. Thrust-type earthquakes along subduction zones that cause vertical movement of the ocean floor are tsunamigenic. Many offshore faults could generate tsunami due to earthquake and undersea landslide associated with the earthquake. Most tsunamis to the tune of approximately 70%–80% are mainly caused by large earthquakes that occur in the marine realm (Tappin 2010, Kremer et al. 2012). The subduction zones and zones of compression where tsunamis usually occur are found on both western and eastern parts of the Northern Indian Ocean. These zones are Sunda Arc, Andaman–Nicobar islands, and Burma–Bangladesh region in the eastern side, while Makran accretion zone and Kutch–Saurashtra region are in the western part of the Northern Indian Ocean.

In order to emphasize the factors responsible for the generations of tsunamis, some of the well-documented tsunamis that occurred in human history due to earthquakes, volcanic eruptions, and landslides are discussed below.

The December 26, 2004, Sumatra–Andaman earthquake of magnitude 9.3 generated at the subduction plate boundary where the Indian and Australian plates converge and plunge below the Sunda Plate led to a 15-m slip of the ocean floor along a 1,300 km-long and 160–240-km-wide rupture that generated 30-m-high tsunami (Rastogi and Jaiswal 2006). The earthquake (followed by the tsunami) was felt in Bangladesh, India, Malaysia, Myanmar, Thailand, Sri Lanka, and the Maldives (Løvholt et al. 2006).

The August 27, 1883, volcanic eruption of Krakatau in Indonesia led to the generation of one of the deadliest tsunamis of 35-m-high tsunami in western Java and southern Sumatra. Tsunami waves were observed throughout the Indian Ocean, the Pacific Ocean, the American West Coast, South America, and even as far away as the English Channel. The Krakatau volcanic eruption of 416 AD as recorded in ancient Japanese scriptures generated a series of catastrophic tsunamis, which must have been much greater than those generated in 1883. The time of tsunami with a wave height of several meters that affected Tamil Nadu in India matches with this early Krakatau eruption.

The undersea landslides or submarine mass failures (SMFs) are yet another possible trigger of tsunamis that poses threats to several coastal cities that are particularly built on or near massive layers of sediment put down by major rivers (Lace 2008, Tappin 2010). However, the magnitude of tsunamis is mainly dependent on water depth of failure, SMF volume and failure mechanism, cohesive slump or fragmental landslide, the sediment type, and rate of sedimentation together with its postdepositional alteration, etc. SMFs occur in numerous environments, including the open continental shelf, submarine canyon/fan systems, fjords, active river deltas, and convergent margins. The devastating tsunami of July 1998 at Sissano Lagoon in Papua New Guinea, in which 2,200 people died, was found to be caused by submarine landslides (Tappin et al. 1999). The catastrophic seabed failure on the Gulf Coast of Mississippi delta was caused by Hurricane Camille in 1969 that led to the collapse of three offshore drilling platforms. The seabed sediment failure resulted in a change of seabed relief of up to 12 m (Bea et al. 1983).

IMPACT OF TSUNAMI ON COASTAL OCEANOGRAPHIC AND COASTAL GEOMORPHIC FEATURES

Tsunami waves near the place of their generation disturb a water column of huge volume. When tsunami waves travel from their source, these waves transport seawater of deep ocean to several hundreds and thousands of kilometers to the various coastal regions of the world. When the deep ocean water mixes with the coastal water, the physicochemical properties of coastal water change significantly. This sudden change in coastal water properties affects the coastal marine lives. The tsunami tidal waves transport large volumes of seawater into inland and affect the coastal geomorphology and ecology of inland. Sugawara et al. (2008) noticed the typical current velocity of tsunami runup near the coast to be around several meters per second, and the inundation depth is several tens to hundreds of centimeters. Such currents have a sufficient strength for the removal of the material from the preexisting surface. Manickraj et al. (2005) compared the beach profile of pre-tsunami data with that of the post-tsunami data of Central Tamil Nadu Coast, India, and observed a total change in the beach equilibrium of the different stations manifested only after tsunami. They recorded the highest erosion of beach in post-tsunami at places by 454 M³/M, and this was attributed to the narrow shelf topography of the region, whereas an accretionary trend on the order of 92 M³/M at some other places was attributed to the presence of riverlet and the wider shelf topography, which must have acted as the controlling factors for the changes brought in by tsunami waves in the beach morphology. Murthy et al. (2006) while studying the factors guiding the tsunami surge in the worst-affected coast of Tamil Nadu in east coast of India during the very infamous tsunami of December, 26, 2004, have inferred that the wider shelf, submarine canyons with a concave coastline and concave morphology of the shelf, close proximity to the source of the tsunami, are some of the reasons that have devastated the coast during the tsunami. The submarine canyons, river channels in the tsunami-affected region, and low coastal plains have facilitated the inundation of coastal plains by tsunami waves. On the other hand, mangrove forest, beach sand dunes, coastal plantations, and higher-elevation coastal plains have acted as barriers to absorb the energy of tsunami waves and minimize the damages caused by tsunami waves in other coastal regions.

Chandrasekar et al. (2007) analyzed the pore water chemistry in the beach sands of Central Tamil Nadu, by using atomic absorption spectrometry and graphite furnace, and established a typical trend of variation for the zones of erosion, deposition, and neutral zone in the study area. The elemental concentrations of heavy metals like copper, nickel, cobalt, and lead were found to be extremely high, more than the average seawater. Such higher concentration was surmised to have been the contribution of mixing up of metal-rich waters from deeper portion of the sea. The calcium and potassium distributions were found to be normal, whereas sodium in the pore waters was seen with very high order of enrichment. Devi and Shenoi (2012) provided a comprehensive report of December 26, 2004, Indian Ocean tsunami and its effects on coastal morphology and ecosystems. Tsunami waves destroy the corals in the nearby shallow areas and affect the diversity and population of marine lives, mainly fishes, on which hundreds of millions of coastal-dwellers depend for their livelihood. When giant tsunami waves smash onto the shores, the coastal plains are inundated, and the massive backwash that returns to sea carrying a deadly cargo of sediments results in choking of the live coral reefs of the region and destroys it. The tsunami surges also destroy the mangroves, releasing silt, sediments, nutrients, and pollutants that may cause significant damages to ecosystems having a long-lasting effect. Ramachandran et al. (2005) while studying the ecological impact of the Indian Ocean Tsunami of December, 26, 2004, on Nicobar Islands observed that the mangrove ecosystems got damaged to the extent of 51%–100%, coral reef 41%–100%, and forest ecosystems 6.5%–27%. Considering the severity of damages and their consequences, a definite restoration ecology program was suggested.

Anilkumar et al. (2006) studied the physical oceanographic conditions along the east and west coasts of India immediately after the Indian Ocean Tsunami of December 26, 2004, and observed changes in thermocline, mixed-layer depth, salinity, and turbidity conditions of the coastal water. They observed the mixed-layer deepening (>50 m) along the east coast of India and downward tilt in the thermocline towards north in the southeast coast. They further observed two layers of turbidity: one around 40 m water depth and the other around 250 m water depth. The shallow turbidity layer is associated with high chlorophyll concentration, but the deeper layer noticed at shallow stations off the west and east coasts of India could be due to the resuspension of the sea floor sediments due to turbulence generated by the tsunami.

Marghany (2014) while utilizing Moderate Resolution Imaging Spectroradiometer (MODIS) satellite data on sea surface salinity along Banda Aceh Coastal Waters inferred that during pre-tsunami event, the isohaline contours ranged between 28.5 and 29.0 psu. However, the isohaline contours were found to be increased to 36.7 psu during tsunami event and continued to increase to 37 psu. This clearly indicates that the tsunami 2004 has significant impacts on the sea surface salinity (SSS) because of high sediment deposit concentrations which added more salts and minerals to coastal waters of Banda Aceh.

Conceptual framework for the deposition of tsunami and storm surge sediments: Several coastal regions of the world that are prone to and have been affected by tsunamis are frequently hit by the cyclones and storm surges every year. Hence, the sediments deposited by these two events in onshore and offshore regions need to be broadly categorized in light of hydrodynamic conditions, sea-floor topography, sediment nature, and coastal geomorphology. The deposition of tsunami sediments in onshore and offshore regions depends on magnitude and intensity of tsunami waves, distance of coast from tsunami generation, seafloor configurations, wave energy imparted on the sea floor and coast, availability and nature of sediments, the entrainment of sediments in tsunami waves and their transportation, inundation distance in coastal regions, coastal morphology, etc. Tsunami waves that travel at a very high speed of 500–1,000 km/h with a large wavelength of ~500 km, a low amplitude of <1 m, and a large wave period of about 3,000 seconds in deep sea change their characteristics when they approach to shallow sea. In shallow sea when tsunami wave contacts with the seafloor, the speed of tsunami waves decreases but succeeding waves push the water to the coast, resulting further increase in wave height due to wave compression. The pushing of the succeeding waves towards the coast shortens the interval between the wave trains and amplifies their amplitude; as a result, tsunami energy keeps piling on along with the increase in wave height and volume of water. The tsunami height near the coast can reach several tens of meters (up to 30 m height) in the most severe cases. Further, the duration of the water-level rise increases and the inundation area widens due to large wave characteristics of tsunamis. According to Bryant (2001), about 60% of the increase in wave height takes place in the last 20 m depth change near the coast. When tsunami approaches to shallow sea, the initial kinetic energy decreases and the potential energy increases but the total mechanical energy at the initial and final stages remains equal. Therefore, when the tsunami waves approach the coast and water depths progressively reduce, the wave energy in the water column is concentrated over a much shorter distance horizontally and wave begins to interact significantly with the sea floor. Particle velocities near the coast reach a value of about 7 m/s or ~ 25 km/h (Shetye 2005). It is with such velocities that the waters associated with a tsunami move towards a beach and inundate the coast for several hundred of meters to several kilometers inland.

At its final phase of a runup, a tsunami wave uses almost all of its kinetic energy, which is converted into the potential energy of the water mass. The potential energy is immediately reconverted into kinetic energy; thus, a backwash is induced. The backwash of a tsunami erodes the surface, including previously formed tsunami layers, local soil, and other material, and transports them during backwash. The direction of the backwash is not always opposite to that of the runup, as the backwash often concentrates on topographic depressions, such as channels and rivers, causing severe erosion (Kon'no 1961). Intensity of tsunami backwash varies according to the topography of the current path. As the grain size of runup deposits directly reflects the granulometry of the source area, so that these deposits show poor sorting (ranging from fine sand to large cobbles), whereas tsunami deposits formed by backwash are commonly composed of finer particles such as sandy mud. Sand layers by tsunami runup are often covered by a mud cap, which is associated with the stagnation of water with suspended matters under relatively calm conditions (Gelfenbaum and Jaffe 2003, Sato et al. 1995).

The presence of successive tsunami layers with fining-upward sequences is evidence of the repetition of tsunami incursions. The hydraulic energy of a tsunami generally decreases with each new incursion; thus, the thickness and the mean particle diameter of the sandy runup deposits decrease upward, resulting in multiple fining-upward sequences (e.g., Gelfenbaum and Jaffe 2003, Nishimura and Miyaji 1995, Shi et al. 1995). Clay- and silt-sized particles suspended within the water may begin to settle down during a relatively calm interval between wave trains. Thus, mud layers often cover the previously deposited coarser sediment (Takashimizu and Masuda 2000).

Submarine tsunami deposits typically show evidence of significant hydrodynamic energy and oscillations of runup and backwash. Based on tsunami video clippings, numerical model, textural, geochemical, mineralogical, and magnetic results of two sediment cores collected off Nagapattinam, Tamil Nadu, India, in the water depths of 5 and 10 m, Veerasingam et al. (2014) developed a tsunami sediment depositional model. The deposition model shows four progressive steps: pre-tsunami stage, tsunami stage (wave propagation landward and inundation), depositional stage (backwash flows and related gravity-driven processes seaward), and post-tsunami stage. Figure 10.2a–d shows all the four stages of the depositional model. This model also shows that the backwash current has transported and deposited sediments from a landward source during the tsunami event to the shallow water offshore regions.

Like tsunamis, tropical cyclones impact the coastal regions by bringing strong winds and rain to both the land and sea. Storm surge causes a widespread erosion from the sea and transports the sediment landward or seaward to the open sea. During cyclonic events, the strong storm winds push the seawater towards the coastal inland to a long distance for few hundred of meters to several tens of kilometers consistently for long hours. The 5- to 6-m storm surge during 1999 Odisha supercyclone had brought water up to 35 km inland, carrying along with its coastal debris

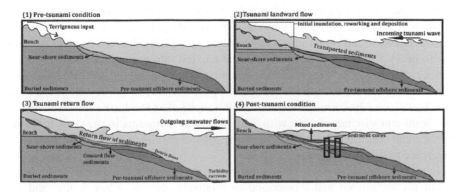

FIGURE 10.2 A tsunami sediment depositional model in the coastal region during pre-tsunami stage, tsunami stage, backwash flow stage, and post-tsunami stage. (Veerasingam et al. 2014.)

and inundated towns and villages (Wikipedia, n.d.-a). Under such circumstances, the strength of backwash in the events may be weak; hence, the sediment transportation during backwash in offshore regions may be less as compared to the tsunami events. Conley and Beach (2003) noticed that during storms, offshore-directed transport of suspended sediment occurs but onshore-directed transport may also occur near the bottom. Hence, seaward transport of sediment from land provides a potential new source of sediment for offshore areas, in addition to sediment inputs from runoff. Under normal condition, low wave energy may result in onshore transport, while higher wave energy is usually associated with offshore transport due to strong offshore currents near the bottom (Warren 2017).

While some sediment from land or nearshore may be advected offshore, it is important to remember that on the shelf overall, most subaqueous sediment is material eroded from and redeposited on the shelf. Storm surge ebb transport of coastal sediment can be key to the inner shelf (Goff et al. 2004, 2010). Where inner shelf materials are transported seaward, sandy deposits may be recognized over muddy strata (Chaumillon et al. 2017).

Tropical cyclones are capable of eroding relatively deeper water sediments and fauna contained within from continental shelves. Chang et al. (2001), Bentley et al. (2002), and Xu et al. (2016) noticed that the strong shear stress generated by hurricane-induced currents at the bottom boundary of the shallow continental shelf can resuspend sediments from as deep as the 100-m isobath. It can mix the water column down to great depths and resuspend massive volumes of sediments on the continental shelves (Bianucci et al. 2018). They observed the highest concentration of suspended matter over the shelves within 50-m isobath from coast during Hurricane Rita that made landfall on September, 24, 2005, in the Gulf of Mexico. They noticed a large increase in surface total suspended matter (TSM) (2.4 g/m^3 or 53%) in the shelf during the hurricanes period followed by further increase in TSM to the tune of additional 42% (1.9 g/m^3) in the following week. However, by the third week, TSM concentrations declined due to settling and nearly approached their pre-storm concentrations. The two main sources of surface water TSM over the shelf during or after hurricane have been noticed, first the sediments resuspended from the seafloor by hurricane-generated bottom shear (Chang et al. 2001, Goñi et al. 2006) and the additional suspended matter eroded from the land by river systems due to heavy rain during the hurricane and carried into the region. Their observation was supported by rain and SSS signatures near the mouth of the Mississippi River. A mean decrease in SSS during hurricane was due to storm-associated rainfall. The second freshening around days 9 and 10 after the hurricanes was due to the freshwater discharge along with the associated sediments in the shelf from the river catchment, after about 2 weeks of the typical residence time of freshwater in watershed and Lower River.

In such high-energy environment during cyclonic events, the deeper water depth sediments are eroded and transported to relatively shallow water depths and get deposited in a low-energy environment. Under such circumstances, the top layer of seafloor sediments representing the time of storm event must have been eroded first from deeper depths followed by the erosion and transportation of progressively older sediments in the subsequent storm events. Conversely, at the deposition site

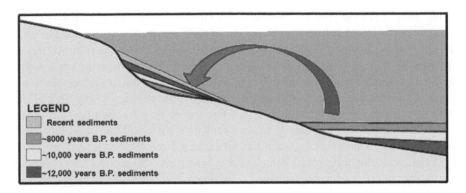

FIGURE 10.3 Erosion and transportation of sediments from deeper water depth under severe storm condition and deposition of the same at shallow water depth in low-energy environment. (After Nigam and Chaturvedi 2006.)

in shallow water depths of calm environment, the younger sediments would have been deposited first at the bottom followed by the progressively older sediments on the top, giving rise to the inverted sequence of deposition. The similar situation of an inverted sequence of deposition in three shallow water sediment cores collected between 10 and 20 m of water depths off Kachchh, Gujarat, India, in the Northern Arabian Sea, has been observed (Chaturvedi 2001, Nigam and Chaturvedi 2006). Figure 10.3 shows the erosion and transportation of sediments from deeper water depth and deposition of the same at shallow water depth in the low-energy environment supporting the hypothesis for the occurrence of inverted sequence of sediment deposition.

TSUNAMI DEPOSITS IN ONSHORE AND OFFSHORE REGIONS

ONSHORE TSUNAMI DEPOSITS

Coastal flats are the best accessible site for investigating the effect of tsunamis. The direct observations and measurement on the process of erosion, transportation, and deposition by the tsunami waves and their impact could be investigated in the coastal regions. Certain coastal geomorphological features such as coastal lakes and lagoons have a high preservation potential of tsunami deposits as such regions have in general a calm and stable sedimentation environment, whereas certain other coastal flats such as beaches are highly dynamic and are the regions of erosion. In addition to it, the major part of tsunami deposits are easily disturbed and removed due to the effect of wind, precipitation, anthropogenic activities, and other processes; hence, they have little chance of preservation potential. The characteristics of onshore tsunami deposits vary from region to region depending upon the local bathymetry and topography, sources of sediment, transport processes, etc.

As the runup of tsunami spreads landward and exhausts the kinetic energy, the mean diameter of the sedimentary particles gradually decreases landward. The lateral changes in thickness and particle size of the tsunami layer are good indicators of the direction of the current. Therefore, the landward thinning and fining of a tsunami layer is the major criterion for the identification of a tsunami runup. The sharp decrease in the velocity of the tsunami-induced bottom current just above the sea floor in the bottom boundary layer results in a very strong shearing on sea floor sediments. This implies that the effect of shearing is at a much greater depth in a tsunami than in violent wind-induced deep-water waves. After lifting-up and during transportation in the main, nonsheared layer, even very fragile materials are not broken. A tsunami loses its energy when it runs up a coast and inundates the coastal area essentially because of the water movement in a direction that is "against" gravity. Friction on the bottom and percolation of water into the ground also contribute to the energy loss of a runup tsunami. The depth (the thickness of water column) and the speed of the water flow decrease until the flow stops and starts to turn over and flow in the opposite direction, towards the sea. Consequently, the runup part of a tsunami site thins out inland and leaves only a very thin sedimentary layer. The thicknesses of tsunami deposits typically decrease inland (Atwater 1987), so that they form a gradually tapering wedge (Dawson et al. 1988), and the mean grain size of these deposits also decreases with distance from the sea (Benson et al. 1997). Tsunami deposits commonly cover coastal flats to several hundreds of meters from the coastline to even few kilometers with sand layers. However, its thickness varies from a few millimeters to several tens of centimeters. The particle size of a tsunami layer often decreases upward, i.e., from coarse sand at the base to fine sand at the top (Benson et al. 1997, Gelfenbaum and Jaffe 2003, Shi et al. 1995). This is considered to be a result of the gradual settling from suspension. At the time of sediment transport, coarser particles are likely to exist near the bottom, while finer ones can be distributed throughout the water column. A structureless (massive or homogeneous) tsunami layer may reflect the severe mixing of sediment and water due to high turbulence and rapid deposition.

Tsunami waves erode, transport, and deposit sediments and fauna contained within. Foraminifera, exclusively marine microorganisms and extremely sensitive towards environmental/sea-level changes (Boltovskoy and Wright 1976), play a significant role in geological, oceanographic, and paleoceanographic studies. Many researchers (Bandy 1964, Wantland 1975, Nigam 1986, Martin and Liddell 1988) proposed that benthic foraminifera are ideal candidates for tracing sediment movement because of their abundance, restricted habitats, and small size. By diagnostic assemblages as well as their size distribution, it could be possible to track sediment movement (Li et al. 1997, Collins et al. 1999, Hippensteel and Martin 1999). After December 26, 2004, Indian Ocean tsunami, the post-tsunami sediments were collected across the beach from the mid-tide to 175 m inland of tsunami-inundated cultivated land of New Wandoor region of Andaman and Nicobar Island, and studied for foraminifera distributions (Khare and Chaturvedi 2006b, Khare et al. 2009). It was observed that the foraminiferal assemblage is dominated by benthic foraminifera. Total foraminiferal numbers decreased from beach towards inland. The percentage of living specimen tends to diminish from the present coastline towards inland. The presence of living specimen only at the stations closer to coast line could be due to the persistent marine influence in that particular region. The inland regions that were subjected to marine influence due to tsunami waves only for a short while witnessed the dead yet modern foraminifera. The foraminiferal diversity in sediment samples suggested the presence of beach-inner shelf realm sediments over the cultivated fields. Kon'no (1961) identified several kinds of marine microfossils in the deposit left by the 1960 Chilean earthquake tsunami and clarified that the sediment was derived from the nearshore areas. He concluded that the tsunami deposits did not originate in the deep sea. In the case of the 1993 southwest Hokkaido earthquake tsunami, Nanayama et al. (1998, 2000) studied the foraminifers from the tsunami sediments and suggested that the tsunami deposit was derived from the sea floor of about 50–60 m water depths.

Chaturvedi and Rajamanickam (2016) studied the pre- and post-tsunami surface sediment samples collected from different geomorphic units along 35-km-long coastal stretch of the Central Tamil Nadu Coast, India, for foraminiferal investigations to trace the tsunami signature. While comparing the foraminifera of the pre-tsunami sediments with those of the post-tsunami sediments, they observed a significant increase in total foraminiferal numbers along with the occurrences of several new genera and species of foraminifera in the post-tsunami sediments in the area where the beach width was wide. The presence of the appreciable number of forams along with the addition of several new species in post-tsunami sediments in the study area is a definite indicator that forams have been brought from the offshore regions and deposited in the study area. The occurrences of inner shelf species of foraminifera such as *Asterorotalia trispinosa*, *Ammonia dentate*, *Ammonia tepida*, *Amphistegina rediata*, *Quinqueloculina seminulum*, *Spiroloculina* sp., and *Nonion* sp., in the post-tsunami sediments of the study area indicate that the sediments have been transported from inner shelf regions. Their study clearly demonstrated that the likelihood chances of preservation of paleotsunami deposits in wider beaches are much more compared to the narrower beach sites. Therefore, the wider beaches may be considered to be the suitable site for paleotsunami investigations.

Chandrasekaran et al. (2005) observed an upshot in the percentage of carbonate content from 3.12% to 6.90% in the post-tsunami samples highlighting the fresh deposition of detritus and molluscan shells from the offshore. Shelf topography has controlled the changes in the sedimentology of the beach sediments as evidenced by the occurrence of medium-sized sands on the beaches facing narrow shelf, whereas fine sands are deposited on the beaches where the shelf is wider. Rajamanickam et al. (2007) observed that the number of polymodalities in the distribution of heavy minerals during pre- and post-tsunami sediments that prevailed is uniform except the abundance of heavy minerals in post-tsunami sediments in the study area. Angusamy et al. (2005) observed an increase in heavy mineral enrichment from the north to the south in the study area. They inferred that the source for the distribution of heavy minerals remains to be the same, that is, the shelf sediment. They concluded that the presence of low-gradient, wider inner shelf must have retarded the tsunami waves, and further, the steep gradient of the beach supported by the dunes in this area must have stalled the tsunami waves impact by dissipating the energy. According to Kon'no (1961), particle sorting in a tsunami deposit is generally poor. The distribution of the particle sizes of tsunami deposits can be unimodal or multimodal. The multimodal grain-size distribution in tsunami deposits is considered to reflect the nature of the source material and the processes of tsunami sedimentation (Shi et al.

1995). For example, a bimodal grain-size distribution of a tsunami deposit is attributed to the activity of two different transport processes within a single flow, that is, suspended load and bed load (Minoura et al. 1996).

Offshore Tsunami Deposits

Shallow marine environments have the higher sedimentary preservation potential than onshore and wave-dominated nearshore regions; hence, this region can house tsunami records. On the other hand, an erosion surface may develop in some places in this environment also because of the very high-erosion energy of tsunami waves. The depths of sediment sources can obviously vary, depending on the magnitude of the tsunami, the local bathymetry and topography, and the distribution of sediment. A number of researchers have reported that the tsunami deposits are largely derived from the nearest source, such as the local coastal substratum, without transport over a long distance (Sato et al. 1995, Shi et al. 1995, Nanayama et al. 2000, Gelfenbaum and Jaffe 2003). According to the results of Gelfenbaum and Jaffe (2003), the volume of deposition is twice the volume of erosion in nearshore areas, while Sato et al. (1995) suggested that the volume of erosion could account for the volume of tsunami deposits on land. The sediment source and characteristics of a tsunami control the components of a tsunami deposit. The frequency curves of the grain-size particles in tsunami deposits show relatively well to poorly sorted distributions.

Khare and Chaturvedi (2006b) in a post-tsunami sediment sample observed 3-cm-thick clay layer atop coral sand at a shallow depth of 18 m water depth collected off Nancowry in Andaman and Nicobar Islands. The presence of coral sand confirms the coral reef (shallower) environment. Moreover, the present 18 m water depth would have been actually shallower before tsunami event, which caused land subsidence and sea-level rise. Possibly when the tsunami subsided and water started receding, the tsunami waves must have lost the energy gradually. Under high-energy conditions, even larger particles can be moved to quite distant regions, whereas with the gradual reduction in energy, the larger (sandy-silt) particles could not have been carried to much distances and in the process were deposited in onshore to nearshore regions and only finer particles (silt-clay) could be carried with the retreating water up to a depth of the present study area or even beyond. However, the possibility of transportation of finer sediments from relatively deeper regions to the sampling site due to tsunami may not be totally ruled out.

Veerasingam et al. (2014) studied the textural, mineralogical, geochemical, and rock-magnetic characteristics of two sediment cores collected off southeast coast of India at 5 and 10 m water depths. They inferred that the tsunami deposit is relatively rich in sand, quartz, feldspar, carbonate, SiO_2, TiO_2, K_2O, and CaO, and depleted of clay and iron oxides. They suggested that most of the quartz-rich coarse sediments have been transported from beach area through the backwash towards offshore by the tsunami wave.

During the transportation of sediments from relatively deeper water depth by such high-energy environment, the fauna along with the sediments can also be displaced and get deposited in low-energy regime in shallow regions (Schiebel et al. 1995, Li et al. 1997, Nigam and Chaturvedi 2006, Chaturvedi 2018). Chaturvedi

(2001) and Nigam and Chaturvedi 2006 studied the three nearshore sediment cores collected at water depths between 10 and 20 m from the cyclone-prone coastal area of Kachchh, Gujarat, to reconstruct the comprehensive overview of the paleoclimatic events of the study area. The foraminiferal study and radiocarbon dating of these three sediment cores revealed an interesting result. They observed 70 cm thick of ~10,000 to ~12,000 years BP fine-grained sediment deposits sandwiched between ~8,000 years BP, sediments at the bottom, and the normal detrital sediments on the top. Further, a second similar deposition of ~10,000-year-old sediments over a normally deposited ~7,000-year-old sequence was observed in the core. The second inverted sequence was subsequently overlain by a normal sequence (Figure 10.4). These older sediments aged between 10,000 and 12,000 years BP, which raised an interesting question about their origin. Various possibilities about the origin of older sediments deposited on the younger sediments in the study area were discussed below that have led to develop a hypothesis and the model to explain the deposition of inverted sequences in the study area (Figure 10.3).

Possibility 1: Transportation of older sediments from land to the study area: If sediments would have devoid of foraminifera, it could have been concluded that they were originally deposited in lacustrine environment and later transported to the present site. But this is not the case. These sediments contain smaller-sized foraminifera with clay, and therefore, they are sediments of marine origin. Among various possibilities, let us consider that these marine sediments were deposited on land and later eroded and transported to the present site. But the sea-level position during

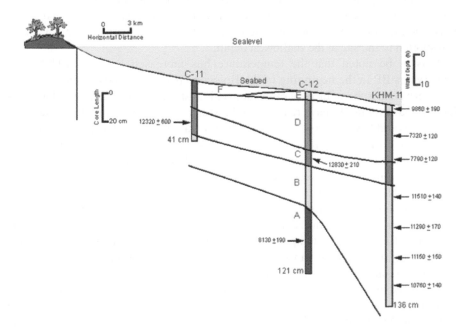

FIGURE 10.4 Schematic diagram showing the location of cores, radiocarbon dating, and zonation sequences in the cores based on foraminiferal cluster analysis. (Nigam and Chaturvedi 2006.)

~10,000 years BP was about 60 m lower than the present sea level (Hashimi et al. 1995) and has crossed the present level about 6,000 years BP in Arabian Sea and other places (Katupotha and Fujiwara 1988, Pirazzoli 1991). Therefore, a pool of older marine sediments on land is ruled out.

Possibility 2: Transportation of older sediments of comparatively deeper sea to the study area: The second possibility is that originally, sediments were deposited at greater depth and later transported to the present site at shallow depth. This means sediment transportation is against slope that requires high energy. Studies documenting the onshore transport of sediment by storm are common (Graus et al. 1984, Ball et al. 1987, Macintyre et al. 1987, Jones and Hunter 1992, Scoffin 1993, Hernandez-Avila et al. 1997, Li et al. 1997). However from Indian region, such reports are limited. In view of this, for the inverted sequence of deposition in the study area, it may be hypothesized that the older sediments might have been transported from the deeper region and deposited at the present site under the storm condition. The study area is known to be a storm-prone zone. Therefore, the present results may reflect the paleostorm surges in this area. Alternatively, the inverted sequence might have been resulted from neotectonic activity. It can be hypothesized that due to faulting, area covering core locations (i.e., nearshore area) subsided and deeper side has become upthrow site. Thus, sediments eroded from the upthrow site might have deposited in downthrow site and gave rise to inverted sequence. In this case, the sediments are transported against the gravity, which requires a lot of energy. Such energy is possible through storm surges. The storm can churn the sediments along with fauna that can go into suspension. When the suspended sediments travel towards the land, the coarse sediments may not travel much distance due to large particle size and heavy mass; on the other hand, fine material along with the small fauna might have been transported to the site at the shallowest depth.

It is to be noted that the temperature has increased between 10,000 and 8,000 years BP in the beginning of the Holocene (Coughan and Neyenzi 1991) in Northern Hemisphere. In view of the above, it is surmised that around 8,000 years BP the sea level might have reached to the study area, and during the same time, the area would have undergone through tectonic activity, which caused depression around the coring sites. This depression was later filled with the reworked sediments eroded from the deeper region. It is well known that the coiling direction in many foraminiferal species is temperature-controlled (Nigam 1988, Nigam and Rao 1989, Nigam and Khare 1992). Chaturvedi (2018) studied the coiling direction of the foraminiferal species *Rotalidium annectens* of one of the cores collected at 20 m water depth from the study area. *Rotalidium annectens* prefers the dextral coiling with high temperature. Since the temperature increased with time during early Holocene, it can be deduced that the original older sediments contained lesser number of dextral forms followed by an increase in the younger sediments in the upper portion of the cores. But it is observed that the dextrally coiled specimens of this species gradually decrease upward within inverted sequence though, under normal conditions, an upward increase was expected. Thus, this anomaly provides a clue for reasoning, because in early Holocene the temperature rose gradually with time. Therefore, when these sediments were eroded, transported, and redeposited as inverted sequence at the core location, higher values of the dextral forms were observed at the bottom of the core (which corresponds to originally younger sediments) and less values in upper portion of the sediment core. This observation supports the conclusion that ~8,000 years back, some physical forces such as storms must have played a major role. Hence, it is hypothesized that this inverted sequence formed due to high-energy events such as storms/tsunami action that eroded the original normal sequence of paleoshelf sediments of early Holocene (~8,000 years BP), lifted, transported, and redeposited them at the present-day shallower depth.

The rate of sedimentation is very slow in this area, and one interesting observation that can be noted in the present study is the missing (or very thin layer) of sedimentary sequence after 6,000 years BP to the present. This could be due to the upliftment of land; most of the small rivers that used to flow in this part of the sea either have dried or have changed their course and therefore terrigenous supply stopped/reduced. The drying of rivers and disappearing of Saraswati River once flowing through this region and discharging its water into the Arabian Sea are well documented (Valdiya 1996). Though the biogenic materials might have been deposited, either they are too less to have any signature or they are removed from the site of deposition. The published report revealed that sea level was higher by few meters (2–6 m) about 6,000 years BP than the present (Hashimi et al. 1995). It is also believed that ~3,500 years BP city of Dwarka existed (on reclaimed land as mentioned in Bhagavat Gita) which later drowned. Now remnants of this city have been discovered at ~10 m water depth (Rao 1999). Coupling these phenomena with our observations leads us to conclude that ~3,500 years BP relative sea level reached to the lowest (at least up to ~10 m water depth) after a high at ~6,000 years BP. This lowering must have exposed a large part of the present shallow water and thus eroded the loose sediments.

Distinction between tsunami and storm deposits: Tsunami deposits are not identifiable on the basis of exclusive and diagnostic criteria because other kinds of sediment deposits by storm surges, gravity flows, tidal waves, and sea-level changes share some of their characteristics (Pinegina and Bourgeois 2001). Particularly, storm surges are a major candidate for an alternative interpretation because of the area that they affect and the order of wave energy can be comparable with those of tsunamis, and this holds also for the resulting deposits. Distinguishing between tsunami and storm deposits has therefore been a study object for a long time already (Dawson et al. 1991, Dawson and Shi 2000). Although the distinction between tsunami and storm deposits is difficult, the characteristics of tsunamis offer possible criteria for the identification. Tsunamis are characterized by long-wave properties. This is responsible for the relatively long period of wave oscillations with strong, enduring currents around coastal areas. It results in repetitive runup and backwash, as well as in larger runup distances and heights. On the other hand, storm surges do not have long-wave properties; thus, the periodicity of wave repetition is shorter, and the runup distance and height are smaller (Nott 2004, Pinegina and Bourgeois 2001). During cyclonic events, the traction and bed load transport of sediment are dominant, whereas the suspension mode of transport is dominant during tsunami. Cyclonic events are associated with heavy rainfall; hence, the suspended materials brought from the river catchments by rivers and streams to the offshore regions are much more than during the tsunami events. Dawson et al. (1991) suggested that erosion is more significant than deposition during storms and suggested that there is a low probability of deposition by storms. In the case study of the 1993 southwest Hokkaido earthquake tsunami and the 1959 Miyakojima typhoon in northern Japan, Nanayama et al. (2000) showed that the tsunami deposits were composed of four layers of sand by landward and seaward-directed flow; the storm deposits, on the other hand, comprised only one layer from a flow that was exclusively landward. They reported that the tsunami and storm deposits were similar in that their thickness decreased landward in both cases. According to paleontological analysis, the tsunami deposits were derived from the sea bottom at the depths of 50–60 m, whereas the storm deposits were composed of well-sorted marine sand derived from coastal areas. Their grain-size analyses illustrated that the storm deposits were better sorted than any of the tsunami deposits. Goff et al. (2004) reported significant differences between ancient tsunami and modern storm deposits in New Zealand. The storm deposits are not extensive and show no relationship between the inundation distance and their thickness. They noted that the sorting of the storm deposit was better than that of the tsunami deposit, and that the mean grain size of the storm deposit was larger than that of the tsunami deposit.

The widespread deposition of sand by a major tropical cyclone has been described by Nott (2004). The storm deposit was composed of tabular medium- to coarse-grained cross-bedded sands, shells of various sizes, and coral fragments. Most remarkable is that the sand sheet extends inland for 300 m. The major storm did not affect a boulder accumulation, which is interpreted as having been deposited by prehistoric tsunamis. There is no doubt that major storms are capable of forming sand layers that somewhat resemble tsunami deposits. The wave energy of major storm surges seems to be locally comparable to tsunamis. However, sand sheets are a common feature in tsunami-devastated lowlands such as coastal marshes and lagoons. In general, onshore tsunami deposits commonly exhibit (1) relatively coarser sediments such as sandy or gravelly layers within a peat succession or within lacustrine sediments, (2) the presence of fining-upward sequences or graded structure, and (3) a landward thinning of wedge-shaped sediment sheets (Dawson et al. 1988). Minoura and Nakaya (1991) also pointed out that the seawater that floods onto land as a result of a tsunami incursion causes distinctive processes of sedimentation in coastal marshes and lagoons.

SUMMARY

Tsunami waves cause severe damage to life and properties in coastal regions. Such waves not only destroy the coastal habitats and original coastal landforms but change substantially the physicochemical and biological conditions of coastal water. Hence, the prediction of the occurrences of tsunami events is one of the challenging tasks before the geoscientist. The Northern Indian Ocean is one of the tsunami-prone regions of the world due to the presence of the number of active subduction zones and faults; hence, it would have preserved several episodes of tsunami deposits in onshore and offshore regions. Several micropaleontological, sedimentological, mineralogical, and geochemical signatures distinguish tsunami deposits from the sediments deposited in normal condition. However, the effect of sea-level changes, shifting of river channels, neotectonic activities, sedimentation during storm, etc., should be taken into considerations while interpreting the tsunami deposits. A multiproxy approach is suggested to study the paleotsunami deposits.

ACKNOWLEDGMENTS

The authors are grateful to the President of the Arba Minch University, Ethiopia, and the Secretary, Ministry of Earth Sciences, New Delhi, for giving their permission to publish this paper. We are thankful to Dr. A. K. Chaubey for his help in the preparation of the North Indian Ocean map.

REFERENCES

Angusamy, N., Soosai Manickaraj, D., Chandrasekaran, R., Chandrasekar, N., Loveson, V. J., Gujar, A. R. and Victor Rajamanickam, G. 2005. Heavy mineral distribution in the beaches of Nagapattinam District, Tamilnadu, India. In: *Development Planning of Coastal Placer Minerals*, eds. V. J. Loveson, N. Chadrasekar and A. Sinha, Allied Publishers Pvt. Ltd., New Delhi, 80–92.

Anilkumar, N., Sarma Y. V. B., Babu, K. N., Sudhakar, M. and Pandey, P. C. 2006. Post-tsunami oceanographic conditions in southern Arabian Sea and Bay of Bengal. *Current Science*, 90:421–427.

Atwater, B. F. 1987. Evidence for great Holocene earthquakes along the outer coast of Washington State. *Science*, 236:942–944.

Ball, M. M., Shinne, E. A. and Stockman, K. W. 1987. The geological effects of Hurricane Donna in South Florida. *Geology*, 75:583–597.

Bandy, O. L. 1964. Foraminiferal biofacies in sediments of the Gulf of Batabona, Cuba and their geologic significance. American. *Association of Petroleum Geology Bulletin*, 48:1666–1676.

Bea, R. G., Wright, S. G., Sicar, P. and Niedoroda, A. W. 1983. Wave-induced slides in South Pass Block 70, Mississippi delta. *Journal of Geotechnical Engineering*, 109:619–644. doi: 10.1061/(ASCE)07339410(1983)109:4(619.

Benson, B. E., Grimm, K. A. and Clague, J. J. 1997. Tsunami deposits beneath tidal marshes on northwestern Vancouver Island, British Columbia. *Quaternary Research*, 48:192–204.

Bentley, S. J., Keen, T. R., Blain, C. A. and Vaughan, W. C. 2002. The origin and preservation of a major hurricane event bed in the northern Gulf of Mexico: Hurricane Camille, 1969. *Marine Geology*, 186:423–446.

Bianucci, L., Balaguru, K., Smith, R. W., Leung, L. R. and Moriarty, J. M. 2018. Contribution of hurricane-induced sediment resuspension to coastal oxygen dynamics. *Nature*, 8:15740. doi: 10.1038/s41598-018-33640-3.

Boltovskoy, E. and Wright, R. 1976. *Recent Foraminifera*, Dr. W. Junk B.V. Publishers, The Hague, p. 515.

Bryant, E. 2001. *Tsunami: The Underrated Hazard*, Cambridge University Press, Cambridge, UK.

Chandrasekar, N., Gujar, A. R. and Loveson, V. J., et al. 2007. Pore water chemistry in the beach sands of Central Tamil Nadu, India. In: *Exploration, Exploitation, Enrichment and Environment of Coastal Placer Minerals (PLACER 2007)*, eds. V. J. Loveson, P. K. Sen and A. Sinha, Macmillan India Ltd., New Delhi, pp. 278–288.

Chandrasekaran, R., Angusamy, N., Manickaraj, D. S., Loveson, V. J., Gujar, A. R., Chandrasekar, N. and Victor Rajamanickam, G. 2005. Grain size distribution and annual variation along the beaches from Poompuhar to Nagoor, Tamilnadu, India. *In: Development Planning of Coastal Placer Minerals*, eds. V. J. Loveson, N. Chadrasekar and A. Sinha, Allied Publishers Pvt. Ltd., New Delhi, pp. 156–170.

Chang, G. C., Dickey, T. D. and Williams III, A. J. 2001. Sediment resuspension over a continental shelf during Hurricanes Edouard and Hortense. *Journal of Geophysical Research*, 106:9517–9531.

Chaturvedi, S. K. 2001. Distribution and ecology of foraminifera in Kharo creek and adjoining shelf area off Kachchh, Gujarat, Unpublished PhD Thesis, Goa University, p. 366.

Chaturvedi, S. K. 2018. Differential coiling in Rotalidium annectens - foraminiferal testimony for Paleostorms/Paleotsunami: A case study from Kachchh, Gujarat, India. *Indian Journal of Geo Marine Sciences*, 47(07):1455–1459.

Chaturvedi, S. K. and Rajamanickam, G. V. 2016. Implication of foraminifera in tracing the suitable site for Palaeo-Tsunami impressions: A case study from Central Tamil Nadu Coast. *International Journal of Earth Sciences and Engineering*, 9(6):2390–2394.

Chaumillon, E., Bertina, X., Fortunato, A. B., et al. 2017. Storm-induced marine flooding: Lessons from a multidisciplinary approach. *Earth-Science Reviews*, 165:151–184.

Collins, E. S., Scott, D. B. and Gayes, P. T. 1999. Hurricane records on the South Carolina coast: Can they be detected in the sediment record? *Quaternary International*, 56:15–26.

Conley, D. C. and Beach, R. A. 2003. Cross-shore sediment transport partitioning in the nearshore during a storm event. *Journal of Geophysical Research*, 108(C3):3065.

Coughlan, M. and Nyenzi, B.S. 1991: Climate Trends and Variability. *Proceedings of the Second World Climate Conference*. Cambridge University Press, 71–82.

Dawson, A. G., Foster, I. D. L., Shi, S., Smith, D. E. and Long, D. 1991. The identification of tsunami deposits in coastal sediment sequences. *Science of Tsunami Hazards*, 9(1):73–82.

Dawson, A. G., Long, D. and Smith, D. E. 1988. The Storegga slides: Evidence from eastern Scotland for a possible tsunami. *Marine Geology*, 82:271–276.

Dawson, A. G. and Shi, S. 2000. Tsunami deposits. *Pure and Applied Geophysics*, 17:875–897.

Devi, E. U. and Shenoi, S. S. C. 2012. Tsunami and the effects on coastal morphology and ecosystems: A report. *Proceedings of the Indian National Science Academy*, 78:513–521.

Draper, I. 1967. Wave activity at the sea bed around north-western Europe. *Marine Geology*, 5:133–140.

Encyclopaedia Britannica. n.d. Indian Ocean. https://www.britannica.com/place/Indian-Ocean, accessed on 09 July 2020.

Gandhi, S. M., Jisha, K. and Rajeshwara Rao, N. 2014. Recent benthic foraminifera and its ecological condition along the surface samples of Pichavaram and Muthupet Mangroves, Tamil Nadu, East Coast of India. *International Journal of Current Research and Academic Review*, 2:252–259.

Gelfenbaum, G. and Jaffe, B. 2003. Erosion and sedimentation from the 17 July, 1998 Papua New Guinea tsunami. *Pure and Applied Geophysics*, 160:1969–1999.

Goff, J., McFadgen, B. G. and Chague-Goff, C. 2004. Sedimentary differences between the 2002 Easter storm and the 15th-century Okoropunga tsunami, southeastern North Island, New Zealand. *Marine Geology*, 204:235–250.

Goff, J., Allison, M. A. and Sean, P. S. G. 2010. Offshore transport of sediment during cyclonic storms: Hurricane Ike (2008), Texas Gulf Coast, USA. *Geology* 38(4):351–354. doi: 10.1130/G30632.1.

Goni, M. A., Gordon, E. S., Monacci, N. M., Clinton, R., Gisewhite, R., Allison, M. A. and Kineke, G. 2006. The effect of Hurricane Lili on the distribution of organic matter along the inner Louisiana shelf (Gulf of Mexico, USA). *Continental Shelf Research*, 26:2260–2280.

Graus, R. R., Mcintyre, I. G., Herchenroder, B. E. 1984. Computer simulation of the reef zonation at Discovery Bay, Jamaica. Hurricane disruption and long-term physical oceanography controls. *Coral Reefs*, 3:59–68.

Hadley, M. 1964. Wave-induced bottom currents in the Celtic Sea. *Marine Geology*, 2:164–167.

Hashimi, N. H., Nigam, R., Nair, R. R. and Rajagopalan, G. (1995). Holocene sea level fluctuation on Western Indian continental margin: An update. *Journal of the Geological Society of India*, 46:157–162.

Hernandez-Avila, M. L., Roberts, H. H. and Rouse, L. J. 1997. Hurricane generated waves and coastal boulder rampart formation. *Proceedings of Third International Coral Reef Symposium, Miami*, 2:71–78.

Hippensteel, S. P. and Martin, E. 1999. Foraminifera as an indicator of overwash deposits, Barrier Island sediment supply, and Barrier Island evolution: Folly Island, South Carolina. *Palaeogeography, Palaeoclimatology, Palaeoecology*, 149:115–125.

Hussain, S. M., Krishnamurthy, R., Gandhi, M. S., Ilayaraja, K., Ganesan, P. and Mohan, S. P. (2006). Micropalaeontological investigations on tsunamigenic sediments of Andaman Islands, *Current Science*, 91:1655–1667.

Ingle, J. and Ames, C. Jr. 1966. *The Movement of Beach Sand*, Elsevier, Amsterdam. 221 p.

Jaiswal, R. K., Rastogi, B. K. and Murty Tad, S. 2008. Tsunamigenic sources in the Indian Ocean. *Science of Tsunami Hazards*, 27(2):33–53.

Jillett, J. B. 1969. Seasonal hydrology of waters off Otago Peninsula, south-eastern New Zealand. *New Zealand Journal of Marine and Freshwater Research*, 3:349–375.

Jones, B. and Hunter, I. G. 1992. Very large boulders on the coast of Grand Cayman: The effects of giant waves on rocky coastlines. *Journal of Coastal Research*, 8:763–774.

Katupotha, J. and Fujiwara, K. 1988. Holocene sea level changes on the southwest and south coast of Sri Lanka. *Palaeogeography, Palaeoclimatology, Palaeoecology*, 68:189–203.

Khare, N. and Chaturvedi, S. K. 2005. Need to investigate coastal marine sediments to study the impact of tsunami. *Journal Geological Society of India*, 66:768–771.

Khare, N. and Chaturvedi, S. K. 2006a. Presence of marine microfossils (foraminifera) in distant onshore region of New Wandoor (south andaman island) – An indication of tsunami induced marine transgression. *Indian Journal of Earth Sciences*, 33:70–75.

Khare, N. and Chaturvedi, S. K. 2006b. A note on the clay layer atop coral sand at shallow depth off Nancowry (Andaman & Nicobar). *Bulletin of Environmental Science*, 4:53–56.

Khare, N., Chaturvedi, S. K. and Ingole, B. 2009. Tsunami induced transportation of the coastal marine sediments to distant onshore regions: Some indications from foraminiferal and microbenthic studies of New Wandoor region (Andaman & Nicobar). *Disaster Advances*, 2:7–13.

Kon'no, E. (ed.) 1961. Geological observation of the Sanriku coastal region damaged by the tsunami due to the Chile earthquake in 1960. *Contribution of Institute of Geology and Paleontology, Tohoku University*, 52:1–45 (in Japanese with English abstract).

Kremer, K., Simpson, G. and Girardclos, S. 2012. Giant Lake Geneva tsunami in AD 563. *Nature Geoscience*, 5:756–757.

Lace, W. W. 2008. *Great Historic Disaster: The Indian Ocean Tsunami of 2004*, Chelsea House Publishers, New York.

LaFond, E. C. 1957. Oceanographic studies in the Bay of Bengal. *Proceedings of the Indian Academy of Sciences–Section B*, 46:1–47.

Li, C., Jones, B. and Blanchon, P. 1997. Lagoon shelf sediment exchange by storms - Evidence from Foraminiferal assemblages east coast of Grand Cayman, British West Indies. *Journal of Sedimentary Research*, 67:17–25.

Loveson, V. J., Gujar, A. R., Rajamanickam, G. V., Chandrasekar, N., Manickaraj, D.S.,
    Chandrasekaran, R., Chaturvedi, S.K., Mahesh, R., Josephine, P.J., Deepa, V., Sudha,
    V., and Sunderasen, D. 2007. Post tsunami rebuilding of beaches and the texture of sedi-
    ments. In *Exploration, Exploitation, Enrichment and Environment of Coastal Placer
    Minerals (PLACER 2007)*, eds. V. J. Loveson, P. K. Sen and A. Sinha, Macmillan India
    Ltd., New Delhi, pp. 131–146.
Løvholt, F., Bungum, H., Harbitz, C. B., Glimsdal, S., Lindholm, C. D. and Pedersen, G.
    2006. Earthquake related tsunami hazard along the western coast of Thailand. *Natural
    Hazards and Earth System Sciences*, 6:979–997. doi: 10.5194/nhess-6-979-2006.
Macintyre, I. G., Graus, P. R., Reinthal, P. N., Litter, M. M. and Litter, D. S. 1987. The barrier
    reef sediment apron: Tobacco reef, Belize. *Coral Reefs*, 6:1–12.
Manickraj, D. S., Chandrasekaran, R., Gujar, A. R., Loveson, V. J., Angusamy, N.,
    Chandrasekar, N. and Rajamanickam, G. V. 2005. Stability of the beaches in
    Nagapattinam District, Tamil Nadu, India. In *Development Planning of Coastal Placer
    Minerals*, eds. V. J. Loveson, N. Chadrasekar and A. Sinha, Allied Publishers Pvt. Ltd.,
    New Delhi, pp. 171–179.
Marghany, M. 2014. Simulation of Tsunami impact on sea surface salinity along Banda
    Aceh Coastal Waters, Indonesia. Intech, Open Science, pp. 229–251. Open access
    peer-reviewed   chapter.   https://www.intechopen.com/books/advanced-geoscience-
    remote-sensing/simulation-of-tsunami-impact-on-sea-surface-salinity-along-banda-
    aceh-coastal-waters-indonesia
Martin, R. E. and Liddell, W. D. 1988. Foraminifera biofacies on a north coast fringing reef
    (1-75), Discovery Bay, Jamaica. *Palaios*, 3:298–314.
Minoura, K., Gusiakov, V. G., Kurbatov, A., Takeuti, S., Svendsen, J. I., Bondevik, S. and
    Oda, T. 1996. Tsunami sedimentation associated with the 1923 Kamchatka earthquake.
    Sedimentary Geology, 106:145–154.
Minoura, K. and Nakaya, S. 1991. Traces of tsunami preserved in inter-tidal lacustrine and
    marsh deposits: Some examples from northeast Japan. *Journal of Geology*, 99:265–287.
Murthy, K. S. R., Subrahmanyam, A. S., Murty, G. P. S., Sarma, K. V. L. N. S., Subrahmanyam,
    V., Mohana Rao, K., Suneetha Rani, P., Anuradha, A., Adilakshmi, B. and Sri Devi,
    T. 2006. Factors guiding tsunami surge at the Nagapattinam–Cuddalore shelf, Tamil
    Nadu, east coast of India. *Current Science*, 90:1535–1538.
Nair, R. R., Hashimi, N. H., Kidwai, R. M., Gupta, M. V. S., Paropkari, A. L., Ambre, N. V.,
    Muralinath, A. S., Mascarenhas, A. and D'Costa, G. P. 1978. Topography and sediments
    of the western continental shelf of India-Vengurla to Mangalore. *Indian Journal of
    Marine Sciences*, 7:224–230.
Nanayama, F., Shigeno, K., Satake, K., Shimokawa, K., Koitabashi, S., Miyasaka, S. and
    Ishii, M. 2000. Sedimentary differences between the 1993 Hokkaido-nansei-oki
    tsunami and the 1959 Miyakojima typhoon at Taisei, southwestern Hokkaido, northern
    Japan. *Sedimentary Geology*, 135:255–264.
Nanayama, H., Satake, K., Shimokawa, K., Shigeno, K., Koitabashi, S., Miyasaka, S. and
    Ishii, M. 1998. Sedimentary facies and sedimentation process of invading tsunami
    deposit—example from the 1993 southwest Hokkaido earthquake tsunami. *Kaiyo
    Monthly*, 15:140–146 (in Japanese).
Nigam, R. 1986. Foraminiferal assemblages and their use as indicators of sediment movement:
    A study in the shelf region off Navapur, India. *Continental Shelf Research*, 5:421–430.
Nigam, R. 1988. Reproductive behaviour of benthic foraminifera: A key to palaeoclimate.
    Proceedings of the Indian National Science Academy, 54:585–594.
Nigam, R. and Chaturvedi, S. K. 2006. Does inverted depositional sequences and alloch-
    thonous foraminifers in sediments along the coast of Kachchh, NW India, indicate
    palaeostorm and/or tsunami effects? *Geo Marine Letters*, 26:42–50. doi: 10.1007/
    s00367-005-0014-y.

Nigam, R. and Khare, N. 1992. The reciprocity between coiling direction and dimorphic reproduction in benthic foraminifera. *Journal of Micropalaeontology*, 11:221–228.

Nigam R. and Rao A. S. 1989. The intriguing relationship between coiling direction and reproductive mode in benthic foraminifera. *Journal of the Palaeontological Society of India*, 34:1–6.

Nishimura, Y. and Miyaji, N. 1995. Tsunami deposits from the 1993 southwest Hokkaido earthquake and the 1640 Komagatake eruption, northern Japan. *Pure and Applied Geophysics*, 144:719–733.

Nott, J. 2004. The tsunami hypothesis—Comparisons of the field evidence against the effects, on the Western Australian coast, of some of the most powerful storms on Earth. *Marine Geology*, 208:1–12.

Pinegina, T. K. and Bourgeois, J. 2001. Historical and paleo-tsunami deposits on Kamchatka, Russia: Long-term chronologies and long-distance correlations. *Natural Hazards and Earth System Sciences*, 1:177–185.

Pirazzoli, P. A. 1991. *World Atlas of Holocene Sea-level Changes*, Elsevier, Amsterdam, 300 pp.

Rajamanickam, G. V., Chandrasekaran, R., Soosai Manickaraj, D., Gujar, A. R., Loveson, V. J., Chaturvedi, S. K., Chandrashekhar, N. and Mahesh, R. 2007. Source rock indication from the heavy mineral weight percentages, central Tamil Nadu, India. In *Exploration, Exploitation, Enrichment and Environment of Coastal Placer Minerals (PLACER 2007)*, eds. V. J. Loveson, P. K. Sen and A. Sinha, Macmillan India Ltd., New Delhi, pp. 75–86.

Rajendran, C. P., Rajendran, K., Srinivasalu, S., Andrade, V., Aravazhi, P. and Sanwal, J. 2011. Geoarchaeological evidence of a Chola period tsunami from an ancient port at Kaveripattinam on the south-eastern coast of India. *Geoarchaeology*, 26:867–887. doi: 10.1002/gea.20376.

Ramachandran, S., Anitha, S., Balamurugan, V., Dharanirajan, K., Ezhil Vendhan, K., Preeti Divien, M. I., Senthil Vel, A., Hussain, I. S. and Udayaraj, A. 2005. Ecological impact of tsunami on Nicobar Islands (Camorta, Katchal, Nancowry and Trinkat). *Current Science*, 89(1):195–200.

Rao, S.R. 1999. *The lost city of Dvaraka*, Aditya Prakashan, New Delhi, 157 pp.

Rao, V. P. and Kessarkar, P. M. 2001. Geomorphology and geology of the Bay of Bengal and the Andaman Sea. In *The Indian Ocean: A perspective*, eds. R. Sen Gupta and E. Desa, Oxford & IBH, New Delhi, 2:817–868.

Rastogi, B. K. and Jaiswal, R. K. 2006. A catalog of tsunamis in the Indian Ocean. *Science of Tsunami Hazards*, 25:128–143.

Salama, A., Meghraoui, M., El Gabry, M., Maouche, S., Hussein, M. H. and Korrat, I. 2018. Paleotsunami deposits along the coast of Egypt correlate with historical earthquake records of eastern Mediterranean. *Natural Hazards and Earth System Sciences*, 18:2203–2219. doi: 10.5194/nhess-18-2203-2018.

Sato, H., Shimamoto, T., Tsutsumi, A. and Kawamoto, E. 1995. Onshore tsunami deposits caused by the 1993 Southwest Hokkaido and 1983 Japan Sea earthquakes. *Pure and Applied Geophysics*, 144:693–717.

Schiebel, R., Hiller, B. and Hemleben, Ch. 1995. Impacts of storms on Recent planktic foraminiferal test production and CaCO3flux in the North Atlantic at 47 °N, 20 °W (JGOFS). *Marine Micropaleontology*, 26:115– 129.

Schofield, J. C. 1967. Sand movement at Mangatawhiri Spit and Little Omaha Bay. With appendices by P. E. Hyde, W. R. Ponder, H. R. Thompson, and J. C. Schofield. *New Zealand Journal of Geology and Geophysics*, 10:697–731.

Schofield, J. C. 1975. Sea-level fluctuations cause periodic, post-glacial progradation, South Kaipara Barrier, North Island, New Zealand. *New Zealand Journal of Geology and Geophysics*, 18:295–316.

Schofield, J. C. 1976. Sediment transport on the continental shelf, east of Otago—A reinterpretation of so-called reuct features. *New Zealand Journal of Geology and Geophysics*, 19(4):513–525. doi: 10.1080/00288306.1976.10423542.

Scoffin, T. P. 1993. The geological effects of hurricanes on coral reefs and the interpretation of storm deposits. *Coral Reefs*, 12:203–221.

Shetye, S. R. 2005. Tsunamis: A large-scale earth and ocean phenomenon. *Resonance*, 10, 8–19.

Shi, S., Dawson, A. G. and Smith, D. E. 1995. Coastal sedimentation associated with the December 12th, 1992 tsunami in Flores, Indonesia. *Pure and Applied Geophysics*, 144:525–536.

Silvester, R. and Mogridge, G. R. 1970. Reach of waves to the bed of the continental shelf. *Proceedings of the 12th Coastal Engineering Conference*, Washington, DC, Vol. 2, 651–665. (Published by the American chapter 1 Society of Civil Engineers, New York.)

Srinivasalu, S., Rajeshwara Rao, N., Thangadurai, N., Jonathan, M. P., Roy, P. D., Ram Mohan, V. and Saravanan, P. 2009. Characteristics of 2004 tsunami deposits of the northern Tamil Nadu coast, southeastern India, *Boletín de la Sociedad Geológica Mexicana*, 61:111–118.

Stahl, L., Koczan, J. and Swift, D. 1974. Anatomy of a shoreface-connected sand ridge on the New Jersey Shelf: Implications for the genesis of the shelf surficial sand sheet. *Geology*, 2:117–120.

Sugawara, D., Minoura, K. and Imamura, F. 2008. Tsunamis and tsunami sedimentology. In *Tsunamiites: Features and Implications*, eds. T. Shiki, Y. Tsuji, T. Yamazaki and K. Minoura, 1st edition. Elsevier, London, UK, pp. 10–45.

Swift, D. J. P., Duane, D. B. and Pilkey, O. H. (Eds.). 1972. *Shelf Sediment Transport: Process and Pattern*. Dowden, Hutchinson and Ross, Stroudsburg, PA.

Takashimizu, T. and Masuda, F. 2000. Depositional facies and sedimentary successions of earthquake-induced tsunami deposits in Upper Pleistocene incised valley fills, central Japan. *Sedimentary Geology*, 135:231–239.

Tappin, D. R. 2010. Submarine mass failures as tsunami sources: their climate control. *Philosophical Transactions of the Royal Society*, 368:2417–2434.

Tappin, D.R., Matsumoto, T., Watts, P., Satake, K., McMurtry, G. M., Lafoy, Y., Tsuji, Y., Kanamastu, T., Lus, W., Lwabuchi, Y., Yeh, H., Matsumotu, Y., Nakamura, M., Mahoi, M., Hill, P., Crook, K., Anton, L. and Walsh, J.P. 1999. Sediment slump likely caused 1998 Papua New Guinea Tsunami. *Eos, Transactions American Geophysical Union*, 80:329.

Valdiya, K. S. 1996. Saraswati that disappeared. *Resonance*, 1:19–28.

Veerasingam, S., Venkatachalapathy, R., Basavaiah, N., Ramkumar, T., Venkatramanan, S. and Deenadayalan, K. 2014. Identification and characterization of tsunami deposits off southeast coast of India from the 2004 Indian Ocean tsunami: Rock magnetic and geochemical approach. *Journal of Earth System Science*, 123:905–921.

Wantland, K. F. 1975. Distribution of Holocene benthonic foraminifera on the Belize Shelf. In *Belize shelf: Carbonate sediments, Ecology*, eds. K. F. Wantland and W. C. Pusey, American Association of Petroleum Geologists Studies in Geology, Tulsa, OK, pp. 322–399.

Warren, W. F. 2017. Storm-Induced Nearshore Sediment Transport. Unpublished Dissertation, Florida Atlantic University, 62pp.

Wikipedia. n.d.-a. 1999 Odisha cyclone. en.wikipedia.org/wiki/1999_Odisha_cyclone, accessed on 02 August, 2020.

Wikipedia. n.d.-b. Lost city. en.wikipedia.org/wiki/Lost_city, accessed on 02 August, 2020.

Xu, K., Mickey, R.C., Chen, Q., Harris, C.K., Hetland, R.D., Hu, K. and Wang, J. 2016. Shelf sediment transport during hurricanes Katrina and Rita. *Computers & Geosciences*, 90:24–39.

11 Ostracod Diversity from Continental Slope Sediments of Gulf of Mannar, India
Bathymetry and Ecological Variations

A. Rajkumar, S. M. Hussain, S. Maniyarasan,
Mohammed Noohu Nazeer, and K. Radhakrishnan
University of Madras

CONTENTS

INTRODUCTION

In recent years, the importance given to environmental and climatic changes in the scientific literature is being increased; organisms that provide a proxy record of changes are particularly valuable. The living assemblages in the marine realm exhibit changes in their ecological characteristics concerning the environmental changes. It appears that evidently, microfossils can be very well correlated to bring out these changes ensued. Ostracoda such as tiny crustaceans are exceptional among calcareous microfossils in that they are commonly found in Quaternary deposits of marine, transitional, and nonmarine environments. Thus, they can be very successfully used as paleoenvironmental and paleoclimate proxies (Julio and Francisco, 2012). Ostracoda forms help in ascertaining the paleoenvironmental and ecological characteristics because of their minute size, abundance, wide geographic distribution, and sensitiveness to the fluctuations that happened in a sedimentary environment of different ages (Mohammed Nishath et al., 2015). The composition and structure of microorganisms are influenced by sediment characteristics (Snelgrove and Butman, 1994).

Ostracoda population and distribution vary with the environmental settings. An increase in diversity in Ostracoda with time was not steady (Sara and Robin, 2012), because of the varying factors influencing the Ostracod growth. Ostracod size projects the macroevolutionary as well as the macroecological behaviors (Gene et al., 2010). These crustaceans have been studied in the past decades for their distribution, ecology, and use in paleoenvironmental studies; the reader may refer to the following contributions (Hong et al., 2019; Jeremy et al., 2019; Brautovic et al., 2018; Kulkoyluoglu et al., 2017; Anne-Sophie et al., 2016; Valls et al., 2015; Frezza and Bella, 2015; Hokuto et al., 2014; Kresna, 2014; Cabral and Loureiro, 2013; Nicole et al., 2013; Martin et al., 2013; Lili et al., 2012; Munif and Dietmar, 2012; Andre et al., 2012; Theodora, 2012; Mazdygan and Chavtur, 2011).

In the present work, we have discussed the distribution of ostracods from continental slope sediments of Gulf of Mannar, although this region has recorded a brief scientific data on taxonomy of shelf ostracods (Hussain, 1998; Sridhar et al., 2019; Rajkumar et al., 2020). This study provides the detailed account of deep-sea Ostracod diversity with bathymetric and ecological variations.

STUDY AREA

The Gulf of Mannar is a large shallow bay-forming part of the Laccadive Sea in the Indian Ocean. It lies between the southeastern tip of India and the west coast of Sri Lanka, in the Coromandel Coast region. It has been announced as India's first Marine Biosphere Reserve and covers an area of about $10,500\,km^2$ that serves as the home for a wide range of organisms nurturing over 3,600 species (Rao et al., 2008; Thanikachalam and Ramachandran, 2003; Kumaraguru, 1999). The Gulf of Mannar is enriched with three definite marine ecosystems, namely, the coral ecosystem, seagrass ecosystem, and mangrove ecosystem. The occurrence of these specialized ecosystems makes Gulf of Mannar a unique large marine ecosystem in the Indian subcontinent (ICMAM, 2001). The prevalence of monsoons is well experienced in the Gulf of Mannar; especially, the intensity of northeast monsoon is higher. There are two courses of drift in water currents seen in Gulf of Mannar: between April and September, the movement is from south to north, and vice versa in October to December (Chacko, 1950). Two cores, namely, MC-2 and MC-60, were collected in the geographical coordinates 08°25′40.50″N 78°43′46.06″E and 08°01′08.77″N 79°05′53.90″E, respectively, from the Gulf of Mannar. Bathymetry map with core location is shown in Figure 11.1.

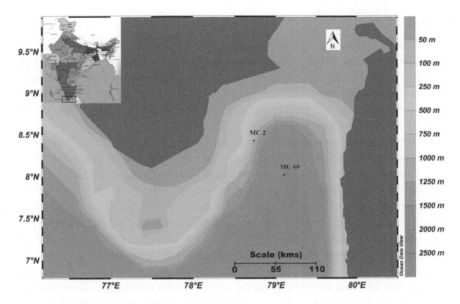

FIGURE 11.1   Bathymetry map showing the sample locations from Gulf of Mannar.

MATERIALS AND METHODS

The Oceanographic Research Vessel Sindhu Sadhana was aided to collect the two short core samples under the cruise program SSD004. Two short core samples were collected at the water depths of 1,235 and 1,887 m (MC-2 and MC-60) from the continental slope region of the Gulf of Mannar. The sediment core samples with a length of 45 and 36 cm were retrieved from the ocean bottom and subsampled at an interval of 3 cm. Thus, 27 subsamples were obtained and subjected to standard micropaleontological studies (Ostracod analysis).

The classification proposed by Hartmann and Puri (1974) has been followed for the Ostracod taxonomy. For retrieving Ostracoda from the sediment matrix, 10 g of dried sediment samples was soaked overnight with 0.025 normality of $(NaPO_3)_6$ solution for disintegration and washed through a 0.063-mm sieve. The nature of substrate, calcium carbonate, and organic matter in the sediment samples was determined by the methodology suggested by Krumbein and Pettijohn (1938), Piper (1947), and Gaudette et al. (1974), respectively.

Multivariate statistical analyses have been performed by using the PAST software (version 3.0) to measure species diversity and to group the samples. Cluster analyses (unweighted pair-group method with arithmetic mean (UPGMA) algorithm, Bray–Curtis similarity index for abundance data) have been applied to group the samples (Angela Garcia et al., 2017), and dendrograms are prepared to arrange the samples based on species diversity. Species richness, Shannon–Weaver index, dominance, and evenness are calculated to examine the species diversity.

Ostracod Population and Its Ecology

A total of 237 Ostracod specimens belonging to 33 species of 23 genera were present in MC-2 core samples, and 52 specimens belonging to 10 species of 8 genera were present in MC-60 core samples. The population and relative abundance of the Ostracod species identified in the study area are given in Table 11.1. Selected scanning electron microscope (SEM) images of Ostracods depicting different views are illustrated in Plate 11.1.

Ostracoda distribution depends on environmental factors such as substrates, water temperature, water depth, salinity, and hydrochemical gradients (Smith, 1993; Yasuhara and Irizuki, 2001; Rumes et al., 2016). Ecologically, silty sand and sand are the most favorable substrates for Ostracod growth (Sridhar et al., 2002). Silt is the only substrate found in all the subsamples of MC-2 and MC-60, and Ostracods are found to accommodate subsamples in the silty substrate (Table 11.2). The result of calcium carbonate shows that the amount of $CaCO_3$ in the silty substrate may be due to the occurrence of calcareous microfauna noticed in the sediment samples. The result of organic matter shows the presence of mud-sized particles that accumulate organic matter content in the sediments. Comparatively, the percentage of $CaCO_3$ and organic matter shows a slight increase in MC-60 than in MC-2 core samples. The result indicates that the values are directly proportional to the water depth and inversely proportional to the Ostracod population, found in the study area.

Dominant Species

The distributional patterns of all species, which show their relative abundance more than 5% of the total population in each core, are considered as dominant species for the present study. *Krithe* sp.1, *Krithe kroemmelbeini*, *Krithe* sp.2, *Bradleya japonica*, and *Acanthocythereis* sp. are dominant in MC-2 samples and hold a count of 82 specimens in the study area. The dominant forms in MC-60 samples are *Krithe* sp. *Krithe* sp.2, *Krithe kroemmelbeini*, *Xestoleberis* sp., *Parakrithella* sp., *Bradleya japonica*, and *Paijenborchella* sp. The vertical distributions of dominant species for MC-2 and MC-60 samples are given in Figures 11.2 and 11.3, respectively.

The genus *Krithe* is one of the most ubiquitous deep-sea benthic Ostracoda; the dominance of this genus is generally observed after 1,000 m water depth (Ayress et al., 1999; Tanaka, 2016; Yasuhara et al., 2008). Different species of genus *Krithe* are used as a marker for cooler water environment and also an indicative genus for glacial–interglacial cycles (Elewa, 2004; Stepanova and Lyle, 2014; Mohammed Noohu et al., 2019). *Genus* Krithe holds the maximum population, and it is dominantly present in both MC-2 and MC-60 samples in the study area. Next to that, *Bradleya japonica* is a dominant species present in both the water depths. The distribution of *Bradleya japonica* is recorded in deeper as well as shallower water region, i.e., upper abyssal to outer shelf (Mohammed Nishath et al., 2017). The count of *Krithe* sp.2 and *Bradleya japonica* is similar, which may indicate that both the forms prefer a similar ecological condition for their growth and sustainability.

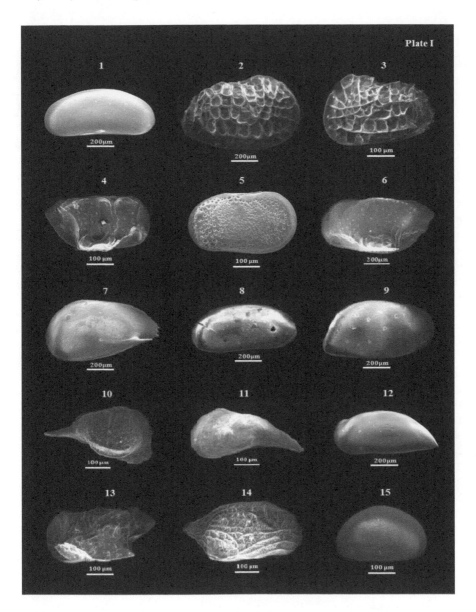

PLATE 11.1 (1) *Argilloecia* sp., (RV external view), (2) *Bradleya andamanae* (RV external view), (3) *Bradleya japonica* (LV external view), (4) *Bythoceratina mandviensis* (RV external view), (5) *Cytherella hemipuncta* (RV external view), (6) *Cytheropteron* sp.1 (LV external view) (7) *Keijella karwarensis* (LV external view), (8) *Krithe kroemmelbeini* (LV external view), (9) *Krithe* sp.1 (RV external view), (10) *Paijenborchella* sp., (RV external view), (11) *Paijenborchellina prona* (LV external view), (12) *Paracypris* sp., (LV external view), (13) *Paracytheridea pseudoremanei* (LV external view), (14) *Semicytherura contraria* (RV external view), and (15) *Xestoleberis* sp., (LV external view), (LV, left valve), (RV, right valve).

TABLE 11.1
Distribution of Ostracoda in the Gulf of Mannar

Sl. No	Species Name	MC-2 (1,235 m) A	MC-2 (1,235 m) R	MC-60 (1,887 m) AI	MC-60 (1,887 m) R
1	*Acanthocythereis* sp.	12	5.1	0	0.0
2	*Actinocythereis* sp.	9	3.8	0	0.0
3	*Argilloecia* sp.	7	3.0	0	0.0
4	*Bairdoppilata* sp.	3	1.3	0	0.0
5	*Bradleya andamanae* Benson, 1972	3	1.3	0	0.0
6	*Bradleya japonica* Benson, 1972	14	5.9	6	11.5
7	*Bythoceratina mandviensis* Jain, 1978	7	3.0	0	0.0
8	*Cytherella hemipuncta* Swanson, 1969	3	1.3	0	0.0
9	*Cytherella semitalis* Brady, 1868	4	1.7	0	0.0
10	*Cytheropteron* sp.1	5	2.1	0	0.0
11	*Cytheropteron* sp.2	5	2.1	0	0.0
12	*Cytheropteron volantium* Whatley and Masson, 1979	9	3.8	0	0.0
13	*Echinocythereis* sp.	3	1.3	0	0.0
14	*Hemicytheridea ornata* Mostafawi, 1992	4	1.7	0	0.0
15	*Henryhowella* sp.	11	4.6	0	0.0
16	*Keijella karwarensis* Bhatia and Kumar, 1979	2	0.8	0	0.0
17	*Krithe kroemmelbeini* Jain, 1978	18	7.6	7	13.5
18	*Krithe* sp.1	24	10.1	11	21.2
19	*Krithe* sp.2	14	5.9	7	13.5
20	*Macrocyprina* sp.	6	2.5	0	0.0
21	*Pacombocythere* sp.	9	3.8	1	1.9
22	*Paijenborchella malaiensis* Kingma, 1948	11	4.6	0	0.0
23	*Paijenborchella* sp.	6	2.5	4	7.7
24	*Paijenborchellina prona* Lyubimova et al., 1960	4	1.7	0	0.0
25	*Paracypris* sp.	7	3.0	1	1.9
26	*Paracytheridea pseudoremanei* Bonaduce et al., 1980	5	2.1	0	0.0
27	*Parakrithella* sp.	9	3.8	6	11.5
28	*Pterygocythereis chennaiensis* Mohan et al., 2001	5	2.1	0	0.0
29	*Pterygocythereis* sp.	3	1.3	2	3.8
30	*Semicytherura contraria* Zhao and Whatley, 1989	10	4.2	0	0.0
31	*Semicytherura* sp.	2	0.8	0	0.0
32	*Xestoleberis* sp.	2	0.8	7	13.5
33	*Xestoleberis variegata* Brady, 1880	1	0.4	0	0.0
Total		237		52	

A, absolute abundance; R, relative abundance.

TABLE 11.2

Estimated Values of Environmental Parameters with the Ostracod Population for MC-2 and MC-60 in Gulf of Mannar

Sl. No	Depth (cm)	MC-2						MC-60					
		CaCO$_3$ (%)	OM (%)	Sediment Type	Carapace	Open Valves	Total Population	CaCO$_3$ (%)	OM (%)	Sediment Type	Carapace	Open Valves	Total Population
1	0–3	27.5	9.3	Silt	4	29	33	25.5	10	Silt	1	2	3
2	3–6	27.5	8	Silt	3	8	9	31	9.6	Silt	0	4	4
3	6–9	28.5	9.2	Silt	3	13	17	29.5	9.4	Silt	1	3	4
4	9–12	27.5	9.3	Silt	1	11	13	30	10.3	Silt	1	4	5
5	12–15	29	9.4	Silt	3	16	20	28	10.6	Silt	0	1	1
6	15–18	27.5	9.2	Silt	3	13	16	32	9.2	Silt	0	2	1
7	18–21	28.5	9.4	Silt	4	16	21	32.5	9.9	Silt	1	6	8
8	21–24	29	9.7	Silt	3	18	22	33	10.5	Silt	0	3	3
9	24–27	27.5	8.7	Silt	2	13	13	33.5	10.4	Silt	0	3	3
10	27–30	27.5	8.2	Silt	2	9	12	32.5	10.3	Silt	1	5	6
11	30–33	28.5	8.3	Silt	3	16	20	28	10.3	Silt	1	6	7
12	33–36	27.5	9.3	Silt	1	9	7	27	9.2	Silt	1	6	7
13	36–39	29.5	8.9	Silt	3	7	11	-	-	-	-	-	-
14	39–42	29	9.4	Silt	2	10	12	-	-	-	-	-	-
15	42–45	28.5	8.1	Silt	1	11	11	-	-	-	-	-	-
				Total	38	199	237			Total	7	45	52

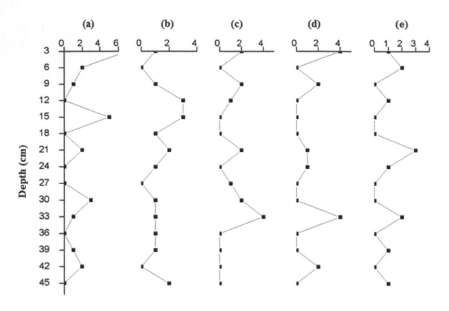

FIGURE 11.2 Vertical distributions of (a) *Krithe* sp.1, (b) *Krithe kroemmelbeini*, (c) *Krithe* sp.2, (d) *Bradleya japonica*, and (e) *Acanthocythereis* sp. in MC-2 of Gulf of Mannar.

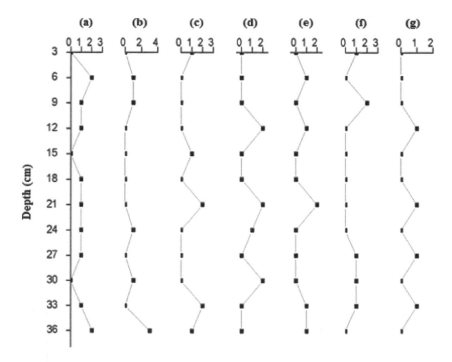

FIGURE 11.3 Vertical distributions of (a) *Krithe* sp., (b) *Krithe* sp.2, (c) *Krithe kroemmelbeini*, (d) *Xestoleberis* sp., (e) *Parakrithella* sp., (f) *Bradleya japonica*, and (g) *Paijenborchella* sp. in MC-60 of Gulf of Mannar.

Xestoleberis is a cosmopolitan genus; it is relatively easy to identify due to its subovate and smooth carapace nature (Titterton and Whatley (2005). Genus *Xestoleberis* is generally found in the shallow and intermediate marine environments, but this genus has also been recorded from the deep-marine environment (Rajkumar et al., 2020).

OSTRACOD CARAPACE/VALVE RATIO

Ostracod microfauna are enclosed in bivalve carapace which is covered by an epicuticle. Hussain et al. (2002) studied the carapace/valve ratio and observed the faster rate of sedimentation in the inner shelf of the Gulf of Mannar, off Tuticorin, southeast coast of India. The number of preserved carapaces is 38, and there are 199 open valves that are present in MC-2 samples, with a total number of 237. In MC-60 samples, the total open-valve counts are 45 and the number of carapaces is 7, thus making the total count to 52 (Table 11.2). For the present study, the number of open valves is higher than that of carapaces encountered in both MC-2 and MC-60 samples. This indicates a gradual and slower rate of sedimentation in the continental slope region. This is due to less terrigenous input by the rivers from the inner continental shelf to the slope region, in the Gulf of Mannar.

SPECIES DIVERSITY INDICES

Multivariate statistical analysis is the easiest of the methods available, to understand the Ostracod species diversity in an area (Hammer et al., 2001). Species diversity characterizes the biodiversity/its relation with the number of various species in an area. It indicates the maturity in communities (higher diversities being related to higher maturity) and ecological trends because it can be assumed that areas supporting diverse species have, generally, better living conditions. Many researchers have worked on Ostracod species diversity with respect to the sediment nature and other environmental parameters such as pH, temperature, salinity, and oxy-redox conditions (Zhao et al., 1985; Mostafawi, 1992; Hussain et al., 2009). Species diversity includes species richness, Shannon–Weaver index, dominance, and evenness (Tables 11.3 and 11.4).

SPECIES RICHNESS AND SHANNON–WEAVER INDEX (H)

Species richness is the total number of species (S), which is dependent on the sample size (the bigger the sample, the more species). In MC-2, a maximum of 18 species (33 specimens) and a minimum of 5 species (9 specimens) were found at the depth of 0–3 and 3–6 cm, respectively. Species ranging from 1 to 6 are observed in MC-60 samples, of which the maximum taxa (seven specimens) were found at the depth of 30–33 cm and the minimum of one taxon was found at the depths of 15–15 and 15–18 cm, which have been recorded in the study area.

Shannon's diversity index has been a popular diversity index used in ecological-related studies. A diversity index, taking into account the number of individuals as well as several taxa, varies from 0 for communities with only a single

TABLE 11.3
Species Diversity Values for MC-2 of Gulf of Mannar

Depth (cm)	Taxa (S)	Individuals	Dominance (D)	Simpson (1-D)	Shannon (H)	Evenness ($e^{H/S}$)
0–3	18	33	0.1	0.9	2.62	0.76
3–6	5	9	0.23	0.77	1.52	0.92
6–9	12	17	0.1	0.9	2.4	0.91
9–12	8	13	0.16	0.84	1.95	0.88
12–15	10	20	0.14	0.87	2.15	0.86
15–18	9	16	0.16	0.84	2.01	0.83
18–21	13	21	0.09	0.91	2.49	0.93
21–24	15	22	0.09	0.91	2.59	0.89
24–27	10	13	0.11	0.89	2.25	0.94
27–30	8	12	0.15	0.85	1.98	0.9
30–33	8	20	0.15	0.85	1.97	0.9
33–36	6	7	0.18	0.82	1.75	0.96
36–39	9	11	0.12	0.88	2.15	0.95
39–42	10	12	0.11	0.89	2.25	0.95
42–45	10	11	0.11	0.89	2.27	0.97

TABLE 11.4
Species Diversity Values for MC-60 of Gulf of Mannar

Depth (cm)	Taxa (S)	Individuals	Dominance (D)	Simpson (1-D)	Shannon (H)	Evenness ($e^{H/S}$)
0–3	3	3	0.33	0.67	1.1	1
3–6	3	4	0.38	0.63	1.04	0.94
6–9	3	4	0.38	0.63	1.04	0.94
9–12	4	5	0.28	0.72	1.33	0.95
12–15	1	1	1	0	0	1
15–18	1	1	1	0	0	1
18–21	5	8	0.22	0.78	1.56	0.95
21–24	3	3	0.33	0.67	1.1	1
24–27	3	3	0.33	0.67	1.1	1
27–30	5	6	0.22	0.78	1.56	0.95
30–33	6	7	0.18	0.82	1.75	0.96
33–35	4	7	0.31	0.69	1.28	0.9

taxon to high values (up to about 5.0) for communities with multiple taxa, each with few individuals (Shannon and Weaver, 1949). In MC-2 samples, the diversity value ranges from 1.52 to 2.62, and in MC-60 samples, the diversity values are confined in between 0 and 1.75. Shannon's diversity index values are higher in MC-2 samples, and those are comparatively lower in MC-60 samples (Figure 11.4).

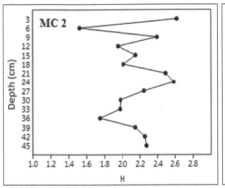 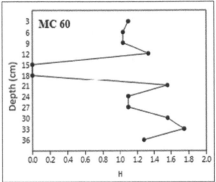

FIGURE 11.4 Shannon–Weaver diversity index for MC-2 and MC-60 of Gulf of Mannar.

DOMINANCE AND EVENNESS

The Simpson dominance index is a measure of both the species richness and the proportion of each species, which means a community with high diversity results in low dominance value. The Simpson dominance index ranges from 0 (if all species are equally present) to 1 (if one species dominates the community completely). Dominance (D) value ranges from 0.09 to 0.23 in MC-2 core samples. The result shows that the study area is having low dominance, which means there is no such species that dominates the community completely.

Evenness expresses how evenly the individuals are distributed among the different species, and it is often termed equitability. It makes sense to consider species richness and species evenness as two independent characteristics of biological communities that together constitute their diversity (Heip, 1974). From the results, MC-2 and MC-60 samples are showing high evenness values, which shows that species present in these stations are more or less evenly distributed.

CLUSTER ANALYSIS

Dendrograms are prepared to group the samples according to similarity with depth for both the MC-2 and MC-60 cores (Figure 11.5). Cluster analyses are used to classify the species and to analyze them into small interpretable categories. The relationships among the subsamples in each core are determined based on the species richness, population, diversity index, dominance, and evenness values. For convenience, subsamples are numbered as 1–15, irrespective of the depths for the two multicore samples. Four biotopes are obtained for 15 subsamples (0–45 cm) of MC-2. Biotope I includes subsamples 2 and 12; Biotope II comprises subsamples 3, 6, 5, 11, 7, and 8; and Biotope III holds subsamples 4, 10, 9, 14, 13, and 15 – all these subsamples are having close ecological similarities. Biotope IV of MC-2 exhibits a slight variation compared to other subsamples. Three major biotypes are identified from the cluster dendrogram of 12 subsamples (0–35 cm) of MC-60. Biotope I has subsamples 8, 1, 9, 2, and 3. Biotope II has subsamples 4, 12, 7, 10, and 11. Biotope III has subsamples 5 and 6.

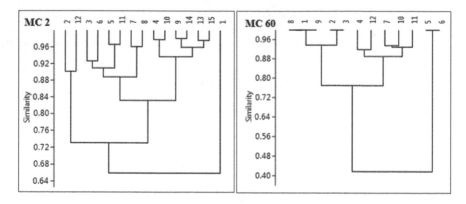

FIGURE 11.5   Cluster dendrogram of species similarity for MC 2 and MC 60 of Gulf of Mannar.

CONCLUSION

The terrigenous finer sediment influx associated with the tropical climate of the region supplies the muddy substrate to the continental slope, which is the prime factor favoring the Ostracod growth. The temperature of bottom waters also affects the diversity and distribution of Ostracoda. The percentage of $CaCO_3$ and organic matter shows a slight increase in MC-60 samples than in MC-2 samples, which indicates that the values are directly proportional to the water depth. The carapace/open-valve ratio of Ostracoda reveals a slower and gradual rate of sedimentation that has been observed for both the cores MC-2 and MC-60 from the area. Ecologically, both the cores exhibit similar dominant species. Both the cores have the maximum relative abundance of *Krithe* spp. Species diversity indices indicate no individual species is dominating the diversity and the species are evenly distributed in both the cores. Shannon index values are higher in MC-2 compared to MC-60 samples, indicating the reduction of species diversity towards increasing depth.

ACKNOWLEDGMENTS

The authors are thankful to the University Grant Commission for providing financial support under UGC-BSR scheme (GCCO/A-2/UGC-Meritorious/2014/2100, Dated: 24 January 2014). The authors are very much thankful to Dr. Rajeev Saraswat, the Scientist, National Institute of Oceanography, Goa, and Dr. Mohammed Nishath for their help in sample collection. The authors also thank Mrs. Kresna Tri Dewi, Centre of Geological Survey, Bandung, for help in Ostracod identification. Thanks are also due to Sophisticated Analytical Instrument Facility, IIT Madras, for providing permission to take SEM images of Ostracoda.

REFERENCES

Andre, B., Pirkenseer, C.M., De Deckker, P., and Speijer, R.P. 2012. Oxygen and carbon isotope fractionation of marine ostracod calcite from the eastern Mediterranean Sea. *Chemical Geology* 310:114–125.

Angela, G.G, Patrick, G., Marlies, V.S., Francisco, J.S., Francisco, J.J., Carlos, A.A.Z., and Werner, E.P. 2017. Benthic foraminifera-based reconstruction of the first Mediterranean-Atlantic exchange in the early Pliocene Gulf of Cadiz. *Palaeogeography, Palaeoclimatology, Palaeoecology* 472:93–107. doi: 10.1016/j.palaeo.2017.02.009.

Anne-Sophie, F., Maria-Angela, B., Christophe, F., Alina, T., and Serge, B. 2016. Sedimentary archives of climate and sea-level changes during the Holocene in the Rhone prodelta (NW Mediterranean Sea). *Climate of the Past* 12:2161–2179. doi: 10.5194/cp-2016-57.

Ayress, M.A., Timothy, B., Vicki, P., and Whatley, R. 1999. Neogene to recent species of Krithe (Crustacea: Ostracoda) from the Tasman Sea and off southern Australia with description of five new species. *Records of Australian Museum* 51(1):1–22.

Brautovic, I., Bojanic, N., Vidjak, O., Grbec, B., and Gangai, Z.B. 2018. Composition and distribution patterns of marine planktonic ostracods (Crustacea, Ostracoda) in the Adriatic Sea - A historical perspective. *Acta Adriatica* 59(1):71–90.

Cabral, M.C., and Loureiro, I.M. 2013. Overview of recent and Holocene ostracods (Crustacea) from brackish and marine environments of Portugal. *Journal of Micropalaeontology* 32:135–159.

Chacko, P.I. 1950. Marine plankton from waters around the Krusadai Island. *Proceedings: Plant Sciences* 31(3):162–174.

Elewa, A.M.T. 2004. Quantitative analysis and paleoecology of Eocene Ostracoda and benthonic Foraminifera from Gebel Mokattam, Cairo, Egypt. *Paleogeography, Paleoclimatology, Paleoecology* 211:309–323.

Frezza, V., and Bella, L.D. 2015. Distribution of recent ostracods near the Ombrone River mouth (Northern Tyrrhenian Sea, Italy). *Micropaleontology* 61:101–114.

Gaudette, H.E., Flight, W.R., Toner, L., and Folger, D.W. 1974. An inexpensive titration method for determination of organic carbon in recent sediments. *Journal of Sedimentary Petrology* 44:249–253.

Gene. H., Satrio A., W, Julia E. B., and Kenneth G.M. 2010. Climate-driven body-size trends in the Ostracod fauna of the deep Indian ocean. *Palaeontology* 53(6):1255–1268.

Hammer, O., Harper, D.A.T., and Ryan, P.D. 2001. Past: Palaeontological statistics software package for education and data analysis. *Palaeontologia Electronica* 4:9.

Hartmann, G., and Puri, H.S. 1974. Summary of neontological and paleontological classification of Ostracoda. *Mitteilungen aus dem Hamburgischen Zoologischen Museum und Institut* 70:7–73.

Heip, C. 1974. A new index measuring evenness. *Journal of the Marine Biological Association of the United Kingdom* 54(3):555–557.

Hokuto, I., Young, M.Y., Irizuki, T., Sampei, Y., and Ishiga, H. 2014. Spatial variations in recent ostracode assemblages and bottom environments in Trincomalee Bay, northeast coast of Sri Lanka. *Micropaleontology* 60(6):509–518.

Hong, Y., Yasuhara, M., Iwatani, H., and Mamo, B. 2019. Baseline for ostracod-based northwestern Pacific and Indo-Pacific shallow-marine paleoenvironmental reconstructions: ecological modeling of species distributions. *Biogeosciences* 16:585–604.

Hussain, S.M. 1998. Recent benthic Ostracoda from the Gulf of Mannar, off Tuticorin, southeast coast of India. *Journal of the Paleontological Society of India* 43:1–22.

Hussain, S.M., Mohan, S.P., and Manivannan, V. 2002. Microenvironmental inference of Recent Benthic Ostracoda from the Gulf of Mannar, off Tuticorin, Southeast coast of India. *National Seminar on Management of Natural Resources*, Dept. of Geology, Nagarjuna University, 23–43.

Hussain, S.M., Ravi, G., Mohan, S.P., and Rajeshwara Rao, N. 2009. Distribution of recent Ostracoda in the Bay of Bengal, off Karikkattukuppam (near Chennai), southeast coast of India - Implication for biodiversity. *Gondwana Geological Magazine* 24(1):35–39.

ICMAM. 2001. Critical habitat information system of Kadamat island-Lakshadweep. Technical report. Chennai: DOD, GOI.

Jeremy, M.C., Finn, V., Derya, A., Immenhauser, A., and Ola, K. 2019. Ostracods as ecological and isotopic indicators of lake water salinity changes: the Lake Van example. *Biogeosciences* 16:2095–2114. doi: 10.5194/bg-16-2095-2019.

Julio, R.L., and Francisco, R.M. 2012. A general introduction to ostracods: Morphology, distribution, fossil record and applications. *Developments in Quaternary Science* 17:1–14.

Kresna, T.D. 2014. Ostracoda from subsurface sediments of Karimata strait as indicator of environmental changes. *Bulletin of Marine Geology* 29(1):1–10.

Krumbein, W.C., and Pettijohn, F.J. 1938. *Manual of Sedimentary Petrography*, New York: Appleton-Century-Crofts, 549.

Kulkoyluoglu, O., Yılmaz, S., and Yavuzatmaca, M. 2017. Comparison of Ostracoda (Crustacea) species diversity, distribution and ecological characteristics among habitat types. *Fundamental and Applied Limnology* 190(1):63–86.

Kumaraguru, A.K. 1999. Monitoring coral reefs environment of Gulf of Mannar – A pilot study. Report submitted to IOC/UNESCO.

Lili, F., Irizuki, T., and Sampei, Y. 2012. Spatial distribution of recent ostracode assemblages and depositional environments in Jakarta Bay, Indonesia, with relation to environmental factors. *Paleontological Research* 16(4):267–281.

Martin, G., Maria, I.R., Marco, C., and Werner, E.P. 2013. Ostracods (Crustacea) and their palaeoenvironmental implication for the Solimões formation (Late Miocene; Western Amazonia/Brazil). *Journal of South American Earth Sciences* 42:216–241.

Mazdygan, E.R., and Chavtur, V.G. 2011. The composition and distribution of pelagic ostracods (Ostracoda: Myodocopa) in antarctic waters adjacent to the d'Urville Sea. *Russian Journal of Marine Biology* 37(4):263–271.

Mohammed Nishath, N., Hussain, S.M., and Rajkumar, A. 2015. Distribution of ostracoda in the sediments of the northwestern part of the Bay of Bengal, India - Implications for microenvironment. *Journal of the Palaeontological Society of India* 60(2):27–33.

Mohammed Nishath, N., Hussain, S.M., Neelavnnan, K., Thejasino, S., Saalim, S. and Rajkumar, A. 2017. Ostracod biodiversity from shelf to slope oceanic conditions, off central Bay of Bengal, India. *Paleogeography, Paleoclimatology, Paleoecology* 483:70–82. doi: 10.1016/j.palaeo.2017.05.004.

Mohammed Noohu, N., Radhakrishnan, K., Hussain, S.M., Sivapriya, V., and Rajkumar, A. 2019. Genus *Krithe* (Ostracoda) as a proxy to decipher paleoceanography: A global review of the genus. *Oceanography and Fisheries Open Access Journal* 10(3):555789.

Mostafawi, N. 1992. Rezente Ostracoden aus dem mittleren Sunda-Schelf, Zwischen der malaiischen Halbinsel und Borneo. *Senckenberg Leth* 72:129–168.

Munif, M., and Dietmar, K. 2012. Recent ostracods from the tidal flats of the coast of Aden City, Yemen. *Marine Biodiversity* 42:247–280. doi: 10.1007/s12526-012-0112-9.

Nicole, B., Bart, D.B., Qichao, Y., Klaus, P.J., Peter, F., Meinrat, O.A., and Antje, S. 2013. Ostracod shell chemistry as proxy for paleoenvironmental change. *Quaternary International* 313–314:17–37.

Piper, F.B. 1947. Soil and plant analysis. University of Adelaide Press, Adelaide, 368.

Rajkumar, A., Hussain, S. M., Nishath, N.M., Dewi, K.T., Sivapriya, V., and Radhakrishnan, K. 2020. Recent ostracod biodiversity from shelf to slope sediments of Gulf of Mannar, India: Ecologic and Bathymetric implications. *Journal of the Palaeontological Society of India* 65(1):73–80.

Rao, V.J., Usman, P.K., and Bharat Kumar, J. 2008. Larvicidal and insecticidal properties of some marine sponges collected in Palk Bay and Gulf of Mannar. *African Journal of Biotechnology* 7(2):109–113.

Rumes, B., Van der Meeren, T., Martens, K., and Verschuren, D. 2016. Distribution and community structure of Ostracoda (Crustacea) in shallow waterbodies of southern Kenya. *African Journal of Aquatic Science* 41(4):377–387. doi: 10.2989/16085914.2016.1241174.

Sara, C.B., and Robin, W. 2012. New perspectives on the stratigraphical and temporal distribution patterns of Argentinian Jurassic marine Ostracoda. *Revue de Paléobiologie, Genève* 11:169–185.

Shannon, C. E. and Weaver, W. 1949. *The Mathematical Theory of Communication.* Univ. Illinois Press.

Smith, A.J. 1993. Lacustrine ostracodes as hydrochemical indicators in lakes of the north-central United States. *Journal of Paleolimnology* 8:121–134.

Snelgrove, P.V., and Butman, C.A. 1994. Animal sediment relationships revisited, cause vs. effect. *Oceanography and Marine Biology Annual Review* 32:111–177.

Sridhar, S.G.D., Deepali, K., Prabhu, V.S., Hussain, S.M., and Maniyarasan, S. 2019. Distribution of recent Benthic ostracoda, around Pullivasal and Poomarichan islands, off Rameswaram, Gulf of Mannar, Southeast coast of India. *Journal of the Palaeontological Society of India* 64(1):27–38.

Sridhar, S.G.D., Hussain, S.M., Kumar, V., and Periakali, P. 2002. Recent ostracoda from Palk Bay, off Rameswaram, southeast coast of India. *Journal of the Paleontological Society of India* 47:17–39.

Stepanova, A., and Lyle, M. 2014. Deep sea ostracoda from the Eastern Equatorial Pacific (ODP Site 1238) over the last 460 ka. *Marine Micropaleontology* 111:100–117.

Tanaka, G. 2016. Redescription of two krithid species (Crustacea, Ostracoda) from the Sea of Japan, with a comment on the taxonomic characters of Krithidae. *Paleontological Research* 20(1):31–47.

Thanikachalam, M., and Ramachandran, S. 2003. Shoreline and coral reef ecosystem changes in Gulf of Mannar, Southeast coast of India. *Journal of the Indian Society of Remote Sensing* 31(3):157–173.

Theodora, T. 2012. Composition and distribution of recent marine ostracod assemblages in the bottom sediments of Central Aegean Sea (SE Andros Island, Greece). *International Review of Hydrobiology* 97(4):276–300.

Titterton, R., and Whatley, R. 2005. Recent marine ostracoda from the Solomon Islands. Part 2: Cytheracea, Xestoleberididae. *Revista Española de Micropaleontogía* 37:291–313.

Valls, L., Zamora, L., Rueda, J., and Mesquita-Joanes, F. 2015. Living and dead ostracod assemblages in a coastal Mediterranean wetland, *Wetlands* 36:1–9. doi: 10.1007/s13157-015-0709-4.

Yasuhara, M., and Irizuki, T. 2001. Recent ostracoda from the northeastern part of Osaka Bay, southwestern Japan. *Journal of Geosciences* 44:57–95.

Yasuhara, M., Cronin, T. M., de Menocal, P. B., Okahashi, H., and Linsley, B. K. 2008. Abrupt climate change and collapse of deepsea ecosystems. *Proceedings of the National Academy of Sciences of the United States of America* 105:1556–1560.

Zhao, Q., Wang, P., and Zhang, Q. 1985. Ostracod in bottom sediments of the South China Sea, off Guangdong Province, China. Their taxonomy and distribution. *Marine Micropaleontology of China* 296–317.

12 Ecological Assessment of Recent Benthic Ostracoda, Off Kurusadai Island, Gulf of Mannar, India

S. Maniyarasan, Kumari Deepali,
S. G. D. Sridhar, S. M. Hussain, and
A. Rajkumar
University of Madras

CONTENTS

INTRODUCTION

Ostracods are one of the best documented groups within the whole of the animal kingdom due to a wealth of characteristic features and well-calcified, tiny bivalve carapaces (Jain and Bhatia, 1978; Jain, 1981). Ostracods are known to inhabit in a wide variety of aquatic environments such as marine, brackish, freshwater, and even terrestrial environments, and their shells are widely dispersed in the fossil record (Puri, 1960). They are important in evolutionary (Cohen and Johnston, 1987; Chaplin and Ayre, 1997), ecological (De Deckker, 1981; Külköylüoğlu et al., 2010), and biological (Kesling and Crafts, 1962; Lopez et al., 2002) studies as indicators of water quality (Benson, 1990) and habitat preferences (Külköylüoğlu and Vinyard, 2000). Taking into account the data of ecological factors, it has been possible to recognize that ostracod groups indicate the relationship with depth and substrate (Montenegro et al., 2004). Ostracods can be good markers of the environmental variability, and they are useful tools for the monitoring of a coastal area showing particular conditions linked to strong river influence and human activity (Pugliese et al., 2006).

There are many factors affecting the survival of Ostracods, such as temperature, bottom topography, depth, salinity, pH, dissolved oxygen, food supply, substrate, $CaCO_3$, and organic matter content, but salinity represents an important factor that controls Ostracods (Bate, 1971; Evans et al., 1973). The major controlling factors governing the distribution of Ostracods in estuarine environments and continental shelf zones are salinity, water temperature, and substrate (Yassini and Jones, 1995). Jain (1978, 1981), Bhatia and Kumar (1979), Khosla et al. (1982), Varma et al. (1993), Kumar and Hussain (1997), Hussain and Mohan (2000, 2001), Sridhar et al. (1998, 2002, 2018), Gopalakrishna et al. (2007), Hussain et al., (2007), Elumalai et al. (2010), Baskar et al. (2013), Maniyarasan (2016), and Kumari Deepali et al. (2019) have done very conspicuous studies on recent marine and marginal water bodies of the Indian coast.

Most habitats are occupied by several species, each with its own ecological niche, defined by the physical, chemical, and biological limits of the organisms. The environmental factors have a lower and upper limit within which a species lives, and these are the maximum stretchable tolerance for the survival of a species. There are, however, critical threshold values beyond which the environment is stressful and an optimum value most favorable for the species at which the abundance is likely to be high (Saraswati and Srinivasan, 2016). The objective of the present study is to assess the ecology of recent benthic Ostracoda based on the environmental parameters related to water and sediment utilizing its total population in this environment, where corals as well as mangroves coexist.

STUDY AREA

The study area, Kurusadai island, is one of the 21 islands in the Gulf of Mannar Biosphere Reserve (GOMBR), located in the southeast part of Tamil Nadu. The area under investigation is off the coast of Mandapam (Lat. from 9°13′ N to 9°15′ N and Long. from 79°07′ E to 79°12′ E) that comes under Ramanathapuram district of Tamil Nadu, along the southeast coast of India. It falls in 57 O/7 and 57 O/8 topo sheets of Survey of India. Kurusadai island is about 3.5 km south off Pamban bridge as shown in Figure 12.1. Known traditionally as a paradise for zoological collections, the fauna around this island has been extensively depleting since many decades. There is an old marine biological laboratory with few other buildings and a dilapidated old museum. The entry being restricted, presently, good vegetation and good coral reef growth are seen around this island. Fringing reefs encircle the study area that is seen at a depth of about 4 m, and the reef is seen in the adjoining islands namely, Pullivasal and Poomarichan islands. Mangrove plants are seen on the northern shore of the study area as well as in the adjoining islands.

The general drift of water in the Gulf of Mannar is from the south to north from about the end of April to the end of September and reverse direction during the height of the northeast monsoon (Chacko, 1950). The climatic conditions of the study area are tropical for which the southwest monsoon has little contribution, while the

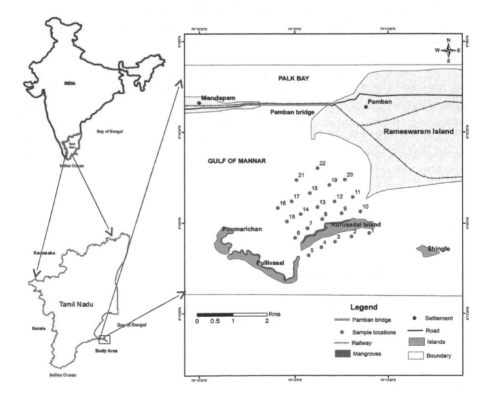

FIGURE 12.1   Location map of the study area.

northeast monsoon during October–December contributes much to the annual rainfall, making the mean annual rainfall that range from 762 to 1,270 mm. The climate is hot during January to May, and the sea is rough from April to August, stormy between June and August, and calm during September; the coldest climate is experienced in December with an average temperature of 25°C.

MATERIALS AND METHODS

In order to study the various environmental aspects of recent Benthic Ostracoda, bottom surface sediments (BSS) and bottom water samples (BWS) were collected from the coast of Kurusadai island, Gulf of Mannar, with the help of a motor launch using a van Veen grab sampler for bottom sediments and Aqua trap sampler for bottom water samples. The samples were collected during three different seasons in a year representing all the seasons that exist in the study area, namely, (1) fair weather (May 2013), (2) southwest monsoon (August 2013), and (3) northeast monsoon (December 2013). Totally, 66 BWS and 66 BSS have been collected from 22 stations representing three different seasons of the year. The sample collections were made 3 km from the shore of Kurusadai island, and it covers an area of 12 km².

At each sampling location, a portion (25 mL) of wet sediment sample was preserved in a mixture of one part of buffered formalin in nine parts of water (4% solution) with a pinch of calcium carbonate to achieve neutrality. The preserved sediment samples were subjected to *rose Bengal* staining technique as described by Walton (1952), to differentiate living and dead Ostracods. Further, species-level identification has been done under stereo zoom microscope, and their respective population was counted for each and every station. Temperature and pH of the collected bottom water samples were measured using portable multiparameter hand-held probe in situ. Salinity was determined using the standard titration method and equation proposed by Knudsen (1901). Dissolved oxygen was determined as proposed by Strickland and Parsons (1968). Calcium carbonate and organic matter were determined from collected bottom sediment samples adopting the methodology after Loring and Rantala (1992) and Gaudette et al. (1974), respectively. Sand, silt, and clay percentages were computed by the pipette method, in accordance with Krumbein and Pettijohn (1938). Trilinear plots were prepared, and Trefethen's (1950) textural nomenclature has been utilized for finding out the nature of the substrate.

RESULTS AND DISCUSSION

WATER AND SEDIMENT PARAMETERS

The present study area comprises 22 stations off Kurusadai island. The results of ecological parameters of bottom water samples (BWS) and bottom sediment samples (BSS) during the different three seasons such as fair weather (May 2013), southwest monsoon (August 2013), and northeast monsoon (December 2013) transects are given in Tables 12.1–12.3. From Bottom Water Samples (BWS), the depth of the water column, temperature, pH, dissolved oxygen, and salinity were found, and

TABLE 12.1

Ecological Parameters off Kurusadai Island during Fair Weather Season with Population of Ostracods

S. No	Depth (m)	Water Parameters				Sediment Parameters						Ostracoda Population	
		Temp (°C)	pH	DO (mg/L)	Salinity (‰)	$CaCO_3$ (%)	Org. Matter (%)	Sand (%)	Silt (%)	Clay (%)	Substrate	Living	Total
1	1.0	31.50	8.10	3.35	33.04	83.5	0.58	97.9	2.0	0.1	Sand	27	372
2	1.0	31.40	8.10	4.46	33.50	79.0	0.64	98.2	1.6	0.2	Sand	25	340
3	1.0	31.10	8.09	3.90	33.19	80.5	0.58	88	11.9	0.1	Sand	32	414
4	1.0	31.10	8.05	4.18	33.95	74.0	0.64	70.4	29.4	0.2	Sand	26	430
5	1.0	30.90	8.13	4.46	33.80	47.5	0.47	57.2	42.7	0.1	Siltysand	42	484
6	1.0	30.70	8.09	4.46	32.28	88.5	0.88	97.3	2.5	0.2	Siltysand	34	427
7	1.0	31.20	8.01	4.74	33.04	79.5	1.12	96.5	3.2	0.3	Sand	29	469
8	1.0	31.50	8.00	3.90	33.65	86.5	0.88	67.1	32.6	0.3	Siltysand	37	503
9	1.0	31.30	8.05	4.74	33.95	78.5	0.53	98.7	1.0	0.3	Sand	35	476
10	1.0	31.30	8.05	4.18	34.40	87.0	0.47	86.8	12.9	0.3	Sand	28	441
11	2.5	31.40	8.06	4.74	34.56	87.5	0.94	65.4	34.4	0.2	Siltysand	53	490
12	2.5	31.30	8.10	3.90	33.95	87.0	1.06	77.7	22.1	0.2	Siltysand	40	518
13	3.0	31.00	8.16	4.18	34.56	67.0	0.64	60.3	39.4	0.3	Siltysand	56	504
14	4.0	30.90	8.13	3.35	33.65	75.5	1.29	87.1	12.5	0.4	Sand	44	580
15	3.0	31.00	8.12	3.35	33.95	79.5	0.58	80.5	19.2	0.3	Sand	37	573
16	3.0	31.10	8.09	3.62	33.80	70.0	1.41	82.9	16.8	0.3	Siltysand	72	620
17	3.0	31.20	8.13	4.18	33.95	72.0	1.17	80.3	19.4	0.3	Siltysand	65	590
18	5.0	31.30	8.07	3.35	33.80	69.5	1.06	81.1	18.7	0.2	Siltysand	87	644
19	5.0	31.20	8.12	3.62	33.80	70.5	1.23	75.4	24.3	0.3	Siltysand	49	575

(Continued)

TABLE 12.1 (*Continued*)
Ecological Parameters off Kurusadai Island during Fair Weather Season with Population of Ostracods

		Water Parameters				Sediment Parameters						Ostracoda Population	
S. No	Depth (m)	Temp (°C)	pH	DO (mg/L)	Salinity (‰)	CaCO$_3$ (%)	Org. Matter (%)	Sand (%)	Silt (%)	Clay (%)	Substrate	Living	Total
20	4.0	30.80	8.09	3.90	33.34	69.0	0.94	78.7	21.0	0.3	Sand	56	540
21	4.0	30.60	8.10	3.62	33.65	71.5	1.06	61.1	38.7	0.2	Siltysand	50	501
22	3.0	31.30	8.13	3.07	33.80	70.5	0.94	60.6	39.1	0.3	Siltysand	44	515
Max	5.0	31.50	8.16	4.74	34.56	88.5	1.41	98.7	42.7	0.4	Max	87	644
Min	1.0	30.60	8.00	3.07	32.28	47.5	0.47	57.2	1.0	0.1	Min	25	340
Avg	2.4	31.14	8.09	3.97	33.71	76.1	0.87	79.5	20.2	0.2	Avg	44	500

TABLE 12.2

Ecological Parameters off Kurusadai Island during Southwest Monsoon with Population of Ostracods

		Water Parameters				Sediment Parameters						Ostracoda Population	
S. No	Depth (m)	Temp (°C)	pH	DO (mg/L)	Salinity (‰)	CaCO$_3$ (%)	Org. Matter (%)	Sand (%)	Silt (%)	Clay (%)	Substrate	Living	Total
1	1.0	30.1	8.11	2.23	32.30	83.0	0.548	82.8	17.1	0.1	Sand	34	355
2	1.0	30.2	8.12	2.51	32.20	60.5	0.947	97.2	2.6	0.2	Sand	27	370
3	1.0	29.6	8.18	2.51	32.20	84.0	0.533	98.9	0.9	0.2	Sand	38	409
4	1.0	29.5	8.14	2.79	32.30	88.5	0.339	97	2.9	0.1	Sand	41	385
5	1.0	31.1	8.19	2.23	32.40	75.0	0.824	72	27.7	0.3	Siltysand	39	411
6	1.0	30.9	8.22	2.79	32.50	78.5	0.291	89.4	10.4	0.2	Sand	56	540
7	1.0	30.9	8.27	2.79	32.70	79.5	0.581	90.4	9.4	0.2	Sand	32	315
8	1.0	30.7	8.16	2.23	32.60	82.5	0.388	97.4	2.5	0.1	Sand	46	388
9	1.0	30.7	8.19	3.07	32.60	70.0	0.185	98.7	1.1	0.2	Sand	30	330
10	1.0	30.2	8.23	3.07	32.60	78.5	0.231	89.8	10.1	0.1	Sand	37	394
11	2.5	30.7	8.27	2.79	32.20	88.5	0.598	95.4	4.4	0.2	Sand	55	460
12	2.5	30.8	8.20	2.23	32.50	89.0	0.748	97.7	2.2	0.1	Sand	27	430
13	3.0	29.4	8.20	3.63	32.20	81.5	0.199	97.4	2.5	0.1	Sand	46	590
14	4.0	29.2	8.20	3.91	32.30	69.5	0.698	88.4	11.4	0.2	Sand	34	547
15	3.0	29.3	8.19	3.91	32.30	50.0	1.296	55	44.9	0.1	Siltysand	38	586
16	3.0	29.6	8.20	2.23	32.50	41.0	1.246	52	47.9	0.1	Siltysand	61	607
17	3.0	28.6	8.13	3.91	32.60	43.0	0.833	45	54.7	0.3	Sandysilt	30	450

(Continued)

TABLE 12.2 (*Continued*)
Ecological Parameters off Kurusadai Island during Southwest Monsoon with Population of Ostracods

S. No	Depth (m)	Water Parameters				Sediment Parameters						Ostracoda Population	
		Temp (°C)	pH	DO (mg/L)	Salinity (‰)	CaCO$_3$ (%)	Org. Matter (%)	Sand (%)	Silt (%)	Clay (%)	Substrate	Living	Total
18	5.0	29.9	8.13	3.35	32.60	42.5	0.787	32.4	67.3	0.3	Sandysilt	24	490
19	5.0	29.9	8.13	3.07	32.40	37.0	1.246	40	59.8	0.2	Sandysilt	55	680
20	4.0	29.9	8.14	3.63	32.20	37.5	1.446	37.7	62.2	0.1	Sandysilt	48	560
21	4.0	29.3	8.16	3.35	32.50	38.5	1.595	38.5	61.3	0.2	Sandysilt	31	479
22	3.0	29.3	8.14	3.07	32.70	36.5	1.795	39.4	60.4	0.2	Sandysilt	26	405
Max	5.0	31.1	8.27	3.91	32.70	89.0	1.795	98.9	67.3	0.3	Max	61	680
Min	1.0	28.6	8.11	2.23	32.20	36.5	0.185	32.4	0.9	0.1	Min	24	315
Avg	2.4	30.0	8.18	2.97	32.43	65.2	0.789	74.2	25.6	0.2	Avg	39	463

TABLE 12.3

Ecological Parameters off Kurusadai Island during Northeast Monsoon with Population of Ostracods

S. No	Depth (m)	Water Parameters				Sediment Parameters						Ostracoda Population	
		Temp (°C)	pH	DO (mg/L)	Salinity (‰)	CaCO$_3$ (%)	Org. Matter (%)	Sand (%)	Silt (%)	Clay (%)	Substrate	Living	Total
1	1.0	28.4	8.04	2.23	31.70	81.0	0.748	83.2	16.6	0.2	Sand	29	390
2	1.0	27.7	8.07	3.07	31.70	78.5	0.698	95.4	4.4	0.2	Sand	36	412
3	1.0	28.1	8.01	2.51	31.80	80.5	0.872	96.4	3.5	0.1	Sand	21	378
4	1.0	28.0	8.00	3.35	31.90	85.0	0.727	98	1.8	0.2	Sand	33	345
5	1.0	28.1	8.01	2.23	31.90	77.5	0.872	86	13.7	0.3	Sand	18	460
6	1.0	28.5	8.03	2.51	32.00	76.5	0.775	91.2	8.6	0.2	Sand	42	530
7	1.0	28.3	8.01	3.35	31.90	75.0	0.630	88.5	11.4	0.1	Sand	37	556
8	1.0	27.9	8.00	2.79	31.90	82.5	1.163	95	4.9	0.1	Sand	54	522
9	1.0	27.9	7.99	2.23	31.70	79.0	0.417	93.4	6.3	0.3	Sand	35	497
10	1.0	28.3	8.03	3.63	31.90	81.0	0.787	95.7	4.2	0.1	Sand	65	570
11	2.5	28.6	8.05	3.35	32.00	84.0	0.748	96.8	2.9	0.3	Sand	30	544
12	2.5	29.3	8.10	2.51	31.80	86.5	0.449	91.3	8.5	0.2	Sand	27	603
13	3.0	29.9	8.13	4.19	32.00	83.0	0.698	92.6	7.1	0.3	Sand	52	580
14	4.0	31.2	8.15	3.91	32.50	76.0	0.798	84.3	15.5	0.2	Sand	39	475
15	3.0	30.9	8.21	3.35	32.00	56.0	0.548	65.2	34.6	0.2	Siltysand	47	490
16	3.0	30.4	8.21	2.79	32.50	52.0	1.147	72	27.8	0.2	Siltysand	61	567
17	3.0	29.8	8.16	3.07	32.50	47.5	1.573	63.4	36.3	0.3	Siltysand	50	614
18	5.0	30.7	8.17	3.91	31.80	53.5	1.018	51.6	48	0.4	Siltysand	43	658

(*Continued*)

TABLE 12.3 (*Continued*)

Ecological Parameters off Kurusadai Island during Northeast Monsoon with Population of Ostracods

		Water Parameters				Sediment Parameters						Ostracoda Population	
S. No	Depth (m)	Temp (°C)	pH	DO (mg/L)	Salinity (‰)	$CaCO_3$ (%)	Org. Matter (%)	Sand (%)	Silt (%)	Clay (%)	Substrate	Living	Total
19	5.0	30.3	8.11	3.63	31.80	36.0	1.246	48.7	51.1	0.2	Sandysilt	64	586
20	4.0	29.1	8.00	3.35	31.80	39.0	0.848	55.7	44.2	0.1	Siltysand	56	519
21	4.0	29.5	8.10	3.91	32.00	35.5	0.997	60.3	39.4	0.3	Siltysand	61	606
22	3.0	29.8	8.12	3.63	31.90	38.0	1.147	55.9	43.9	0.2	Siltysand	48	530
Max	5.0	31.2	8.21	4.19	32.50	86.5	1.573	98.0	51.1	0.4	Max	65	658
Min	1.0	27.7	7.99	2.23	31.70	35.5	0.417	48.7	1.8	0.1	Min	18	345
Avg	2.4	29.1	8.08	3.16	31.95	67.4	0.859	80.0	19.8	0.2	Avg	43	520

from Bottom Sediment Samples (BSS), the following parameters namely, $CaCO_3$, organic matter, and the type of substrate were determined. All these parameters are discussed station-wise and season-wise. Their spatial distribution has been presented and compared with the population of Ostracoda that were found in the study area.

DEPTH

For the study area, water depth ranges from 1 to 5 m as shown in spatial distribution in Figure 12.2. During fair weather season, the highest population of Ostracoda was recorded as 644 specimens at the 18th station where the depth was 3 m, and lowest of 340 specimens were recorded at station 2 where the water depth was 1 m only. This is further evidenced during SW monsoon by the recording of 680 specimens at the 19th station where the water depth was 5 m and the lowest of 315 specimens that were recorded at the 7th station where the depth is 1 m only. NE monsoon recorded the highest of 658 specimens at the 18th station where the depth was 5 m and lowest of 345 specimens that were recorded at the 4th station where the depth is 1 m only. An attempt that relates to the depth factor with the distribution of Ostracoda is made. Living as well as the total population of Ostracods is found to be directly proportional to water depth. The stations from 1 to 5 that are devoid of mangroves recorded lower population compared to other stations, namely from 6 to 22. Spatially, the depth has a positive correlation with the total population of Ostracoda during all the seasons.

Depth in itself does not affect Ostracod distribution; however, a number of important ecological factors, including hydrostatic pressure, temperature, salinity, and dissolved oxygen have their influence in the abundance of Ostracod fauna and its

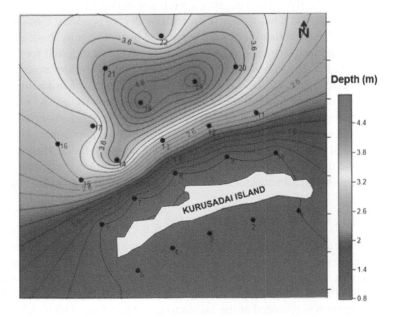

FIGURE 12.2 Spatial distribution of water depth.

diversity (De Deckker and Forester, 1988). Sridhar et al. (2011) observed that the population of Ostracods is directly proportional to water depth along eastern transect off Rameswaram island, and they (op cit.) observed no correlation with the total population along northeast and north off Rameswaram. Baskar et al. (2013) reported northeast and east off Rameswaram that the total population of Ostracods is directly proportional to water depth. Ostracods are, therefore, sensitive indicators of bottom-water conditions, and their geographical distributions are effective tracers of different benthic environments and distinct water masses. The relationship between marine Ostracods and their chemical/physical environment and morphological responses to changing environments have been reviewed by Carbonel et al. (1988).

TEMPERATURE

For the study area, bottom water temperature (BWT) ranges from 27.7°C (Northeast monsoon) to 31.5°C (fair weather season), as shown in the spatial distribution diagrams from Figure 12.3a–c. The fair weather season recorded the temperature that ranges from 30.6°C to 31.5°C, having a mean value of 31.1°C. Higher values have been observed during fair weather season at the 1st and the 8th station, and a lower value was recorded at the 21st station. During southwest monsoon, the temperature ranges from 28.6°C to 31.1°C, having a mean value of 30°C. Higher values have been observed at station 5, and a lower value was recorded at the 17th station. During northeast monsoon, the temperature ranges from 27.7°C to 31.2°C. A higher value has been recorded at the 14th station, and the lowest value was observed at the 2nd station. BWT is influenced by the intensity of solar radiation, evaporation, freshwater influx, and cooling and mix up with ebb and flow from adjoining neritic waters. The present study reveals that the BWT has a negative correlation during fair weather season and southwest monsoon, and it has a positive correlation during northeast monsoon.

Hutchins (1947) stated that species are controlled by temperature, basically in two ways, that is, continued existence and optimum temperatures required for reproduction and population. Hussain et al. (2002) observed that the temperature condition was within the tolerance for the thriving of Ostracoda fauna at Karikkattukuppam, Southeast coast of India. Temperature is one of the controlling factors of Ostracods in Pichavaram mangroves (Arul et al., 2003). Baskar et al. (2013) reported that the temperature of bottom water shows inverse proportion to the total population of Ostracods in northeast and east off Rameswaram, Southeast coast of India. Hussain (1992) reports that the differences in depths between the stations in his study area are only marginal; the rather small temperature variations (fractions of degree) are justified.

HYDROGEN ION CONCENTRATION (PH)

In the present study, the pH ranges from 8.0 to 8.16 with an average of 8.09 during fair weather season. Higher values have been observed at the 13th station, and a lower value was recorded at the 8th station. It is from 8.11 to 8.27 with an average of 8.18 during southwest monsoon; higher values have been observed at the 7th station, and a lower value was recorded at the 1st station, and it is from 7.99 to 8.21 with an average of 8.08 during northeast monsoon season. Higher values have been observed

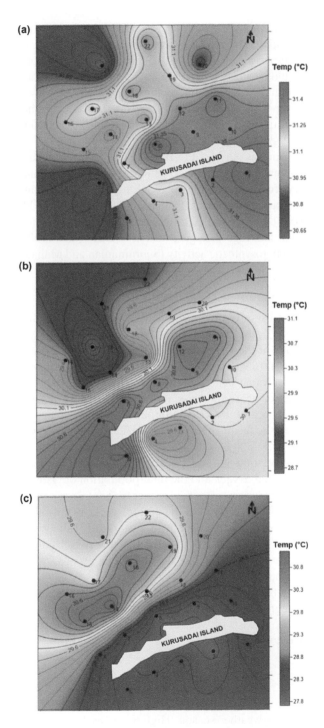

FIGURE 12.3 (a) Spatial distribution of BWT for FW. (b) Spatial distribution of BWT for SW. (c) Spatial distribution of BWT for NE.

at the 15th station, and a lower value was recorded at the 9th station. However, the pH has similar variation during all the seasons of the study area, as shown in spatial distribution diagrams in Figure 12.4a–c. Insignificant variation in pH has been observed, and it could be attributed to the relatively low land drainage and terrestrial runoff at this location and also due to the absence of freshwater discharge from any other perennial river. Considering all the seasons, the fair weather and northeast monsoon have a positive correlation, and southwest monsoon has a negative correlation with the total population of Ostracoda.

Its seasonal variation is attributed to factors such as the removal of CO_2 by photosynthesis throughout bicarbonate degradation, dilution of seawater by freshwater influx, low primary productivity, reduction of salinity, temperature, and decomposition of organic matter (Karuppasamy and Perumal, 2000; Rajasegar, 2003; Paramasivam and Kannan, 2005). The recorded high summer pH might be due to the influence of seawater penetration and high biological activity (Das et al., 1997) and due to the occurrence of high photosynthetic activity (Subramanian and Mahadevan, 1999).

Dissolved Oxygen

In the present study, dissolved oxygen recorded in the fair weather season ranges from 3.07 to 4.74 mg/L, the lowest being observed at the 22nd station and a higher value being recorded at the 7th station, and the mean value is 3.97 mg/L. During southwest monsoon, it ranges from 2.23 to 3.91 mg/L, lower being recorded at 1st station, and a higher value being observed at 7th station, and the mean value is 2.97 mg/L. During northeast monsoon, it ranges from 2.23 to 4.19 mg/L, the lowest value being observed at the 1st station, and a higher value being recorded at the 13th station, and the mean value is 3.16 mg/L, as shown in spatial distribution diagrams (Figure 12.5a–c). Based on the facts established, during fair weather season, negative correlation and southwest and northeast monsoon positive correlation between dissolved oxygen and the total population of Ostracods are inferred for the study area.

An increase in dissolved oxygen in the bottom water is favorable for a comparative abundance of Ostracoda population in waters off Tuticorin, southeast coast of India (Hussain et al., 1996). It is inferred that the dissolved oxygen content of water has a positive correlation with the total population of Ostracods, as reported off Rameswaram, southeast coast of India (Sridhar et al., 1998, 2011). Arul et al. (2003) reported that the dissolved oxygen is directly proportional to the population of Ostracods in mangrove environment at Pichavaram. Mohan et al. (2002) observed that a direct relationship exists between dissolved oxygen content of the bottom water and the living Ostracod population.

Salinity

In the present study, the salinity ranges from 32.28‰ to 34.56‰ with an average of 33.71‰ during fair weather season. Higher values have been observed at the 13th station, and a lower value was recorded at 6th station. It is from 32.20‰ to 32.70‰ with an average of 32.43‰ during southwest monsoon, higher values have

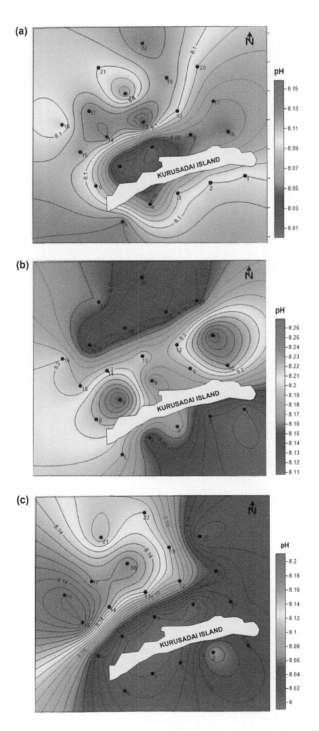

FIGURE 12.4 (a) Spatial distribution of pH for FW. (b) Spatial distribution of pH for SW. (c) Spatial distribution of pH for NE.

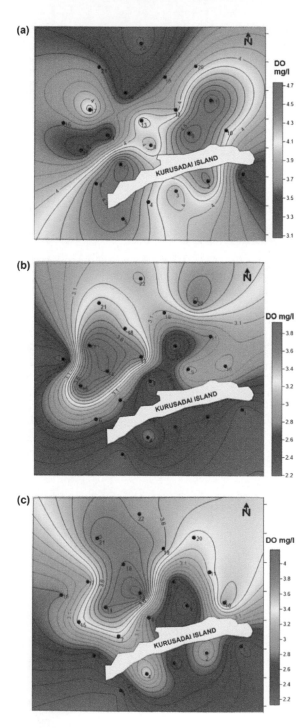

FIGURE 12.5   (a) Spatial distribution of DO for FW. (b) Spatial distribution of DO for SW. (c) Spatial distribution of DO for NE.

been observed at 7th station, and a lower value was recorded at 2nd station, and it is from 31.70‰ to 32.50‰ with an average of 31.95‰ during northeast monsoon season. Higher values have been observed at the 14th station, and a lower value was recorded at the 1st station, as shown in spatial distribution diagrams in Figure 12.6a–c. Salinity is one of the factors that were high during the fair weather season, and the higher values of salinity may be due to the low amount of rainfall, higher rate of evaporation, and the dominancy of neritic water; while the lower concentration of salinity was recorded in the northeast monsoon, which may be due to heavy rainfall and large quantity of freshwater inflow. In the present study, positive correlation was observed during fair weather and northeast monsoon and negative correlation during southwest monsoon with the population of Ostracoda. Sridhar et al. (1998) reported that increase in salinity correlates with increase in total population of Ostracods in Palk Bay, off Rameswaram, that supports the results of Hussain et al. (1996) in Gulf of Mannar.

CALCIUM CARBONATE

The occurrence of corals in the study area might have facilitated the recording of higher $CaCO_3$ in the sediments, the calcium carbonate ranges from 47.5% to 88.5% with an average of 76.0% during fair weather season. Higher values have been observed at the 6th station, and a lower value was recorded at the 5th station. It is from 36.5% to 89.0% with an average of 65.2% during southwest monsoon, and higher values have been observed at 12th station, and a lower value was recorded at 22nd station. It is from 35.5% to 86.5% with an average of 67.4% during northeast monsoon season, where higher values have been observed at 12th station, and a lower value was recorded at 21st station, as shown in spatial distribution diagrams from Figure 12.7a–c. The present study reveals that calcium carbonate has a negative correlation with the total population of Ostracoda during all the seasons. $CaCO_3$ in the study area cannot be compared with the population of Ostracods excepting that the population has an influence over the occurrence of corals and mangroves.

Jonathan and Ram Mohan (2003) reported that the Gulf of Mannar has a high content of calcium carbonate (75%) in the southern part of the study area attributed to carbonate materials of skeletal fragments that act as a nondetrital carrier phase. Baskar et al. (2013) reported that the coral reef that are present in the middle segment east off Rameswaram facilitates the recording of higher $CaCO_3$ in the sediments, and the total population of Ostracods was also recorded to be more in the middle segment. Sridhar et al. (2011) reported that the $CaCO_3$ has no significant correlation with the total population of Ostracods in east, northeast, and north off Rameswaram except that the coral reef might have facilitated the recording of higher $CaCO_3$ in the sediments. Manivannan et al. (1996) found that $CaCO_3$ is the major factor that controls the foraminiferal population in the Gulf of Mannar. Mohan et al. (2002) reported that the $CaCO_3$ content of the sediments is one of the important factors that govern the population of Ostracods and was found to be directly proportional to the population of Ostracods off Karikkattukuppam, Southeast coast of India.

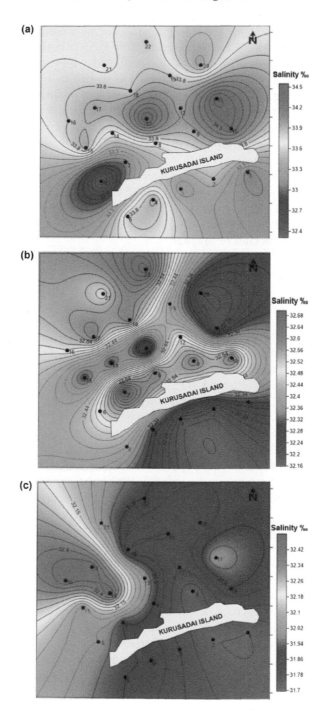

FIGURE 12.6 (a) Spatial distribution of salinity for FW. (b) Spatial distribution of salinity for SW. (c) Spatial distribution of salinity for FW.

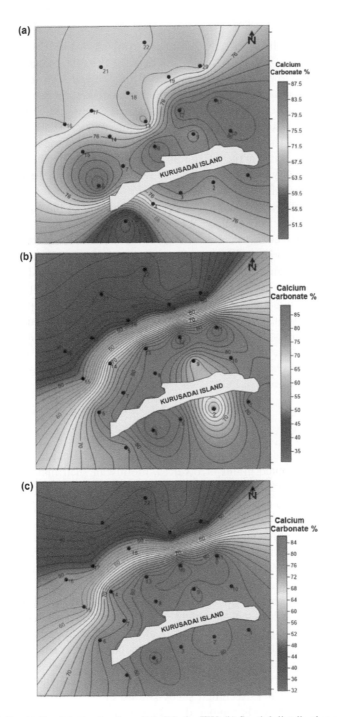

FIGURE 12.7   (a) Spatial distribution of $CaCO_3$ for FW. (b) Spatial distribution of $CaCO_3$ for SW. (c) Spatial distribution of $CaCO_3$ for NE.

Organic Matter

In the present study, the organic matter ranges from 0.47% to 1.41% with an average of 0.87% during fair weather season, where higher values have been observed at the 16th station, and a lower value was recorded at the 5th and 10th stations. It is from 0.18% to 1.79% with an average of 0.78% during southwest monsoon, where higher values have been observed at the 22nd station, and a lower value was recorded at the 9th station. It is from 0.41% to 1.57% with an average of 0.85% during northeast monsoon season, where higher values have been observed at the 17th station, and a lower value was recorded at the 9th station, as shown in spatial distribution diagrams in Figure 12.8a–c. Organic matter has a positive correlation with the total population of Ostracoda during all the seasons.

Whatley and Zhao (1987, 1988) considered the nature of the substrate to be the main controlling factor in Ostracod abundance; the highest values occur in association with medium to coarse sand rich in organic debris and lower values with gravel and those sands poor in carbonate and dominated by quartz. Sridhar et al. (2011) reported that the organic matter is inversely proportional to the total Ostracoda population in east, northeast, and north off Rameswaram. Hussain et al. (1997) state that the lower the organic matter of the sediments, the higher the population abundance off Tuticorin in the Gulf of Mannar. Baskar et al. (2013) reported that the organic matter is inversely proportional to the total population of Ostracods east off Rameswaram. But increase in organic matter has been correlated with increase in total population of Ostracods by Mohan et al. (2002) off Karikattukuppam, (near Chennai), southeast coast of India.

Nature of Substrate

The present study area, the terrestrial sand and coral sand consist of coral fragments mixed with comminuted grains of bivalves, molluscan shell materials, calcareous algae, and foraminifera that are heaped up on the coral reef platform forming Reef island. The results show that a significant amount of sand and silt in the samples consists of carbonate material, probably comminuted. Trefethen (1950) nomenclature is being followed to find out the substrate of different locations of the present study. Figure 12.9a–c show the type of substrate during different seasons for the study area. During fair weather season, sand and silty-sand substrate were identified. The most dominant substrate was found to be silty sand at stations 5, 6, 8, 11–13, 16–19, and 12–22. Sand was recorded in stations 1–4, 7, 9, 10, 14, 15, and 20. During southwest monsoon, it is sand, silty-sand, and sandy-silt substrate. The most dominant substrate was found to be sand (stations 1–14) except at station 5. Silty sand was recorded at station 5, 15, and 16. The stations from 17 to 22 recorded sandy silt. During northeast monsoon, it is sand, silty-sand, and sandy-silt substrate. The most dominant substrate was to be sand at stations 1–14. Silty-sand was recorded at stations 15–22 (except station 19). The station 19 recorded sandy-silt substrate. Silty sand and sandy silt recorded more of Ostracods compared to the substrate sand in the study area. It is observed that the finer the grain size of the sediment, the higher the population of Ostracods from sand to silty-sand substrate.

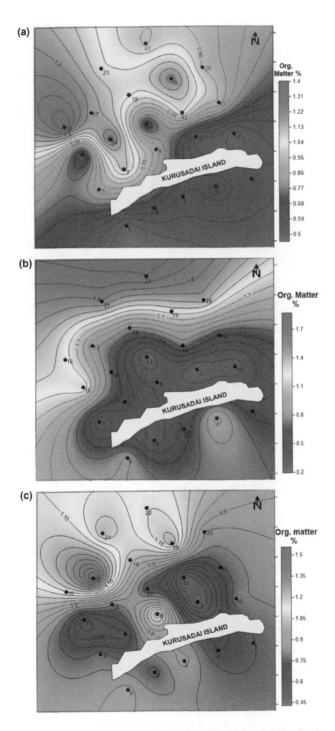

FIGURE 12.8 (a) Spatial distribution of Org.M for FW. (b) Spatial distribution of Org.M for SW. (c) Spatial distribution of Org.M for NE.

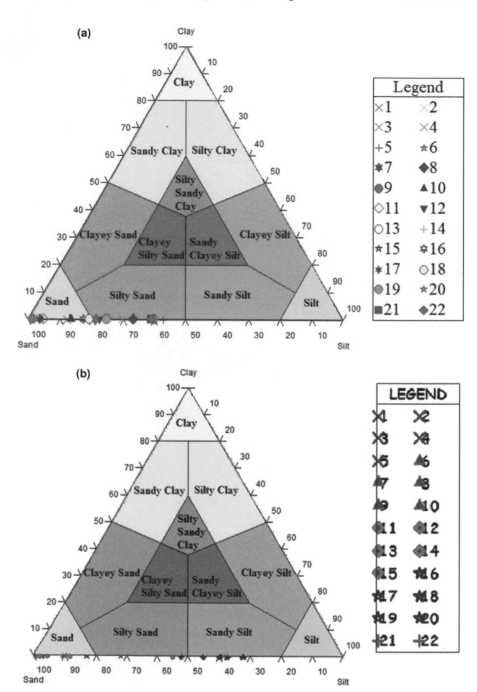

FIGURE 12.9 (a) Types of substrates during fair weather season. (Based on Trefethen, 1950.) (b) Types of substrates during southwest monsoon. (Based on Trefethen, 1950.) (c) Types of substrates during northeast monsoon. (Based on Trefethen, 1950.)

(Continued)

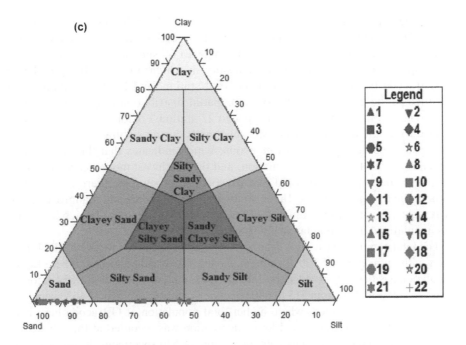

FIGURE 12.9 (CONTINUED) (a) Types of substrates during fair weather season. (Based on Trefethen, 1950.) (b) Types of substrates during southwest monsoon. (Based on Trefethen, 1950.) (c) Types of substrates during northeast monsoon. (Based on Trefethen, 1950.)

Sridhar et al. (1998) reported from Palk Bay, off Rameswaram that the substrate sand and silty sand are more congenial substrate for the abundance of the fauna. Baskar et al. (2013) observed that the substrate sand with higher silt content is found to be more congenial substrate for Ostracod population in east off Rameswaram. Vaidya et al. (1995) reported a higher number of Ostracods in substrate of medium to fine grained sand, whereas poor presence was noticed in coarse grained sand. The substrate texture has a control on the kind of Ostracod fauna that can colonize a particular sediment type (Brasier, 1980). The texture stability of the sediment composing the substrate exerts a strong influence on marine Ostracod, just as it does on brackish water and freshwater forms. Smooth shelled forms are predominant in fine-grained mud, whereas more ornamental forms are being found in coarser or in more calcareous sediments (Benson, 1961; Brasier, 1980).

POPULATION OF OSTRACODA

The living population of Ostracoda station-wise is observed from 25 to 87 specimens with an average of 44 specimens during fair weather season, in which higher population has been observed at the 18th station, and lower population was recorded at the 2nd station. It is from 24 to 61 specimens with an average

of 39 specimens during southwest monsoon, in which higher living population of Ostracoda has been observed at the 16th station, and a lower population
was recorded at the 18th station. It is from 18 to 65 specimens with an average
of 43 specimens during northeast monsoon, in which higher living population
of Ostracoda has been observed at the 16th station, and lower population was
recorded at the 5th station, as shown in spatial distribution diagrams in Figure
12.10a–c. The living population (from 1 to 22 stations) is found to be 968 specimens during fair weather season, 855 specimens during southwest monsoon, and
948 specimens during northeast monsoon, which indicates that the fair weather
season recorded the highest population, and the southwest monsoon recorded the
lowest population. It is observed that irrespective of seasons, there is no station
without the living specimens. In the study area where corals and mangroves coexist is a clear record of the influence of mangroves that supports Ostracoda to have
a congenial and accommodative environment.

The total population (living + dead) of Ostracoda is observed to be 340–644 specimen with an average of 500 specimens during fair weather season, where higher
population has been observed at the 18th station, and lower population was recorded
at the 2nd station. It is from 315 to 680 specimens with an average of 463 specimens
during southwest monsoon, where higher total population of Ostracoda have been
observed at the 19th station, and lower population was recorded at the 7th station.
It is from 345 to 658 specimens with an average of 520 specimens during northeast
monsoon, where higher total population of Ostracoda has been observed at the 18th
station, and lower population was recorded at the 4th station, as shown in spatial
distribution diagrams in Figure 12.11a–c. The total population (from 1 to 22 stations)
is found to be 11,006 specimens during fair weather season, 10,181 specimens during
southwest monsoon, and 11,432 specimens during northeast monsoon, which indicates that the northeast monsoon recorded the highest population, and the southwest
monsoon recorded the lowest population.

Quite interesting facts arrived at, comparing the total population of Ostracods
off Rameswaram, are as follows: east off Rameswaram, from near shore to 15 km in
the inner shallow shelf, silty sand and sand are congenial substrate for Ostracods to
thrive. Sridhar et al. (1998) reported the total population as 2,477 during SW monsoon, when the field work was done in the year 1982 in Palk Bay. The middle segment where corals are plenty recorded more population than near-shore and farther
stations. After three decades (28 years), for the same location in Palk Bay, when field
work was done in the year 2010, Baskar et al. (2013) reported that the total population was 2,964 during SW monsoon, which indicates an increase in population that is
being attributed to a better thriving environment. Northeast off Rameswaram, from
near shore to 15 km, where clay being the only substrate, Baskar et al. (2013) reported
total population as 1,099 during SW monsoon, attributed to the absence of corals
when it is compared with eastern transect.

The present study is southwest off Rameswaram, around Kurusadai island where
corals and mangroves coexist; recorded 10,181 specimens during SW monsoon
(August 2013) alone is a clear record of the influence of mangroves that supports
Ostracods to have a congenial and accommodative environment. The population is almost five times more comparing the eastern transect, off Rameswaram,

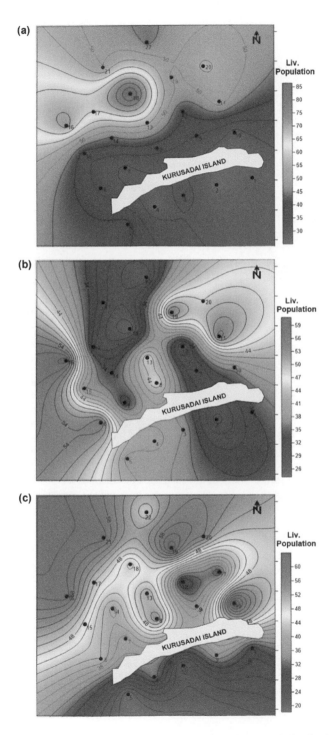

FIGURE 12.10 (a) Spatial distribution of Liv.Pop for FW. (b) Spatial distribution of Liv.Pop for SW. (c) Spatial distribution of Liv.Pop for NE.

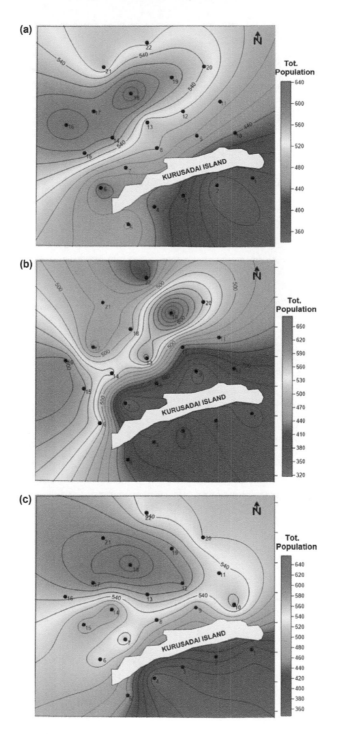

FIGURE 12.11 (a) Spatial distribution of Tot.Pop for FW. (b) Spatial distribution of Tot.Pop for SW. (c) Spatial distribution of Tot.Pop for NE.

where corals are found, and it is almost ten times more comparing the northeastern transect, off Rameswaram, where corals are absent. This reveals very clearly
that the mangroves have a positive influence over the thriving of Ostracods. In the
present study, the following species were found to be widespread and abundant:
(1) *Caudites javana*, (2) *Caudites sublevis*, (3) *Hemicytheridea reticulata*, (4)
Jankeijcythere mckenziei, (5) *Neomonoceratina iniqua*, (6) *Loxocorniculum lilljeborgii*, (7) *Neosinocythere dekrooni*, (8) *Tanella gracilis*, (9) *Hemicytheridea paiki*,
(10) *Loxoconcha megaporaindica*, (11) *Mutilus pentoekensis*, and (12) *Xestoleberis
variegata* which are shown in Plate 12.1. All these widespread and abundant species
have an affinity toward mangroves.

ECOLOGY OF WIDESPREAD AND ABUNDANT SPECIES

In the present study, the ecology of species that occur as "widespread and abundant" is presented. Twelve species were found to be present in all the stations during all seasons as living and total population to be more than ten in number. The
parameters considered for their ecology are tabulated in Table 12.4. For each and
every species with their environmental parameters being considered, their ecology
is assessed.

CAUDITES JAVANA AND CAUDITES SUBLEVIS

Caudites javana and *Caudites sublevis* have higher population, in the study area,
from stations 2, 6 to 18 during all the seasons. They have a living population
higher during fair weather season at the 16th station. Similarly, temporal distribution of the total population size (all the station put together) for *C. javana* and *C.
sublevis* are found to be maximum in fair weather season, which is followed by
northeast monsoon. Considering the ecological parameters of these stations mentioned above, during all the seasons, the ecology of both the species is being fixed
as follows:

> Depth: 1.0–5.0 m
> Temperature: 27.9°C–31.8°C
> pH: 7.99–8.27
> DO: 2.79–4.74 mg/L
> Salinity: 31.70‰–34.56‰
> CaCO$_3$: 41%–89%
> Org. matter: 0.19%–1.41%
> Substrate: Sand and silty sand

HEMICYTHERIDEA RETICULATA

Hemicytheridea reticulata has highest population, in the study area, from stations
6 to 14 during all the seasons. It has a living population higher during fair weather
season at the 9th station. Similarly, temporal distribution of the total population size
(all the station put together) for *Hemicytheridea reticulata* is found to be maximum
in northeast monsoon, which is followed by fair weather season. Considering the

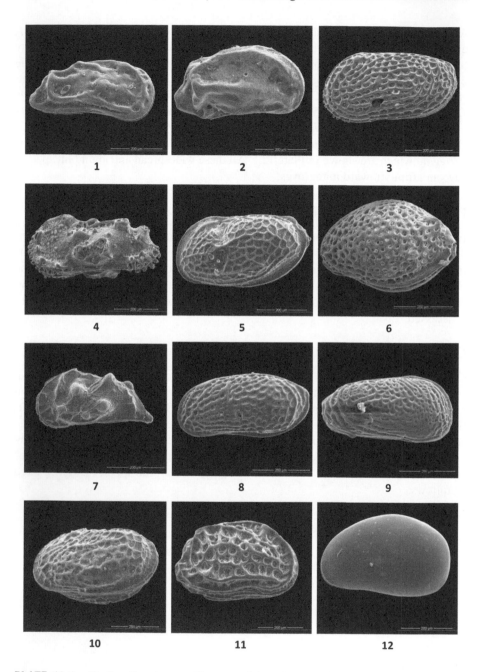

PLATE 12.1   (1) *Caudites javana* (Carapace right valve), (2) *Caudites sublevis* (Carapace right valve), (3) *Hemicytheridea reticulata* (Carapace left valve), (4) *Jankeijcythere mckenziei* (Carapace left valve), (5) *Neomonoceratina iniqua* (Carapace left valve), (6) *Loxocorniculum lilljeborgii* (Carapace left valve), (7) *Neosinocythere dekrooni* (Carapace right valve), (8) *Tanella gracilis* (Carapace left valve), (9) *Hemicytheridea paiki* (Carapace right valve), (10) *Loxoconchamegapora indica* (Carapace right valve), (11) *Mutilus pentoekensis* (Carapace right valve), (12) *Xestoleberis variegata* (Carapace right valve).

TABLE 12.4

Ecological Parameters for Widespread and Abundant Species of the Study Area

S.N	Species/Parameters	Depth (m)		Temp. (°C)		pH		DO (mg/L)		Salinity (‰)		CaCO$_3$ (%)		Org. Matter (%)		Living Popl.		Total Popl.		Substrate
		Min	Max	Min	Max	Min	Max	Min	Max	Min	Max	Min	Max	Min	Max	Min	Max	Min	Max	
1	*Caudites javana*	1.0	5.0	27.9	31.8	7.99	8.27	2.79	4.74	31.70	34.56	41.0	89.0	0.19	1.41	92	186	198	590	Sand/Siltysand
2	*Caudites sublevis*	1.0	5.0	27.9	31.8	7.99	8.27	2.79	4.74	31.70	34.56	41.0	89.0	0.19	1.41	65	137	209	352	Sand/Siltysand
3	*Hemicytheridea Reticulate*	1.0	5.0	29.2	31.5	7.99	8.19	3.07	4.74	31.70	33.95	53.5	69.5	0.38	1.16	50	110	277	322	Sand
4	*Jankeijcythere Mckenziei*	1.0	5.0	30.9	32.2	8.09	8.20	2.23	4.18	31.70	34.56	67.0	83.0	0.19	1.24	31	94	217	263	Sand/Siltysand
5	*Neomonoceratinailniqua*	1.0	5.0	29.4	31.0	8.13	8.20	3.63	4.18	32.00	34.56	67.0	81.5	0.19	0.64	52	102	235	349	Sand/Siltysand
6	*Loxocorniculum tiltjeborgii*	1.0	5.0	28.6	31.2	8.05	8.13	2.23	4.18	32.50	33.95	47.5	79.0	0.41	1.23	146	218	522	601	Siltysand/Sandysilt
7	*Neosinocythere Dekrooni*	1.0	5.0	28.6	31.2	8.05	8.13	2.23	4.18	32.50	33.95	47.5	79.0	0.41	1.23	75	115	210	364	Siltysand/Sandysilt
8	*Tanella gracilis*	1.0	5.0	28.6	31.4	7.99	8.27	2.79	4.74	32.00	34.56	75.0	88.5	0.18	1.57	43	70	395	485	Sand/Siltysand
9	*Hemicytheridea paiki*	1.0	5.0	29.1	30.3	8.00	8.07	3.07	3.91	32.20	33.95	37.0	69.0	0.94	1.23	57	130	298	352	Sand/Siltysand
10	*Loxoconchamegapora Indica*	1.0	5.0	29.3	30.6	8.10	8.16	3.35	3.62	32.50	33.65	35.5	71.5	0.99	1.59	45	87	205	340	Siltysand
11	*Mutilus pentoekensis*	1.0	5.0	29.3	30.6	8.10	8.16	3.35	3.62	32.50	33.65	35.5	71.5	0.99	1.59	34	80	178	453	Siltysand
12	*Xestoleberis variegate*	1.0	5.0	29.9	31.2	8.11	8.13	3.07	3.63	31.80	32.40	36.0	70.0	0.74	1.57	60	96	287	357	Siltysand/Sandysilt

ecological parameters of these stations mentioned above, during all the seasons, the ecology of the species is being fixed as follows:

Depth: 1.0–5.0 m
Temperature: 29.2°C–31.5°C
pH: 7.99–8.19
DO: 3.07–4.74 mg/L
Salinity: 31.70‰–33.95‰
CaCO$_3$: 53.5%–69.5%
Org. matter: 0.38%–1.16%
Substrate: Sand

JANKEIJCYTHERE MCKENZIEI

Jankeijcythere mckenziei has the highest population, in the study area, from stations 1 to 18 during all the seasons. It has a living population higher during fair weather season at the 18th station. Similarly, temporal distribution of the total population size (all the station put together) for *Jankeijcythere mckenziei* is found to be maximum in northeast monsoon, which is followed by fair weather season. Considering the ecological parameters of these stations mentioned above, during all the seasons, the ecology of the species is being fixed as follows:

Depth: 1.0–5.0 m
Temperature: 30.90°C–32.20°C
pH: 8.09–8.20
DO: 2.23–4.18 mg/L
Salinity: 31.70‰–34.56‰
CaCO$_3$: 67.0%–83.0%
Org. matter: 0.19%–1.24%
Substrate: Sand and silty sand

NEOMONOCERATINA INIQUA

Neomonoceratina iniqua has high population, in the study area, for stations 1 and 5 as well as from 13 to 19 during all the seasons. It has a living population higher during fair weather season at the 13th station. Similarly, temporal distribution of the total population size (all the station put together) for *Neomonoceratina iniqua* is found to be maximum in southwest monsoon, which is followed by fair weather season. Considering the ecological parameters of these stations mentioned above, during all the seasons, the ecology of the species is being fixed as follows:

Depth: 1.0–5.0 m
Temperature: 29.4°C–31.0°C
pH: 8.13–8.20
DO: 3.63–4.18 mg/L

Salinity: 32.0‰–34.56‰
CaCO$_3$: 67%–81.5%
Org. matter: 0.19%–0.64%
Substrate: Sand and silty sand

LOXOCORNICULUM LILLJEBORGII AND NEOSINOCYTHERE DEKROONI

Loxocorniculum lilljeborgii and *Neosinocythere dekrooni* has the highest population, in the study area, from stations 6, 9 and from 15 to 20 during all the seasons. They have a living population higher during fair weather season at the 17th station. Similarly, temporal distribution of the total population size (all the station put together) for *Loxocorniculum lilljeborgii* is found to be maximum in fair weather season which is followed by southwest monsoon, and for *Neosinocythere dekrooni* it is found to be maximum in fair weather season, which is followed by northeast monsoon. Considering the ecological parameters of these stations mentioned above, during all the seasons, the ecology of both the species is being fixed as follows:

Depth: 1.0–5.0 m
Temperature: 28.6°C–31.20°C
pH: 8.05–8.13
DO: 2.23–4.18 mg/L
Salinity: 32.50‰–33.95‰
CaCO$_3$: 47.5%–79.0%
Org. matter: 0.41%–1.23%
Substrate: Silty sand and sandy silt

TANELLA GRACILIS

Tanella gracilis has highest population, in the study area, from stations 4 to 8, and 18 during all the seasons. It has a living population higher during southwest monsoon at 11th station. Temporal distribution of the total population size (all the station put together) for *Tanella gracilis* is found to be maximum in southwest monsoon, which is followed by northeast monsoon. Considering the ecological parameters of these stations mentioned above, during all the seasons, the ecology of the species is being fixed as follows:

Depth: 1.0–5.0 m
Temperature: 28.6°C–31.4°C
pH: 7.99–8.27
DO: 2.79–4.74 mg/L
Salinity: 32.0‰–34.56‰
CaCO$_3$: 75%–88.5%
xOrg. matter: 0.18%–1.57%
Substrate: Sand and silty sand

HEMICYTHERIDEA PAIKI

Hemicytheridea paiki has the highest population, in the study area, from stations 7 to 10, 17 and 20, during all the seasons. It has a living population higher during fair weather season at the 20th station. Similarly, the temporal distribution of the total population size (all the station put together) for *Hemicytheridea paiki* is found to be maximum in southwest monsoon, which is followed by fair weather season. Considering the ecological parameters of these stations mentioned above, during all the seasons, the ecology of the species is being fixed as follows:

Depth: 1.0–5.0 m
Temperature: 29.1°C–30.8°C
pH: 8–8.07
DO: 3.07–3.91 mg/L
Salinity: 32.20‰–33.95‰
$CaCO_3$: 37%–69%
Org. matter: 0.94%–1.23%
Substrate: Sand and silty sand

LOXOCONCHA MEGAPORAINDICA AND MUTILUS PENTOEKENSIS

Loxoconcha megaporaindica and *Mutilus pentoekensis* have the highest population, in the study area, from stations 17 and 21, during all the seasons. They have a living population higher during fair weather season at the 21st station. Similarly, the temporal distribution of the total population size (all the station put together) for *Loxoconcha megaporaindica* and *Mutilus pentoekensis* is found to be maximum in fair weather season, which is followed by northeast monsoon. Considering the ecological parameters of these stations mentioned above, during all the seasons, the ecology of both the species is being fixed as follows:

Depth: 1.0–5.0 m
Temperature: 29.3°C–30.6°C
pH: 8.10–8.16
DO: 3.35–3.62 mg/L
Salinity: 32.50‰–33.65‰
$CaCO_3$: 35.5%–71.5%
Org. matter: 0.99%–1.59%
Substrate: Silty sand

XESTOLEBERIS VARIEGATA

Xestoleberis variegata has the highest population, in the study area, from stations 9, 12, 17, and 22, during all the seasons. It has a living population higher during northeast monsoon at the 19th station. Similarly, the temporal distribution of the total population size (all the station put together) for *Xestoleberis variegata* is found to be maximum in northeast monsoon, which is followed by fair weather season. Considering the ecological parameters of these stations mentioned above, during all the seasons, the ecology of the species is being fixed as follows:

Depth: 1.0–5.0 m
Temperature: 29.9°C–31.20°C
pH: 8.11–8.13
DO: 3.07–3.63 mg/L
Salinity: 31.80‰–32.40‰
$CaCO_3$: 36.0%–70%
Org. matter: 0.74%–1.57%
Substrate: Silty sand and sandy silt

CONCLUSION

The water depth has a positive correlation with the total population of Ostracoda during all the seasons. The present study reveals that the bottom water temperature has a negative correlation during fair weather season and southwest monsoon, and it has a positive correlation during northeast monsoon. pH during fair weather and northeast monsoon has a positive correlation, and it has a negative correlation during southwest monsoon with the total population. Based on the facts established, during fair weather season, DO has a negative correlation, and during southwest and northeast monsoons, it has a positive correlation with the total population of Ostracoda. Salinity has a positive correlation during fair weather and northeast monsoon and a negative correlation during southwest monsoon with the population of Ostracoda. $CaCO_3$ in the study area cannot be compared with the Ostracoda population except that the population has an influence over the occurrence of corals. Organic matter has a positive correlation with the total population of Ostracoda during all the seasons. Overall, the most dominant substrate was found to be sand. But the higher population of Ostracoda was recorded in silty sand and sandy silt substrate when compared to sand. It is observed that the finer the grain size of the sediment, the higher the population of Ostracods.

The following 12 species are found to be widespread and abundant: (1) *Caudites javana*, (2) *Caudites sublevis*, (3) *Hemicytheridea reticulata*, (4) *Jankeijcythere mckenziei*, (5) *Neomonoceratina iniqua*, (6) *Loxocorniculum lilljeborgii*, (7) *Neosinocythere dekrooni*, (8) *Tanella gracilis*, (9) *Hemicytheridea paiki*, (10) *Loxoconcha megaporaindica*, (11) *Mutilus pentoekensis*, and (12) *Xestoleberis variegata*, and their ecology is being assessed. In the study area, where corals and mangroves coexist, the population is almost five times more compared to the eastern transect off Rameswaram where corals only are found; it is almost ten times more compared to the northeastern transect off Rameswaram where the region is devoid of corals, which is a clear record of the influence of mangroves along with corals that support Ostracoda to have a congenial and accommodative environment.

ACKNOWLEDGEMENT

The authors place on record their gratitude to UGC, as the research work was carried out under the UGC-CPEPA Project {F.No:8-2/2008 (NS/PE), dt. 14.11.2011} and are grateful to the life support given by the team of fishermen during field work. They are thankful to the Department of Forestry, Coastal Police, and Indian Navy for according permission to carry out the field work off Kurusadai island. They are thankful to the authorities of the University of Madras, and they acknowledge the help rendered by the Department of Applied Geology, University of Madras.

REFERENCES

Arul, B., Sridhar, S.G.D., Hussain, S.M., Darwin Felix, A. and Periakali, P. (2003). Distribution of recent benthic Ostracoda from the sediments of Pichavaram mangroves, Tamil Nadu, Southeast coast of India. *Bull. Pure Appl. Sci.* 22:55–73.

Baskar, K., Sridhar, S.G.D., Sivakumar, T. and Hussain, S.M. (2013). Seasonal comparison of recent benthic ostracoda response to sediment type in the Palk Bay, off Rameswaram island, Tamil Nadu. *Int. J. Earth Sci. Eng.* 6:1047–1059.

Bate, R.H. (1971). The distribution of recent ostracoda in the Abu Dhabi Lagoon, Persian Gulf. In: H.J. Oertli (Ed.), *Paleoecologie des Ostracodes, Colloque Pau (1970)*, Bull. Centre Rech. Pau-SNPA, Suppl. 5:239–256.

Benson, R.H. (1961). Ecology of ostracod assemblages. In: R.C. Moore (Ed.), *Treatise on Invertebrate Paleontology, Part Q, Arthropoda 3, Ostracoda*, Geol. Soc. Amer, Q56–Q63.

Benson, R. (1990). Ostracoda and the discovery of global Cainozoic palaeoceanography events. In: R. Whately and C. Maybury (Eds.), *Ostracoda and Global Events*, Chapman and Hall, London, 41–58.

Bhatia, S.B. and Kumar, S. (1979). Recent Ostracoda from off Karwar, west coast of India, *Proceedings of the VII International Symposium on Ostracodes*, Beograd, 173–178.

Brasier, M.D. (1980). *Microfossils*, George Allen and Unwin Ltd, London, 193p.

Carbonel, P., Colin, J.P., Danielopol, D.L., Löffler, H. and Neustrueva, I. (1988). Paleoecology of limnicostracodes: A review of some major topics. *Palaeogeogr. Palaeoclimatol. Palaeoecol.* 62(1):413–461.

Chacko, P.I. (1950). Marine plankton from waters around the Krusadai Island. *Proc.: Plant Sci.* 31(3):162–174.

Chaplin, J.A. and Ayre, D.J. (1997). Genetic evidence of widespread dispersal in a partheno-genetic freshwater ostracod. *Heredity.* 78(1): 57–67. DOI: 10.1038/hdy.1997.7.

Cohen, A.S. and Johnston, M.R. (1987). Speciation in brooding and poorly dispersing lacustrine organisms. *Palaios.* 5:426–435.

Das, J., Das, S.N. and Sahoo, R.K. (1997). Semidiurnal variation of some physicochemical parameters in the Mahanadi estuary, east coast of India. *Indian J. Mar. Sci.* 26:323–326.

De Deckker, P. (1981). Ostracods of athalassic saline lakes. *Hydrobiologia* 81:131–144. DOI: 10.1007/BF00048710.

De Deckker, P. and Forester, R.M. (1988). The use of ostracods to reconstruct continental palaeoenvironmental records. In: P. De Deckker, J. Colin and J. Peypouquette (Eds.), *Ostraocda in the Earth Sciences*, Elsevier, Amsterdam, 175–199.

Elumalai, K., Hussain, S.M. and Scott Immanuel Dhas, C. (2010). Recent Benthic Ostracoda from the sediments of Ennore Creek, Chennai, Tamil Nadu, India. *J. Palaeo. Soc. India.* 55(1):11–22.

Evans, G., Murray, J.W., Biggs, H.E.J., Bate, R. and Bush, P.R. (1973). The oceanography, ecology, sedimentology and geomorphology of parts of the Trucial Coast barrier island complex, Persian Gulf. In: B.H. Purser (Eds.), *The Persian Gulf*, Springer, Berlin Heidelberg, 233–277. DOI: 10.1007/978-3-642-65545-6_14.

Gaudette, H.E., Flight, W.R., Toner, L. and Folger, D.W. (1974). An inexpensive titration method for determination of organic carbon in recent sediments. *J. Sed. Petrol.* 44:249–253.

Gopalakrishna, K., Hussain, S.M., Mahesh Bilwa, L. and Ayisha, V.A. (2007). Recent benthic Ostracoda from the inner-shelf off the Malabar coast, Kerala, Southeast coast of India. *J. Palaeo. Soc. India.* 52(1):56–68.

Hussain, S. M. (1992). Systematics, ecology and distribution of Recent Ostracoda from the Gulf of Mannar, off Tuticorin, Tamil Nadu. Ph.D. thesis, submitted to the University of Madras.

Hussain, S.M. and Mohan, S.P. (2000). Recent Ostracoda from Adyar river estuary, Chennai, Tamil Nadu. *J. Palaeo. Soc. India.* 45:471–481.

Hussain, S.M. and Mohan, S.P. (2001). Distribution of recent benthic Ostracoda in Adyar river estuary, east coast of India. *Indian J. Mar. Sci.* 30:53–56.

Hussain, S.M., Ragothaman, V. and Manivannan, V. (1996). Distribution of Ostracoda in waters off Tuticorin, south-east coast of India. *Indian J. Mar. Sci.* 25:78–80.

Hussain, S.M., Manivannan, V. and Ragothaman, V. (1997). Sediment-Ostracode relationship in the Gulf of Mannar, off Tuticorin, East Coast of India. *J. Nepal Geol. Soc.* 15:33–37.

Hussain, S.M., Ganesan, P., Ravi, G., Mohan, S.P. and Sridhar, S.G.D. (2007). Distribution of ostracoda in marine and marginal marine habitats off Tamil Nadu and adjoining areas, south east coast of India and Andaman Islands: Environmental implications. *Indian J. Mar. Sci.* 36(4):369–377.

Hussain, S.M., Mohan, S.P. and Manivannan, V. (2002). Microenvironmental inferences of recent benthic ostracoda, from Gulf of Mannar, off Tuticorin, east coast of India, *Proceedings of the National Seminar on Management of Natural Resources*, Nagarjuna University, 23–43.

Hutchins, L. (1947). The bases for temperature zonation in geographical distribution. *Ecol. Monogr.* 17(3), 325–335. DOI: 10.2307/1948663.

Jain, S.P. (1978). Recent Ostracoda from Mandvi Beach, west coast of India. *Bull. Indian Geol. Assoc.* 11 (2):89–139.

Jain, S.P. (1981). Recent Ostracoda from the south-west Kerala Coast, India. *Bull. Indian Geol. Assoc.*, 14(2):107–120.

Jain, S.P. and Bhatia, S.B. (1978). Recent benthonic foraminifera from Mandvi, Kutch. *Proceedings of the VII Indian Colloquium On Micropaleontology & Stratigraphy*, 153–174.

Jonathan, M.P. and Ram Mohan, V. (2003). Heavy metals in sediments of the inner shelf off the Gulf of Mannar, southeast coast of India. *Mar. Pollut. Bull.* 46:263–268.

Karuppasamy, P.K. and Perumal, P. (2000). Biodiversity of zooplankton at Pichavaram mangroves, south India. *Adv. Biosci.* 19(2):23–32.

Kesling, R.V. and Crafts, F.C. (1962).Ontogenetic increase in Archimedean weight of the ostracod Chlamydotheca unispinosa (Baird). *Am. Midl. Nat.* 68:149–153.

Khosla, S.C., Mathur, A.K. and Pant, P.C. (1982). Ecology and distribution of recent ostracods in the Miani lagoon, Saurashtra coast. In: S.S. Mehr (Ed.), *First National Seminar on Quaternary Environment* (Recent Researches in Geology Series), Hindustan Publishing Corp, New Delhi, Vol. 9, 361–371.

Knudsen, M. (1901). *Hydrographical Tables*, G.M. Manufacturing Co., New York, 63p.

Krumbein, W.C. and Pettijohn, F.J. (1938). *Manual of Sedimentary Petrography*, Appleton Century Co. Inc., New York, 549 p.

Külköylüoğlu, O. and Vinyard, G.L. (2000). Distribution and ecology of freshwater Ostracoda (Crustacea) collected from springs of Nevada, Idaho, and Oregon: A preliminary study. *West. N. Am. Nat.* 60:291–303.

Külköylüoğlu, O., Dugel, M., Balci, M., Deveci, A., Avuka, D. and Kilic, M. (2010). Limnoecological relationships between water level fluctuations and Ostracoda (Crustacea) species composition in Lake Sunnet (Bolu, Turkey). *Turk. J. Zool.* 34(4):429–442.

Kumar, V. and Hussain, S.M. (1997). A report on recent ostracoda from Pichavaram mangroves, Tamil Nadu. *Geosci. J.* 18(2):131–139.

Deepali, K., Sridhar, S.G.D., Hussain, S.M. and Maniyarasan S. (2019). Vertical distribution of recent benthic ostracoda at Pullivasal and Kurusadai Islands, Gulf of Mannar, Southeast Coast of India. In: H. Chenchouni, E. Errami, F. Rocha and L. Sabato (Eds.), *Exploring the Nexus of Geoecology, Geography, Geoarcheology and Geotourism: Advances and Applications for Sustainable Development in Environmental Sciences and Agroforestry Research. Proc 1st CAJG 2018.* Advances in Science, Technology & Innovation (IEREK Interdisciplinary Series for Sustainable Development), Springer, Switzerland, 79–84. DOI: 10.1007/978-3-030-01683-8_17.

Lopez, L.C.S., Gonçalves, D.A., Mantovani, A. and Rios, R.I. (2002). Bromeliad ostracods pass through amphibian (Scinaxaxperpusillus) and mammalian guts alive. *Hydrobiologia.* 485(13):209–211.

Loring, D.H. and Rantala, R.T.T. (1992). Manual for the geochemical analyses of marine sediments and suspended particulate matter. *Earth Sci. Rev.* 32:235–283.

Manivannan, V., Kumar, V., Ragothaman, V. and Hussain, S.M. (1996). Calcium carbonate - A major factor in controlling forminiferal population in the Gulf of Mannar, off Tuticorin, Tamil Nadu. *Proceedings XV Indian Colloquiumon Micropalaeontology and Stratigraphy*, Dehradun, 381–385.

Maniyarasan, S. 2016. Systematics, distribution and ecology of recent benthic Ostracoda, off Kurusadai island, Gulf of Mannar, Southeast coast of India. Unpublished Ph.D., Thesis, submitted to the University of Madras, Chennai.

Montenegro, M.E., Pugliese, N. and Sciuto, F. (2004). Shallow water ostracods near the Mea Khlong river mouth (NW Gulf of Thailand). *Boll. Soc. Paleontol. Ital.* 43:225–234.

Mohan, S.P., Ravi, G. and Hussain, S.M. (2002).Distribution of recent benthic Ostracoda off Karikkattukuppam (Near Chennai), southeast coast of India. *Indian J. Mar. Sci.* 31:315–320.

Puglise, N., Montenegro, M.E., Sciuto, F., and Chaimanee, N. (2006). Environmental monitoring through the shallow marine ostracods of Phetchaburi area (NW Gulf of Thailand). *Grazybowki Foundation Special Publication.* 11:85–90.

Paramasivam, S. and Kannan, L. (2005). Physico-chemical characteristics of Muthupettai mangrove environment, Southeast coast of India. *Int. J. Ecol. Environ. Sci.* 31:273–278.

Puri, H.S. (1960). Recent Ostracoda from the west coast of Florida, *Trans. Gulf. Assoc. Soc.*, X:107–149.

Rajasegar, M. (2003). Physico-chemical characteristics of the Vellar estuary in relation to shrimp farming. *J. Environ. Biol.* 24:95–101.

Saraswati, P.K. and Srinivasan, M.S. (2016). *Micropaleontology: Principles and Applications*, Springer International Publishing, Switzerland, DOI: 10.1007/978-3-318–14574-7, 224p.

Sridhar, S.G.D., Hussain, S.M., Kumar, V. and Periakali, P. (1998). Benthic ostracod responses to sediments in the Palk Bay, off Rameswaram, south-east coast of India. *J. Indian Assoc. Sedimentol.* 17(2):187–195.

Sridhar, S.G.D., Hussain, S.M., Kumar, V. and Periakali P. (2002). Recent Ostracoda from Palk Bay, off Rameswaram, southeast coast of India. *J Palaeo. Soc. India.* 47:17–39.

Sridhar, S.G.D., Baskar, K., Hussain, S.M., Solai, A. and Kalaivanan, R. (2011). Distribution and diversity of recent benthic Ostracoda off Rameswaram island, Southeast coast of India, Tamil Nadu. In: O.P. Varma, B.C. Sarkar, A.K. Varma, M.K. Mukherjee and S. Singh (Eds.), *Proceeding 17th Convention of the Geological Congress and International Conference NPESMD 2011*, ISM, Dhanbad, 283–298, ISBN: 978-81-8465-954-2.

Sridhar, S.G.D., Deepali, K., Prabhu, S., Hussain, S.M. and Maniyarasan, S. (2018). Distribution of recent benthic ostracoda, around Pullivasal and Poomarichan islands, off Rameswaram, Gulf of Mannar, Southeast coast of India. *J. Palaeo. Soc. India.* 63(2): 1–11.

Strickland, J.D.H. and Parsons, T.R. (1968). *A Practical Handbook of Seawater Analysis*, Fisheries Research. Board of Canada, Ottawa, Bulletin 167, 311 p.

Subramanian, B. and Mahadevan, A. (1999). Seasonal and diurnal variation of hydrobiological characters of coastal water of Chennai (Madras), Bay of Bengal. *Indian J. Mar. Sci.* 28(4):429–433.

Trefethen, J.M. (1950). Classification of sediments. *Am. J. Sci.* 248:55–62.

Vaidya, A.S., Mannikeri, M.S. and Chavadi, V.C. (1995). Some relationship between the bottom sediments and recent Ostracoda: A case study. *J. Indian Assoc. Sedimentol.* 14:83–88.

Varma, K.U., Shyam Sunder, V.V. and Naidu, T.Y. (1993). Recent Ostracoda of the Tekkali Creek, east coast of India. *J. Geol. Soc. India.* 41(6):551–560.

Walton, W.R. (1952). Techniques for recognition of living foraminifera. *Contr. Cushman Found. Foram. Res.* 3:56–60.

Whatley, R.C. and Zhao, Q. (1988). Recent Ostracoda of the Malacca Straits. Part II. *Rev. Esp. Micropaleontol.* 20(1):5–73.

Whatley, R.C. and Zhao, Q. (1987). Recent Ostracoda of the Malacca straits. Part I. *Rev. Exp. Micropaleontol.* 19(3):327–366.

Yassini, I. and Jones, B.G. (1995). *Foraminifera and Ostracoda from Estuarine and Shelf Environment on the Southern Coasts of Australia*, University of Wollongong Press, Wollongong, 384 p.

13 A Study on Benthic Foraminifera, Texture, and Sediment Geochemistry along the Depositional Environment of Palk Strait, East Coast of India

M. Suresh Gandhi and K. Kasilingam
University of Madras

CONTENTS

INTRODUCTION

Micropaleontology is concerned with the study of microfossils. These microfossils have a large variety of applications in various fields of geology, especially in petroleum exploration. Due to their abundant occurrence and wide geographic distribution in the sediments of all ages and in almost all marine environments, they are being commonly applied in local, regional, interregional correlations and comparisons, primarily in paleogeographic reconstruction. Benthic foraminifers have played a significant role in the reconstruction of the paleoenvironments in the Phanerozoic. The potential of morphogroups of benthic foraminifera is used to reconstruct the paleo-monsoonal precipitations (Nigam et al., 1992; Khare, 1994). Morphogroups are also utilized to discriminate the oxygenated zone from anoxic conditions (Kaiho, 1991). Foraminifera are good indicators for pollution studies (Bhatia and Kumar, 1976; Rao and Rao, 1979; Alve, 1991). In spite of representative studies undertaken in different environments, a study in the strait has not been done so far. The study area of Palk Strait is strategically important one as it is a channel shared by India and Sri Lanka without giving any room for international navigation. Nearly 2,000 years ago, during the period of Sangam, the Palk Strait was widely used for navigation. This strait has been characterized by uninterrupted current movements between Gulf of Mannar and Bay of Bengal. It is necessary to understand the conditions of sediment deposition prevailing in the Palk Strait, as it is being the only strait in the Indian Coastal Zone. In this backdrop, the sedimentological and paleontological studies have been attempted to understand the complexity of the bay and to identify depositional environment prevailing in the bay.

STUDY AREA

The Palk Strait is the narrow channel connecting Gulf of Mannar and Bay of Bengal, and is about 75 km wide between Sri Lanka and India (Figure 13.1). It includes coastal area from Rameshwaram in the southwestern corner to Point Calimere in the

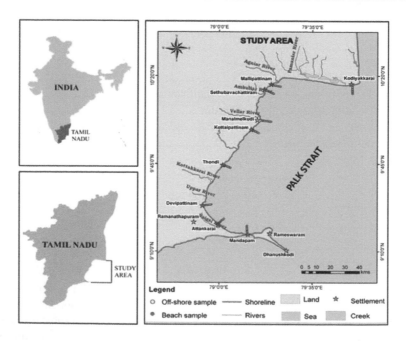

FIGURE 13.1 Location map of the study area.

northeastern corner (Long. 78°50′–79°55′ Lat. 9°15′–10°20′). It falls in the survey of Indian toposheets (No. 58K, 58 J, K, O, and 8 N) prepared in the scale of 1:2, 50,000. The Palk Bay itself is about 110 km long and is surrounded on the northern and western sides by the coastline of the State of Tamil Nadu in the mainland of India. Palk Bay and Gulf of Mannar to its south are connected by a narrow passage called Pamban Strait, which is about 1.2 km wide and 3–5 m deep that separates the Island of Rameshwaram from the mainland.

METHODOLOGY

The study area extends from the coast of Mandapam to Kodiyakarai, Tamil Nadu. Before the commencement of sample collection, a base map in the scale 1:10,000 was prepared using the toposheets (No. 58 K,J,O,N) (No. 58N/3,4 & 8,7,11,15, 58 K/9 to 16, 58O/9, 1–4,7 & 8, 58J/16 scale 1:250,000 and 1:50,000), naval hydrographic charts, and taluk and village maps of Tamil Nadu land surveys. The sediment samples were collected at nine transects in the offshore region, namely, Mandapam, Attankarai, Devipattinam, Thondi, Kottaipattinam, Manamelkudi, Sethubavachatram, Mallipattinam, and Kodiyakarai. All the sediment samples were collected using LaFond Dietz snapper from the sediment–water interface in clean, polythene bags; were immediately preserved in 10% neutralized formaldehyde solution; and then were neatly labeled for further sample processing, analysis, and foraminiferal separation. Calcium carbonate, organic carbon (OC), and organic matter contents in the sediment samples were determined by adopting a methodology suggested by Piper (1947) and Gaudette et al. (1974), and by Walkey-Black method, which is adopted and modified from Jackson (1958), respectively. Sand, silt, and clay percentages were calculated using a combination of sieving and pipette procedures – the latter was done in accordance with Krumbein and Pettijhon (1938). Trilinear plots were prepared, and a description has been given based on Trefethen's (1950). For total digestion, the geochemical analytical procedure suggested by Shapiro and Brannock (1956) was followed. This method was preferred because the sediments essentially consist of detritus silicate minerals, resistant sulfides, and a small quantity of refractory material. Treatment with a mixture of HF, H_2SO_4, and perchloric acid reagent results in a complete dissolution. The solution was finally analyzed for the total Fe, Mn, Cr, Cu, Ni, Co, Pb, and Zn on a PerkinElmer AA 700 AAS equipped with a deuterium background corrector.

RESULTS AND DISCUSSION

SPATIAL DISTRIBUTION AND ECOLOGICAL SIGNIFICANCE

A total of 112 numbers of species belonging to 57 genera have been identified (Table 13.1). When the percentage distribution of superfamilies is taken into account, Miliolacea (39%) in the suborder of Miliolina and Rotaliacea (49%) in the suborder of Rotalina are predominant, while the rest are found to be in the level of less than 100%. Of the 27 benthic families recognized, Hauerinidae, Spiroloculinidae, Elphidiidae, and Rotaliidae are represented by the largest number of species.

TABLE 13.1
Checklist of Benthic Foraminifera

1	*Ammobaculites exiguus* Cushman and Bronnimann. 1948	38	*Quinqueloculina poeyana* d'Orbigny, 1839	75	*Peneroplis planatus* (Fichtel & Moll, 1798)
2	*Textularia agglutinans* d'Orbigny, 1839	39	*Quinqueloculina transversestriata* (Brady, 1881)	76	*Fissurina marginata elegans* (Sidebottom, 1912)
3	*Textularia bocki* Hoglund, 1947	40	*Quinqueloculina echinata* Haake, 1975	77	*Nonionoides boueanum* (d'Orbigny, 1846)
4	*Textularia porrecta* Brady, 1884	41	*Adelosina intricate* (Terquem, 1878)	78	*Nonionoides elongatum* (d'Orbigny, 1826)
5	*Textularia conica* d'Orbigny, 1839	42	*Adelosina laevigata* d'Orbigny, 1826	79	*Nonion Scaphum* (Fichtel & Moll, 1798)
6	*Textularia candeiana* d'Orbigny, 1839	43	*Pyrgo levis* Defrance, 1824	80	*Nonionella labradorica* (Dawson, 1860)
7	*Vertebralina striata* d'Orbigny, 1826	44	*Cycloforina sidebottoni* (Rasheed, 1971)	81	*Neorotalia calcar* (d'Orbigny, 1839)
8	*Rupertianella rupertimta* (Brady, 1881)	45	*Pseudomassilina australis* (Cushman, 1932)	82	*Pararotalia nipponica* Asano, 1936
9	*Edentostomina cultrata* (Brady, 1881)	46	*Pseudomassilina macilenta* (Brady, 1884)	83	*Ammonia beccarii* (Linnaeus, 1758)
10	*Spiroloculina communis* Cushman & Todd, 1944	47	*Pyrgo depressa* (d'Orbigny, 1826)	84	*Ammonia dentata* (Parker and Jones, 1865)
11	*Spiroloculina antillarum* d'Orbigny, 1839	48	*Parrina bradyi* (Millett, 1898)	85	*Ammonia tepida* (Cushman, 1926)
12	*Spiroloculina indica* Cushman & Todd, 1944	49	*Miliolinella circularis* (Bornemann, 1855)	86	*Asterorotalia trispinosa* (Thalmann, 1933)
13	*Spiroloculina orbis* Cushman, 1921	50	*Sigmamiliolinella australis* (Parr, 1932)	87	*Asterorotalia inflata* (Millett, 1904)
14	*Spiroloculina angulata* Cushman, 1917	51	*Miliolinella pyrgoformis* Yassini & Jones, 1995	88	*Elphidium macellum* (Fichtel & Moll, 1798)
15	*Spiroloculina costifera* Cushman, 1917	52	*Flintinoides labiosa (d'Orbigny, 1839)*	89	*Elphidium crispum* (Linnaeus, 1758)
16	*Spiroloculina depressa* d'Orbigny, 1826	53	*Bolivina nobilis* Hantken, 1875	90	*Elphidium discoidale* (d'Orbigny, 1839)
17	*Spiroloculina henbesti* Thabnan, 1955	54	*Bolivian ordinaria* Phleger and Parker, 1952	91	*Elphidium advenum* (Cushman, 1922)

(Continued)

TABLE 13.1 (*Continued*)
Checklist of Benthic Foraminifera

18 *Spirolocidina nitida*
    d'Orbigny, 1826

19 *Spirolocidina robusta*
    Brady, 1884

20 *Spiroloculina affixa*
    Terquem, 1878

21 *Spiroloculina aequa*
    Cushman, 1932

22 *Spiroloculina corrugata*
    Cushman and Todd, 1944

23 *Siphonaperta aqqludnans*
    (d'Orbigny, 1839)

24 *Quinqueloculina costata*
    d'Orbigny, 1878

25 *Quinqueloculina
    lamarckiana* d'Orbigny,
    1839

26 *Quinqueloculina seminulum*
    (Linnaeus, 1758)

27 *Quinqueloculina tropicalis*
    Cushman, 1924

28 *Quinqueloculina elongata*
    Natland, 1938

29 *Quinqueloculina polygona*
    d'Orbigny, *1839*

30 *Quinqueloculina bicostata*
    d'Orbigny, 1839

31 *Quinqueloadina parked*
    (Brady, 1881)

32 *Quinqueloculina tenagos*
    Parker, 1962

33 *Quinqueloculina
    schlumbergeri*
    (Wiesner, 1923)

34 *Quinqueloculina venusta*
    Karrer, 1868

55 *Sagrinella durrandii* (Millett,
    1900)

56 *Boluvina spathulata*
    (Williamson, 1858)

57 *Siphogenerina raphana*
    (Parker & Jones, 1865)

58 *Allassoida virgula* (Brady,
    1879)

59 *Fijiella simplex* (Cushman,
    1929)

60 *Fursenkoina pontoni
    Cushman*

61 *Acemdina inhaerens* Schulze,
    1854

62 *Spiroloxostoma glabra*
    (Millett, 1903)

63 *Rosalina globularis*
    d'Orbigny, 1826

64 *Triloculina tricarinata*
    d'Orbigny, 1826

65 *Triloculina trigonula*
    (Lamarck, 1804)

66 *Pseudotriloculina rotunda*
    (Schlumberger, 1893)

67 *Triloculina insignis* (Brady,
    1881)

68 *Triloculina schreibedana*
    d'Orbigny, 1839

69 *Triloculina terquemiana*
    (Brady, 1884)

70 *Globigerina bulloides*
    d'Orbigny, 1826

71 *Globigerina aequilateralis*
    d'Orbigny, 1826

92 *Cribroelphidium incertum*
    Williamson, 1858

93 *Elphidium craticulatum*
    (Fichtel & Moll, 1798)

94 *Elphidium delicatulum*
    Bermudez, 1949

95 *Discorbinella bertheloti*
    (d'Orbigny, 1839)

96 *Elphidium norvangi* Buzas,
    Smith & Beam, 1977

97 *Elphidium hispidulum*
    Cushman, 1936

98 *Assilina ammonoides*
    (Gronovius, 1781)

99 *Osangularia venusta*
    (Brady, 1884)

100 *Loxostomina limbata*
    (Brady, 1881)

101 *Lagena striata* (d'Orbigny,
    1839)

102 *Lagena perlucida*
    (Montagu, 1803)

103 *Massilina secans*
    (d'Orbigny, 1826)

104 *Eponides repandus* (Fichtel
    & Moll, 1798)

105 *Monalysidium politum*
    (Chapman, 1900)

106 *Sorites marginalis* (Forskal,
    1775)

107 *Lobatula lobatula* (Walker
    c£ Jacob, 1798)

108 *Amphistegina radiata*
    (Fichtel & Moll, 1798)

(Continued)

TABLE 13.1 (*Continued*)
Checklist of Benthic Foraminifera

35	*Quinqueloculina sulcata* Fornasini, 1900	72	*Globigerina ruber* d'Orbigny, 1826	109	*Uvigerina senticosa* Cushman, 1927
36	*Quinqueloculina strigillata* (d'Orbigny, 1850)	73	*Artkulina mayori* Cushman, 1922	110	*Cancris oblanga* Williamson, 1858
37	*Quinqueloculina elegans* d'Orbigny, 1878	74	*Spirolina arietina* (Batsch, 1791)	111	*Hanzawaia concentrica* Cushman, 1918
				112	*Pseudorotalia schroeteriana* Parker, Jones, 1862

Total foraminiferal number (TFN) denotes the total number of recent benthic fora-miniferal specimens recovered from 1 g sediment. The TFN increases away from the shore of the study region (Table 13.2). At Attankarai, a carbonate content was found to be 18.5%–26%, which may be attributable to the broken shell debris – which con-sists of substrate carbonate source – that is dumped through the creek to the seabeds (Figure 13.2).

In all the regions, sediment nature is changing from finer to very finer in terms of texture; i.e., the sediment consists of fine sand/very fine sand and coarse silt. The presence of sandy beds in Thondi and Kottaipattinam in the south of Manamelkudi and in Sethubavachatram in the north of Manamelkudi establishes the current active movement around the spit, which diverts the current activities in a circular direction both in the north and south, enabling the entire area to be filled with suspended fine sediments. The fine sand nature at Kodiyakarai indicates the presence of calm and mixed environment.

LIVING/DEAD FORAMINIFERAL DISTRIBUTION

A total number of 112 foraminiferal species have been identified from 72 sediment samples. The majority of benthic foraminifera are reported to be dead, whereas liv-ing specimens are rarely reported. Kodiyakarai and Attankarai show a high number of dead species. Stations 5 and 7, at Attankarai, record an abrupt decrease in species, while in the remaining stations, there is a gradual increase. Kodiyakarai records more number of dead species in station 8. At Sethubavachatram and Thondi, in all of the stations, the species are more or less uniformly distributed. In all of the stations, the individual species are less than 62 in numbers but the level of total species varies depending upon depths. At Kodiyakarai, even though individual species are less in numbers, the total number of species is much higher. Similarly, at Manamelkudi, in station 3, the total number of species is less despite the presence of more number of species (Figure 13.3).

Out of 112 species, 54 species are distributed only in dead conditions in different stations. Kodiyakarai and Attankarai are found to be highly favorable

TABLE 13.2
Total Foraminifera in Numbers

Station/Depth	Mandapam (1)	Attankarai	Devipattinam	Thondi	Kottaipattinam	Manamelkudi	Sethubavachatram	Mallipattinam	Kodiyakarai (8)
1	199	311	161	68	357	282	35	274	292
2	181	355	441	70	219	280	41	326	338
3	260	433	352	178	265	708	103	272	451
4	287	453	388	130	251	346	72	205	445
5	395	477	580	225	249	185	163	263	565
6	206	563	492	215	234	310	169	246	695
7	160	495	472	154	206	310	100	336	629
8	643	523	497	233	212	128	159	291	757
Total	2,331	3,610	3,383	1,273	1,993	2,549	842	2,213	4,172
Average	291	451	423	159	249	319	105	277	522
Mini	160	311	161	68	206	128	35	205	292
Max	643	563	580	233	357	708	169	336	757

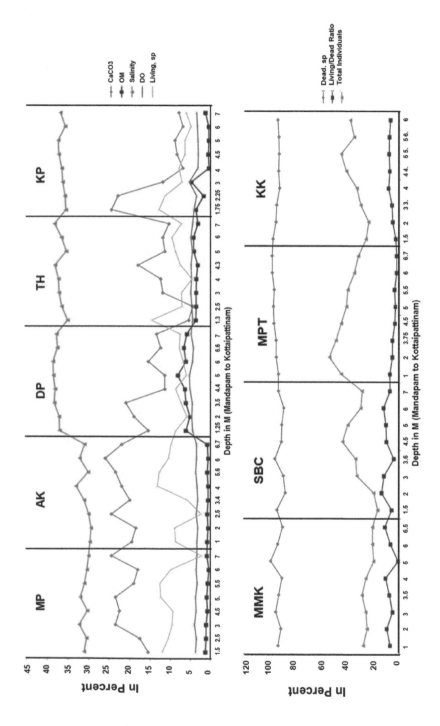

FIGURE 13.2   Results of ecological parameters.

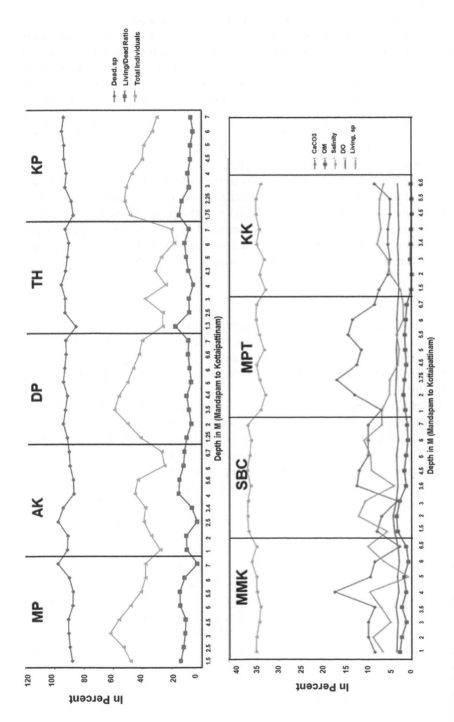

FIGURE 13.3 Results of living/dead ratio.

for the foraminiferal survival; probably, it may be a result of protected environment due to the impact of longshore current. There is a protective environment in Sethubavachatram and Thondi; the absence of more families in Sethubavachatram is attributable to the freshwater influx and frequent turbidity. The total distribution of foraminifera is higher at Kodiyakarai, Attankarai, Devipattinam, Manamelkudi, and Mandapam followed by Mallipattinam, Sethubavachatram, and Thondi. The total number of dead species is comparatively more at Kodiyakarai, Attankarai, Devipattinam, Manamelkudi, and Mandapam. The living species are found more in the south of Manamelkudi because of the high order of churning and mixing of both the monsoons in the vicinity of Devipattinam. The size of the species is comparatively smaller in the north than in the south.

A higher living/dead ratio reflects comparatively higher rate of sedimentation in this strait. The lowest value of 4.14% at Mallipattinam is attributed to the low rate of sedimentation. A maximum living/dead ratio at Mandapam, Attankarai, Thondi, and Sethubavachatram is attributable to the monsoonal push from Point Calimere reaching this coast. As an overview, it can be inferred that more sedimentation takes place in southern regions than in the northern regions of spit growth.

GRAIN SIZE

In the study area, the mean value ranges from 2.296 to 4.306 φ for Mandapam; 2.520 to 2.897 φ for Attankarai; 3.504 to 3.838 φ for Devipattinam; 2.883 to 3.860 φ for Thondi; 3.527 to 4.050 φ for Kottaipattinam; 3.671 to 4.011 φ for Manamelkudi; 3.109 to 3.950 φ for Sethubavachatram; 3.056 to 4.051 φ for Mallipattinam; 2.398 to 3.069 φ for Kodiyakarai.

The textural pattern of sediments present in the study regions shows unimodal to polymodal in nature. The textural pattern of sediments in Attankarai and Kodiyakarai regions is unimodal in nature, which reflects the deposition of sediments primarily by waves and currents, and the textural pattern of sediments in Devipattinam region shows bimodal to polymodal in nature with poorly sorting to moderately sorting. It clearly indicates that whatever sediments that are transported from the river settle in the estuary and do not reach to the coastal region (Angusamy and Rajamacikam, 2006). The textural pattern of sediments present in the study regions shows unimodal to polymodal in nature (Table 13.3).

The sorting nature ranges from well sorted to poorly sorted. The sorting nature of Mandapam region ranges from moderately well sorted to very well sorted, whereas that of Attankarai region is almost well sorted, due to prevailing low wave energy condition. The sorting nature of Devipattinam region is poorly sorted, which may be due to the high energy condition. The sorting nature of Thondi and Kottaipattinam regions is moderately well sorted, which may be due to the addition of sediments of different grain sizes from the reworking of beach ridges or by alluvial action and the prevalence of strong wave convergence throughout the year, and a similar sorting nature may be also due to the prevalence of strong northerly drift. The currents moving from down south region carry the sediments to the northern regions. In this process, the sediments are imparted with a moderate- to well-sorting nature.

TABLE 13.3
The Textural Pattern of Sediments Present in the Study Regions Shows Unimodal to Polymodal in Nature

Textural Parameter	Mean	Sorting	Skewness	Kurtosis
Mandapam	Fine sand to very coarse silt	Moderately sorted to very well sorted	Symmetrical to coarse skewed	Platykurtic to mesokurtic
Attankarai	Fine sand	Well sorted to moderately well sorted	Symmetrical	Mesokurtic to leptokurtic
Devipattinam	Very fine sand	Moderately sorted to poorly sorted	Very coarse skewed	Mesokurtic
Thondi	Fine sand to very fine sand	Moderately sorted	Symmetrical to coarse skewed	Platykurtic to leptokurtic
Kottaipattinam	Very fine sand to very coarse silt	Moderately sorted to moderately well sorted	Coarse skewed to very coarse skewed	Mesokurtic
Manamelkudi	Very fine sand to very coarse silt	Moderately sorted to well sorted	Very coarse skewed	Platykurtic to mesokurtic
Sethubavachatram	Very fine sand	Moderately sorted to poorly sorted	Symmetrical to very coarse skewed	Platykurtic to mesokurtic
Mallipattinam	Very fine sand to very coarse silt	Moderately sorted to moderately well sorted	Coarse skewed to very coarse skewed	Platykurtic to leptokurtic
Kodiyakarai	Fine sand to very fine sand	Moderately well sorted to well sorted	Symmetrical to fine skewed	Mesokurtic to leptokurtic

The Mandapam, Attankarai, and Kodiyakarai regions show an almost positively skewed distribution. The positively skewed distribution indicates a greater quantity of fine sediments and a depositional tendency (Duane, 1964). Devipattinam, Thondi, Kottaipattinam, Manamelkudi, Sethubavachatram, and Mallipattinam regions show negatively skewed distribution. The negatively skewed distribution illustrates the depletion of fine-grained sands and suggests the dominance of erosional processes. The transects like Attankarai and Kodiyakarai display positive skewness. It suggests the possibility of taking sand and slits to nonbeach sediments or a deposit, formed under the conditions of low energy. Attankarai is at the mouth of river confluence; one expects a supply of terrestrial material rather than the beach sediments. The Attankarai transect indicates a mixed sorting (i.e., from well sorted to moderately well sorted). It suggests the prevalence of alternate high and low energy conditions in the seabed. As in the case of Sethubavachatram, the current returning back from Devipattinam is expected to enter through Attankarai where the presence of irregular relief must have conducted the currents to take course through those minor channels to the deeper channel. Such channel banks may be providing a better sorting comparatively to the channel bed, which undergoes the disturbance of the returning currents enabling to get poor sorting in such zone. The complete positive skewness at Kodiyakarai may be attributable to the low energy condition of deposition, or otherwise, the dominant influence of dune/ridge sediments is noticed around the Vedaranyam Ridge.

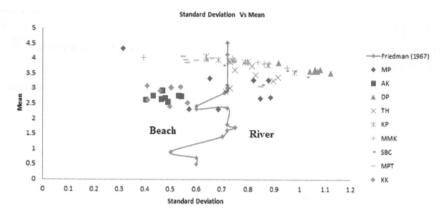

FIGURE 13.4   Results of mean vs standard deviation.

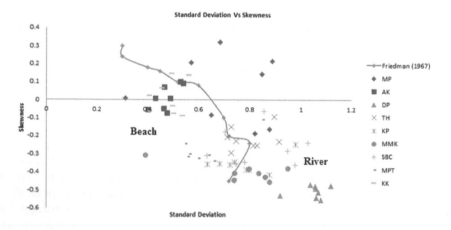

FIGURE 13.5   Skewness vs standard deviation.

The moment kurtosis values are found to vary from 2.429 to 10.492 φ. Mean vs standard deviation is plotted in Figure 13.4. It shows that the stations Attankarai, Kodiyakarai, and Mallipattinam come under the beach environment, whereas at Mandapam, four stations come under beach environment; Devipattinam, Thondi, and Kottaipattinam samples come under the river environment. In Figure 13.5, standard deviation vs skewness is plotted. The following stations Attankarai, Kodiyakarai, Kottaipattinam, and Mallipattinam come under the beach environment.

TRACE ELEMENT GEOCHEMISTRY

In Mandapam, the iron concentration ranges from 4,642 to 6,942 ppm; the average concentration of Fe is 5,550 ppm; the almost similar iron concentration is observed for Thondi, Kottaipattinam, and Manamelkudi, whereas a slight increase in the Fe concentration is observed for Attankarai and Devipattinam regions. In Attankarai

TABLE 13.4
Results of Trace Element Geochemistry

Mandapam	Fe (ppm)	Mn (ppm)	Cr (ppm)	Cu (ppm)	Ni (ppm)	Cd (ppm)	Pb (ppm)	Zn (ppm)
Average	5,550	529	262	73	245	0.247	260	247
Minimum	4,642	424	200	46	215	0.224	238	225
Maximum	6,942	692	326	94	336	0.262	282	264
Attankarai								
Average	7,149	547	369	91	217	0.241	258	264
Minimum	4,484	426	204	81	210	0.21	251	222
Maximum	12,290	690	668	97	225	0.274	270	305
Devipattinam								
Average	10,289	787	380	84	219	0.236	266	262
Minimum	5,448	402	278	51	201	0.224	243	242
Maximum	17,120	1,068	624	96	234	0.252	279	284
Thondi								
Average	4,998	651	274	43	251	0.415	120	238
Minimum	4,358	462	234	27	213	0.22	101	222
Maximum	6,194	864	325	75	307	0.98	154	264
Kottaipattinam								
Average	5,689	571	342	65	247	0.394	135	232
Minimum	4,556	478	235	49	177	0.136	105	223
Maximum	7,636	772	597	95	288	0.63	186	239
Manamelkudi								
Average	5,330	679	303	52	339	0.521	156	233
Minimum	4,158	496	245	27	239	0.414	115	223
Maximum	8,492	1,132	350	84	606	0.642	193	246
Sethubavachatram								
Average	6,560	577	247	69	326	0.507	156	247
Minimum	4,816	422	229	49	235	0.33	107	225
Maximum	9,914	846	280	95	436	0.68	235	266
Mallipattinam								
Average	9,548	867	261	84	339	0.493	210	249
Minimum	6,230	648	213	64	268	0.246	169	237
Maximum	13,114	1,064	317	99	436	0.65	313	257
Kodiyakarai								
Average	7,693	745	282	61	325	0.442	197	226
Minimum	5,172	460	229	43	293	0.192	127	220
Maximum	10,468	922	329	78	351	0.648	249	233

area, the average Fe concentration ranges from 4,484 to 12,290 ppm and the average concentration is 7,149 ppm (Table 13.4). In the Devipattinam samples, the concentration of manganese ranges from 402 to 1,068 ppm and the average concentration is 787 ppm due to the fine substratum and shallow nature of the coast. Except Devipattinam in the southern part of Manamelkudi, the manganese concentration slightly decreases compared to that in the northern part of Manamelkudi. Overall, the result shows that the chromium concentration varies from 200 to 668 ppm with an average concentration of 302 ppm. The Cu concentration is higher in the south of Manamelkudi, which is mainly due to the anthropogenic input. The cadmium concentration slightly increases from the southern part of Manamelkudi (Thondi) to the northern part of Manamelkudi. The zinc concentration in the study regions does not show much variation; almost a similar trend was observed; the overall zinc concentration varies from 220 to 305 ppm with an average of 244 ppm.

The highest concentration was observed in station 2 at Attankarai, and the lowest concentration was observed in station 3 at Kodiyakarai. The contamination factor of the metals in the entire study area is found to be of the order: Pb > Ni > Cr > Cd > Zn > Cu > Mn > Fe. Pb, Ni and Cr, suggesting that the contamination zone with regard to these elements is very high in the study area.

CORRELATION MATRIX

Correlation matrix revealed the existence of a correlation between heavy metal concentration and sediment characteristics (Table 13.5). Lead shows a positive correlation with $CaCO_3$ at 0.479 level, mud at 0.397 level, and also OC at 0.049 level. Libes (1992) reported that finer sediments have a larger surface area, which allows heavy metals and other contaminants to be adsorbed easily. Positive correlations found between OC and studied metals have indicated a high affinity among them (Langston, 1982; Coquery and Welbourn, 1995). In contrast, Cd has presented no significant relationships with OC, presumably due to the fact that Cd does not tend to form stable organic complexes (Campbell et al., 1988). However, all the studied heavy metals showed a negative correlation with sand particles, except nickel and cadmium, which presumably may have low surface area.

OC strongly correlates with mud because mud was the dominant fraction. Fe correlates with other oxides, which have a higher affinity with most elements, especially for trace elements (Stumm and Morgan, 1996), and the organic matter contents are the important controlling factors in the abundance of trace metals (Rubio et al., 2000). As a consequence, the geochemistry of Fe and organic matter could affect the geochemical behavior of trace metals in the aquatic environment. One of the evidences is that concentrations of Fe and OC often correlate with concentrations of other metals in marine environment (Zwolsman et al., 1996; Basaham and El-Sayed, 1998). A significant correlation of Fe with Mn ($r = 0.455$) and Cu ($r = 0.420$) and of Mn with Cr ($r = 0.049$) and Cu ($r = 0.152$) suggests an important associations between the oxide-oxyhydroxides of Fe-Mn and other elements. Notably, the correlation and geochemical associations of trace metals reveal a significant source of contamination reflecting a common origin of similar nature existing from the industrial effluents (Turner, 2000).

TABLE 13.5
Correlation Matrix of the Study Area

Parameters	Fe	Mn	Cr	Cu	Ni	Cd	Pb	Zn	Sand	Mud	CaCO₃	OM	Dead sp	Living sp	Total Pop
Fe	1.000														
Mn	0.455	1.000													
Cr	-0.071	0.049	1.000												
Cu	0.420	0.152	0.108	1.000											
Ni	0.032	0.211	-0.260	-0.120	1.000										
Cd	-0.073	0.192	-0.294	-0.133	0.508	1.000									
Pb	0.334	0.055	0.151	0.554	-0.228	-0.406	1.000								
Zn	0.386	0.098	0.015	0.524	-0.234	-0.215	0.424	1.000							
Sand	-0.420	-0.250	-0.302	-0.257	0.251	0.307	-0.397	-0.373	1.000						
Mud	0.420	0.250	0.302	0.257	-0.251	-0.307	0.397	0.373	-1.000	1.000					
CaCO₃	0.067	-0.154	0.284	0.388	-0.423	-0.595	0.479	0.411	-0.122	0.122	1.000				
OM	0.251	0.176	0.213	-0.013	-0.263	-0.198	0.049	0.224	-0.802	0.802	-0.003	1.000			
Dead sp	0.224	0.371	-0.112	0.011	0.218	0.080	-0.057	-0.134	-0.049	0.048	-0.240	-0.054	1.000		
Living sp	-0.224	-0.371	0.112	-0.011	-0.218	-0.080	0.057	0.134	0.049	-0.049	0.240	0.054	-1.000	1.000	
Total Pop	0.221	0.201	0.227	0.180	0.051	-0.216	0.451	0.112	-0.219	0.219	0.198	-0.089	0.171	-0.171	1.000

DISCUSSION

The textural analysis of the sediments brings out a complexity of the bay with a wide range of sediment nature. The mean size of the offshore samples characterizes a large variation from fine sand to very fine sand and very coarse silty nature. From the southernmost region at Mandapam to Manamelkudi to Kodiyakarai, one can see a change in grain size from coarse to fine nature, leading to an inversely proportional relation to depth; i.e., in general, when the depth increases, the sediment becomes finer. However, in the case of Kodiyakarai region, the offshore sediments consists of predominantly fine sand to very fine sand, which could be attributed to its location being in one of the arms (northern arm) of cusp-shaped basin. Because of the spit across Manamelkudi, the sediments brought by the northerly as well as the southerly drifts are being blocked and deposited there en route due to a clockwise drift to the deeper channel. The existing wave conditions clearly depict the movement of longshore current from the Bay of Bengal in N-S direction, which straightly hit the region at Thondi and Devipattinam during the NE monsoon. Further, the studies of Jena (1997) on oceanographic parameters and sediment movement on this part also support the influence of southwest monsoon in transporting the sediment in Palk Bay. It has been observed till Ponnagaram only. Based on the above inferences, it could be conveniently concluded that the sediments available in the Kodiyakarai area are locally redistributed. During the churning by the waves and currents, whatever brought in suspension must have been diverted to the Kodiyakarai region of calm environment and settled slowly.

The textural pattern of sediments also shows a polymodality nature having one primary dominant mode and rest of secondary modes. The dominant mode is observed in the size range of 2.5–4 φ at Mandapam, 1.5 φ at Devipattinam, 1.5 φ at Kottaipattinam, and 3 φ at Kodiyakarai, which indicates the distribution of finer sediments of recycled nature in the study area (Mohan et al., 2000); in the scatter plots, the offshore samples show two groupings, and this may be due to the differences in the grain sizes. The polymodal nature suggests the existence of many sources for the basinal sediments.

The station-wise distribution of living foraminifera is maximum at Attankarai (334 in number), Kodiyakarai (272 in number), and Devipattinam (244 in number). The lowest distribution is recorded at Sethubavachatram (74 in number), Mallipattinam (87 in number), and Thondi (102 in number). The depth-wise distribution shows that the increasing depth receives more number of living species, compared to shallow depths. In the deeper depth, the channel is the passage for current movements. More water enters in the deeper channel and creates the oxygenated conditions, giving room for more living species. The abundant distribution of TFN indicates a high production activity in deeper depths when compared to the shallower one. But living foraminifera are more in the southern part of Manamelkudi, compared to the northern part. Also, the lowest living foraminifera are found near the spit growth around the region of Manamelkudi.

High turbidity with a lot of suspended matter and poor visibility are attributed to a poor representation of benthic population compared to the rapid sedimentation and high turbidity. The gradual reduction in dead percentage of species from

Manamelkudi to Mandapam in the south and to Kodiyakarai in the north is attributed to the gradual reduction in the size of the dead species.

The following species are distributed only in dead conditions in different stations: *T. bocki, T. candeiana, T. conica, T. porrecta, S. aequa, S. affixa, S. nitida, B. ordinaria, B. spathulata, L. striata, Fissurina marginata, Q. elegans, G. bulloides, G. equilateralis,* and *G. ruber.* Such distribution is possible due to the recurrence of one of the following causes. Murray (1973, 1991) has pointed out that the presence of more dead species is expected to have been due to (1) improper staining during sampling, (2) sudden climatic changes that must have led to the climatic-sensitive species for mass mortality; and (3) transportation by fast currents, from one depth to another or from deeper to shallower conditions, from one latitude to another latitude (lateral transport) having different environments, which must have brought again mass mortality. But in this case, proper staining has been maintained in the field itself. And all the living (Rose Bengal-stained) species which are pink in color were counted. In the present study, improper staining is not a factor for more dead. But, based upon the species distribution, it clearly indicates that most of the species are transported.

The presence of some offshore species like *Lagena, Fissurina, Textularia,* and rare planktonic foraminifera could be attributed to the action of bottom currents and wave action. In the northern portion of Manamelkudi, the following species, namely, *L. perlucida, L. striata, Fissurina marginata, Q. elegans, B. nobilus,* and *Globigerina bulloides,* are seen in dead condition. According to Murray (1973), these species are found in the inner shelf area, a zone somewhat deeper than the present one. Such change from deeper to shallower and that too from open sea to the channel conditions must have caused the mortality here. This is also supported by Quaternary report (Rajamanickam et al., 1999) in such a way that the sediments which are available in Kodiyakarai area are locally distributed by waves and currents. This inference is also in agreement with Setty et al. (1984) and Nigam (1986), who have reported the genus *Bolivina* preferring muddy substrata and restricted to bathyal and marginal marine conditions. Their distribution in the nearshore samples indicates a longer distance of transportation by the fast currents towards the shore from deeper shelf.

Antony (1968) has recognized that larger foraminifera occur in shallower depths that are within 45 m and smaller foraminifera at deeper depths. But in the entire Palk Strait, both the species show a widespread distribution. The species with thick strong walls like *Elphidium, Quinqueloculina, Spiroloculina,* and *Triloculina* are frequently distributed in this strait. It indicates that it can also survive in turbulent water. Boltovskoy and Wright (1976) have found that the attached foraminiferal specimens with thick strong walls are common in the nearshore turbid zone. So, the presence of the above species may be accounted for the presence of the shallow marine turbid environment.

The broken species, abnormality in size, and the absence of the well-grown tests are seen in all the regions (Table 13.6). But in Mandapam, Devipattinam, Thondi, Sethubavachatram, and Mallipattinam, broken and damaged species are more when they are compared to the regions of Kottaipattinam, Manamelkudi, and Kodiyakarai. Setty et al. (1983) have also reported the aberrant features of foraminifera due to pollution along the west coast of India. But in this strait, the broken test is mainly due to the churning action in the substratum by fast current movements. This is also

TABLE 13.6

Foraminifera Size and Morphological Deformative Species in the Study Area

Stations	Size (in mm)	Distortion/Corrosive/Broken Species
Mandapam	L: 0.15 B: 0.13	16%–20% of species are broken, *Lagena*, *Asterorotalia*, *Quinqueloculina*
Attankarai	L: 0.32 B: 0.29	3%–7% of species are broken, *Asterorotalia*
Devipattinam	L: 0.36 B: 0.32	15%–17% of species are broken, *Pararotalia*
Thondi	L: 0.42 B: 0.33	10%–13% of species are broken, *Quinqueloculina*, *Pararotalia*
Kottaipattinam	L: 0.46 B: 0.42	5%–7% of species are broken, *Pararotalia*, *Quinqueloculina*
Manamelkudi	L: 0.52 B: 0.45	2%–3% of species are broken, *Spiroloculina*
Sethubavachatram	L: 0.45 B: 0.40	15%–20% of species are broken; most of the species are *Spiroloculina*, *Pararotalia*
Mallipattinam	L: 0.42 B: 0.40	15%–20% of species are broken; most of the species are *Spiroloculina*, *Pararotalia*
Kodiyakarai	L: 0.13 B: 0.11	8%–12% of species are broken, *Asterorotalia*

supported by Mageau and Walker (1976), and Hickman and Lipps (1983), who have explained that the damaged and distorted tests are used to be seen in the environment where greater turbulence and sediment movement are prevailing because a greater extent of force making the tests smaller.

The living/dead ratio clearly depicts the relative sedimentation possibilities. Based upon the living/dead ratio, the southern portion of Manamelkudi receives more sediments compared to the northern portion of Manamelkudi. Organic matter content also supplements this fact. This is supplemented by the findings of Paropkari (1979), Paropkari et al. (1993), and Setty and Nigam (1982), who have reported low OC content in the sites of low rate of sedimentation and high OC content in the sites of high rate of sedimentation.

The level of enrichment of total trace metals (TTMs) has also increased the levels of most of the elements when compared with all other ecosystems in the world as well as the nearby area. The results also indicate that this region is more heavily contaminated with Cd, Pb, Ni, Cr, Zn, Cu, Mn, and Fe than other regions on the southeast coast of India. The results of the present study suggest the need for a regular monitoring and management program, which will help to improve the quality of Palk Strait pristine regions.

CONCLUSION

A vast stretch of offshore region foraminiferal study has been undertaken. This study highlights the presence of a composite cosmopolitan fauna of foraminifera of 112 benthic species belonging to 27 families, and 57 genera from the 72 sediments samples collected from a depth range of 1–7 m at the Palk Strait. Among them, *Rotalina* (49%) occupies a dominant place, followed by *Miliolina* (39%) in the distribution of suborders.

Out of 112 foraminiferal species, only 55 represent the living species (Rose Bengal-stained). Among them, the genera *Ammonia, Asterorotalia,* and *Quinqueloculina* dominate the living assemblages followed by *Pararotalia* and *Elphidium.*

From the total number of species, *A. beccarii, A. trispinosa, P. nipponica, A. tepida, A. dendata, E. crispum,* and *Q. seminulum* are highly abundant. However, only 13 species, namely, *P. nipponica, Neorotalia calcar, Q. lamarckiana, Q. seminulum, Q. tropicalis, T. trigonula, E. macellum, A. beccarii, E. crispum, A. dentata, A. tepida, S. depressa,* and *S. communis,* keep up a uniform distribution in more than 6 regions.

The Mandapam, Attankarai, and Kodiyakarai regions show fine sand to very fine sand and are positively skewed in nature, which may be due to the prevailing low energy condition indicating the depositional environments and their connection with open sea; irrespective of whatever sediment deposited during the monsoon, the sediments are altered by tidal influence. The sandy nature of the Palk Strait regions outlines the deposition of sediments primarily by northerly moving currents. From the overall study, one can infer that the Palk Strait environment as a whole is a hostile and complicated environment. Here, most of the ecological factors are not suitable for living species as well as dead species, because the bottom is swept by strong current and the unstable and uneven seabed conditions do not permit the growth of living population.

Foraminiferal species are highly tolerant to sediment contamination and disappeared only in extremely polluted conditions. Therefore, they can be used as environmental indicators in highly impacted areas, where other potential bioindicators disappeared due to contamination. The high diversity of foraminiferal assemblages allowed a diversified response of species to different pollutants. The results from the scientific literature evidence are that the benthic foraminifera are excellent indicators in monitoring the marine coastal environments.

The present study has illuminated the fast prograding nature of the shelf as evidenced from living/dead ratios, which show comparatively higher rate of sedimentation taking place in this strait. However, the lowest value with 7.50% at Manamelkudi is attributable to a zone of low sedimentation because it is located at the limbs of the spit, which is acting as a barrier to divert the hitting littoral currents. A depletion of foram species in the present study region reflects the active sedimentation or progradation taking place in the Palk Bay region due to the unfavorable environmental conditions.

ACKNOWLEDGMENTS

The authors are thankful to the Professor and Head, Department of Geology, University of Madras for providing the laboratory facilities to carry out this work. The authors also acknowledge the UGC-CPEPA (F. No. 8-2/2008(NS/PE) Dt 14.12.2011) for providing financial assistance to carry out fieldwork in the mangrove regions at the Palk Strait.

REFERENCES

N. Angusamy and G.V. Rajamanickam, "Depositional environment of sediments along the southern coast of Tamil Nadu, India", *Oceanologia,* Vol. 48, pp. 87–102, 2006.

E. Alve, "Foraminifera, climatic changes and pollution; A study of late Holocene sediment in Drammensfjord, SE Norway", *The Holocene*, Vol. 1, pp. 243–261, 1991.

A. Antony, "Studies of the shell water foraminifera of Kerala coast", *Bulletin of the Department of Marine Biology and Oceanography, University of Kerala*, Vol. 4, pp. 11–54, 1968.

A.S. Basasham and M. A. ELSayed, "Distribution and phase association of some major and trace elements in the Arabian Gulf Sediments, Estuarine, *Coastal Shelf Science*, Vol. 46, pp. 185–194, 1998.

B. Bhatia and S. Kumar, "Recent benthonic foraminifera from the innershelf area around Anjidiv Island, off Binge, West coast of India", *Maritime Sediments Special Publication*, Vol. 1, pp. 239–249, 1976.

E. Boltovskoy and R. Wright, *Recent Foraminifera*, Dr. W. Junk Publishers, The Hague, Netherlands, 515 p, 1976.

A.C. Campbell, M.R. Palmer, G.P. Klinkhammer, T.S. Bowers, J.M. Edmonds, J.R. Lawrence, J.F. Casey, G. Hompson, S. Humphris, P. Rona and J.F. Karson, "Chemistry of hot springs on the Mid-Atlantic Ridge", *Nature*, Vol. 335, pp. 514–518, 1988.

M. Coquery and P. Welbourn, "The relationship between metal concentration and organic matter in sediments and metal concentration in the aquatic macrophyte Eriocaulon septangulare", *Water Research*, Vol. 29, pp. 2094–2102, 1995.

D.B. Duane, 1964, "Significance of skewness in recent sediments, Western Pamlico Sound, North Carolina", *Journal of Sedimentary Research*, Vol. 34, pp. 864–874, 1964.

C.S. Hickman, and J.H. Lipps, "Foraminiferivori-selective injection of foraminifera and test alterations produced by the Neogatropod Olivella", *Journal of Foraminiferal Research*, Vol. 13, pp. 108–114, 1983.

M.L. Jackson, *Soil Chemical Analysis*, Prentice Hall, New York, p. 485, 1958.

B.K. Jena, "Studies on littoral drift sources and sinks along the Indian coast", Unpublished PhD thesis, Berhampur University, p. 204, 1997.

K. Kaiho, "Global changes of paleogene aerobic/anaerobic foraminifera and deep sea circulation", *Palaeogeography, Palaeoclimatology, Palaeoecology*, Vol. 33, pp. 65–85, 1991.

N. Khare, "A note on the *Pavonina flabellaformis* de'Orbigny (Benthic foraminfera) from Arabian sea. Geo", *Science*, Vol. 15, pp. 37–39, 1994.

W.C. Krumbein and F.J. Pettijohn, *Manual of Sedimentary Petrography*, Appleton Century-Crofts, New York, 25, pp. 521–537, 1938.

W.J. Langston, "The distribution of mercury in British estuarine sediments and its availability to deposit-feeding bivalves", *Journal of the Marine Biological Association of the United Kingdom*, Vol. 62, pp. 667–684, 1982.

S.M. Libes, *An Introduction to Biochemistry*, John Wiley & Sons, Inc., New York, p. 73, 1992.

V.J. Loveson and G.V. Rajamanickam, "Coastal geomorphology of southern Tamilnadu", *Special Proceedings of National Symposium on Remote Sensing in Land Transformation and Management at NRSA*, Hyderabad, pp. 115–120, 1987.

V.J. Loveson and G.V. Rajamanickam, "Progradation as evidenced around the submerged port Periapattinam, Tamilnadu India", *Indian Journal of Landscape and Ecolological Studies*, Vol. 12, pp 94–98, 1987.

N.C. Mageau and D.A. Walker, "Effects on ingestion on foraminifera by larger invertebrates", *Maritime Sediments Special Publication*, Vol. 1, pp. 89–105, 1976.

P.M. Mohan, K. Shephard, N. Angusamy, M. Suresh Gandhi and G.V. Rajamianckam, "Evolution of quaternary sediments along the coast between Vedaranyam and Rameshwaram, Tamil Nadu", *Journal of the Geological Society of India*, Vol. 56, pp. 271–283, 2000.

J.W. Murray, *Distribution and Ecology of Living Benthonic Foraminiferids*, Heinemann Educational Books Ltd., London, p. 274, 1973.

J.W. Murray, *Ecology and Paleoecology of Benthic Foraminifera*, Longman Scientific and Technical, England, p. 397, 1991.

R. Nigam, "Foraminiferal assemblages and their use as indicators of sediment movements. A study in the shelf region off Navapur, India". *Continental Shelf Research*, Vol. 5, pp. 421–430, 1986.

R. Nigam, N. Khare and B.V. Borole, "Can benthic foraminiferal morphogroups be used as an indicator for paleomonsoonal precipitation", *Estuarine Coastal and Shelf Science*, Vol. 34, pp. 533–542, 1992.

A.L. Paropkari, A. Mascarenhas and C.P. Babu, "Comment on lack of enhanced preservation of organic matter in sediments under the oxygen minimum on the Oman margin", *Geochimica et Cosmochimica Acta*, Vol. 57, pp. 2399–2401, 1993.

F.B. Phleger, *Ecology and Distribution of Recent Foraminifera*, John Hopkins Press, Baltimore, MD, p. 297, 1960.

C.S. Piper, *Soil and Plant Analysis*, University of Adelaide Press, Adelaide, p. 368, 1947.

B. Rubio, M.A. Nombela and F. Vilas, "Geochemistry of major trace elements in sediments of the Ria de vigo (NW Spain) an assessment of metal pollution", *Marine Pollution Bulletin*, Vol. 40, pp. 968–980, 2000.

M.G.A.P. Setty and R. Nigam, "Foraminiferal assemblages and organic carbon relationship inbenthic marine ecosystem of western Indian continental shelf", *Indian Journal of Marine Sciences*, Vol. 11, pp. 225–232.

M.G.A.P. Setty and R. Nigam and A. Faterpenkar, "An aberrant Spiroloculina Sp. from recent sedmnets of Bombay, Daman, west coast of India", *Mahasagar-Bulletin of the National Institute of Oceanography*, Vol. 16, pp. 77–79, 1983.

M.G.A.P. Setty, S.M. Birajdar and R. Nigam, "Intertidal foraminifera from Miramar Beach, Caranzalam, shoreline Goa", *Indian Journal of Marine Sciences*, Vol. 1, pp. 49–51, 1984.

L. Shapiro and W.W. Brannock, "Rapid analysis of silicate rocks", *United States Geological Survey Bulletin*, Vol. 1036C, pp. 19–55, 1956.

W. Stumm and J.J. Morgan, *Aquatic Chemistry: Chemical Equilibria and Rates in Natural Waters*, 3rd edn., Wiley, New York, 1996.

A. Turner, "Trace metal concentration in sediments from U.K. Estuaries: an empirical evaluation of the role of hydrous iron and manganese oxides", *Estuarine Coastal and Shelf Science*, Vol. 50, pp. 355–371, 2000.

J. Zwolsman, G. Van Eck, and G. Burger, "Spatial and temporal distribution of trace metals in sediments from the Scheldt estuary, south-west Netherlands", *Estuarine Coastal and Shelf Science*, Vol. 43, pp. 55–79, 1996.

14 Quaternary Faunal Records from Upper Reaches of Sina Basin
A Case Study of Math Pimpri – An Early Historic Site

P. D. Sabale, G. L. Badam, and S. D. Kshirsagar
Deccan College Post-Graduate and Research Institute
Deemed University

CONTENTS

INTRODUCTION

Rivers are one of the most significant geomorphic agents, which are responsible for the modification of original surface. Such suitable landforms are developed with human occupations, since prehistoric time to the present. Most of the archaeological settlements are established on ancient fertile flood-borne terrace deposit on both the banks of rivers. Their discovery and documentary evidences emphasize the significant part played by such locations and magnitude of water supplies in lives of human societies in the parts of the world. Such river deposits containing the cultural remains provide a significant information about the age. Landscape, environmental setting of human occupations and the processes that formed the archaeological record. Most of the archaeological data are recorded from sedimentary deposits associated with soil. Artifacts can be considered sediment particles that contributed to the final characteristics of the archaeological record. Accumulation of such sediment takes place by mechanical and chemical processes. The mechanically accumulated sediments include the deposition of fine sand and the mud particles. The level of water and courses of rivers are changed because of high sediment field from drainage basins during the flash flood conditions, which enable to cover the different archaeological sites, such as settlements, temples, mosques, churches, forts, memorials, step-wells, and old river paths of different cultural periods, which are highly affected.

The present paper deals with the Quaternary excavation carried out at the flood-affected the early historic site Math Pimpri, which lies in the upper reaches of Sina River Basin, in the south Ahmednagar district (Figure 14.1).

STUDY AREA

The most awaited Quaternary excavation at Math Pimpri, a prehistory and proto-historic cultural site, is located on top of the island developed in between the past and the present channel on the left bank of obtuse meander of Sina River. The settlement is observed on latitude 17°45.334′ N and longitude 75°15.712′ E and is present 25 km south-south-east from Ahmednagar district headquarter (Figure 14.1). It is accessible from Rui-Chattishi, a weekly market place nearby site, which is present on Ahmednagar–Solapur Highway.

PREVIOUS WORK

A detailed geoarchaeological exploration in Sina River from its source to destination, including the parts of Solapur, Ahmednagar, Bid, and Osmanabad districts of Maharashtra, was carried out, and for the first time, about 54 archaeological sites with different cultural periods were documented, during the field season 2009–2010 (Sabale and Kshirsagar, 2014). Out of these, the sites explored along the channel and few away from it; one of the important *Satavahana* settlement is Math Pimpri. The river flows through the severe drought-prone area, and hence, it is highly seasonal. But the important point is that this basin is also rich in cultural sites similar to the Bhima Basin, a southwestern main river in this area. To study the locations of settlement, availability of raw material and foundation rock, frequency and density of drainage, topographical characteristics and landforms, type and nature of habitat, vegetation, and accessibility of the study area by using Survey of India (SOI), topographical map (47J/13) of 1:50,000 scale, quarter inch map (47/J) of 1:4 miles scale, and satellite data-based map were used.

DETAILS OF THE QUATERNARY EXCAVATION

Math Pimpri is an important settlement and is present on the top of circular hillock covered with flood alluvium, bounded by paleochannel at the east and recent active channel at the west, hence acting as an island (Sabale and Kshirsagar, 2014; Sabale et al., 2015) (Figure 14.2). Here, the continuous stratigraphical sequence up to the prehistorical period was observed. The work carried out at this site has produced fourfold cultural sequence, namely, Period-I: Medieval–Early Medieval; Period-II:

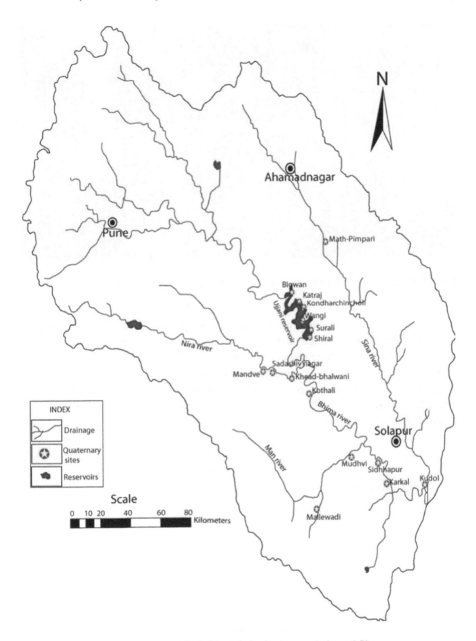

FIGURE 14.1   Location map of Math Pimpri site in upper reaches of Sina.

Early Historic (*Satavahana*); Period-III: flood-logged deposit; and Period-IV: Holocene and Pleistocene deposits with Mesolithic tools which rest on rocky foundation of Deccan Trap basalt.

This habitation deposit rests on the lower variable nature of layers formed by paleo-flood deposit, and finally, both these materials are rest on very solid, hard

FIGURE 14.2 View of Sina River channel and details of each locality.

foundation of Quaternary deposit on Deccan Trap formation. To understand the detail of each deposit which is present in succession, separate three steps like trenches were taken in three respective deposits, and nomenclature for deposits was given in sequence, as per the order of superposition of stratigraphy rule from bottom to top. Therefore, the bottommost Quaternary deposit is referred to as "Locality-I"; similarly, the middle flood deposit "Locality-II"; and the topmost habitation deposit "Locality-III" (Figures 14.2 and 14.3).

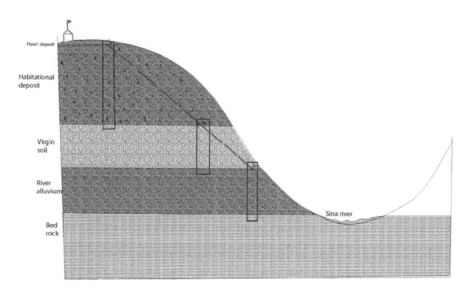

FIGURE 14.3 Vertical section of Math Pimpri site and alignment of trenches at different localities.

This paper is based on the details of the stratigraphy and the faunal remains observed in the lowermost Quaternary deposit at "Locality-I."

This trench shows basically three types of material, namely, (1) top agricultural soil layer, (2) middle Holocene layers, and (3) bottom Pleistocene deposit, which rests on Deccan Trap basalt. The top, thin, single layer of agricultural soil of black cotton soil is intermixed with sediments. The lower set of thin, poorly to moderately stratified layers shows current and cross-bedding of Holocene. This layer covers one-third part of stratigraphy and occupies the upper part of the trench, in which a total of 32 layers (H–L1) were encountered (Table 14.1). These layers are rich in floral remains (such as leaf, stem, root, pollen) and faunal remains Cephalopod, Cerithium, Unio, etc. While the third and bottommost layer of Pleistocene deposit is well cemented with $CaCO_3$ material. Resulting, they are very hard, compact, and show white to brownish color. These layers consist of olive colored clay which is well cemented with the polished pebbly cobbles, which are formed by spheroidal weathering of the bed rock basalt.

This Pleistocene deposit consists of seven (A–G) layers, which rest on Deccan Trap basalt of Upper Cretaceous to Lower Eocene age. The important characteristic of these lowermost layers is that they are rich in animal fossils. Most of the layers show a variation in the size and shape of cemented grains and binding material cement (Plate 14.1). The boundary of most of the layers is very distinct; especially, the unconformity in between Holocene and Pleistocene is very clear. Most of the layers are irregularly deposited on one another.

DETAILS OF PREVIOUS WORK

(1) As stated above, the Math Pimpri, where cultural stratigraphy is found intercalated with Quaternary deposit, is a unique site present in between the two channels of Sina River. Due to the availability of raw materials, some Mesolithic and Paleolithic sites are observed (Sabale and Kshirsagar, 2014; Shivaji et al., 2014, 2016); (2) the excavation was carried up to 2 months in two successive years; (3) to study the different nature of materials, three separate trenches were taken. The sediment samples were collected from each stratigraphic layer from three different localities (trenches). Laboratory experimental analysis of sediments was carried out to understand the depositional sedimentary environment and ultimately to reconstruct the paleoenvironmental conditions of the Sina Basin at Math Pimpri. Therefore, initially according to the required sampling procedure, sufficient quantities of samples were collected. Before sorting of the sample, remains of cultural materials, flora, and fauna were separated and accordingly handed over to the concerned laboratories; (4) after completion of excavation program at the end of March 2014, the various laboratory analytical works were started in the different laboratories; (5) the details of analyses and their proposed aim are given below; (6) sedimentary analysis and size and shape analysis were undertaken to understand the sources of sediment and fossil at the different depositional environment; (7) the qualitative and quantitative geomorphic analytical works were undertaken to understand the suitability of landscape and topographical characteristics. On the basis of the same, one can interpret and understand the interrelationship between landscape and the number of settlements;

TABLE 14.1
Shows the Characteristics of Sedimentary Deposite Observed at Locality-I

Sr. No.	Layer	Thickness (mm)	Lithological Description
1	M1	337–350	Black cotton virgin soil intermixed with subangular to subrounded pebbles, sand and silt.
2	L1	326–337	Silty sand having angular to subrounded pebbles intermixed with fine silty clay.
3	K1	316–326	A homogeneous fine textured black cotton soil layer mixed with fine silt and sand rich in plant remains and roots especially.
4	J1	306–316	Angular to subangular pebbles and subrounded to subangular cobbles cemented with sandy, silty clay.
5	I1	296–306	Sandy, silty compact layer intermixed with oblate to subrounded cobbles which are cremated with fine silty clay.
6	H1	287–296	Broad horizontal layer rich in very coarse subangular to angular rubbles and cobbles as well as subrounded to rounded or oblate pebbles intermixed with very course few sand and fine silty clay.
7	G1	279–287	Squarish or oblate pebbles cemented with course few sand and fine silty clay.
8	F1	274–279	Subrounded to angular cobbles arranged horizontally and mixed with course pebbly sand and poorly cemented with black clay.
9	E1	269–274	Angular to subangular pebbles intermixed with few course subrounded sand grins cemented with silty clay.
10	D1	264–269	Even course angular to subangular loose horizontal, sandy layer poorly cemented with fine silty clay.
11	C1	252–264	Subrounded to oblate shaped pebbles and cobbles intermixed with medium to course sand.
12	B1	244–252	Oblate to subrounded pebbles and course sand mixed with silty clay. At places shell fragments are found.
13	A1	252–244	Even silty sand intermixed with fine silty black cotton soil.
14	Z	216–223	Angular to subrounded pebbles intermixed with fine clayey sand.
15	Y	205–216	Subrounded to subangular grains intermixed with medium sand. At places, shell fragments are found.
16	X	194–205	Subrounded to subangular pebbles as well as cobbles intermixed with course sand.
17	W	190–194	Medium sand intermixed with rounded to subrounded pebbles.
18	V	179–190	Sandy clay layer intermixed with subrounded to rounded pebbles or grains.
19	U	170–179	Rounded to subrounded and angular to subangular pebbles intermixed with sand.
20	T	163–170	Thin bed of medium sand intermixed with angular to subangular and rounded to subrounded pebbles.
21	S	154–163	Medium sized rounded to subrounded pebbles intermixed with clay. At places, shell fragments are found.
22	R	147–154	A thin cross bed having angular grains which are intermixed with course sand and fine silt.

(Continued)

TABLE 14.1 (*Continued*)
Shows the Characteristics of Sedimentary Deposite Observed at Locality-I

Sr. No.	Layer	Thickness (mm)	Lithological Description
23	Q	142–147	Cross bedded thin band of subangular pebbles intermixed with medium sized sand.
24	P	138–142	Cross bedded medium sand intermixed with rounded to subrounded grains.
25	O	127–138	Angular and rounded to subrounded grains intermixed with medium sized sand.
26	N	120–127	Sandy silt intercalated with angular to subangular pebbles.
27	M	114–120	Sity sand having subrounded to rounded grains cemented with black cotton soil.
28	L	105–114	Subrounded to rounded and subangular pebbles intermixed with medium grained sand.
29	K	98–105	Subrounded to rounded and subangular pebbles intermixed with sand.
30	J	91–98	Subangular to subrounded grains intermixed with medium sized sand.
31	I	82–91	A thin band of plane tabular flows in which pebbles intermixed with medium sized sandy silt. Shell fragments are found.
32	H	75–82	Sandy silty calcareous layer intermixed with upper brown silt.
33	G	70–75	Whitish colored fine textured calcareous layer separating upper and lower coarse layer.
34	F	63–70	Angular to subangular brownish to yellowish colored corroded layer of sandy silt.
35	E	48–63	Brownish to reddish silty sand intermixed with angular pebbles and cemented with reddish cement.
36	D	31–48	Angular to subrounded pebbles intermixed with sand.
37	C	18–31	Oblate flat subrounded pebbles cobbles intermixed with sand.
38	B	10–18	Subrounded to subangular pebbles, cobbles intermixed with sand
39	A	0–10	Sandy clay

(8) the remote sensing and aerial photographs of this area are used to understand the suitability of this region; (9) petrographical studies of artifacts were also proposed to understand their source, i.e., whether they belong to local area or are brought out from other areas; (10) for catchment analysis, about 5 km area in a radius from the site was taken for the analysis of the objects; (11) the samples were collected for pollen and archaeobotanical assemblage studies; (12) the faunal assemblages of *Bosindicus* were collected from the middle lower Pleistocene layer of locality and were handed over for the fluorine dating analysis; (13) the assemblages of the cattle burial were handed over to the archaeological laboratory for further analysis; (14) a human skeleton was found at the top of this trench, which was analyzed in the laboratory (Mushrif-Tripathy and Sabale, 2015); and (15) the indurated potteries of the same strata were sent for magnetic analysis.

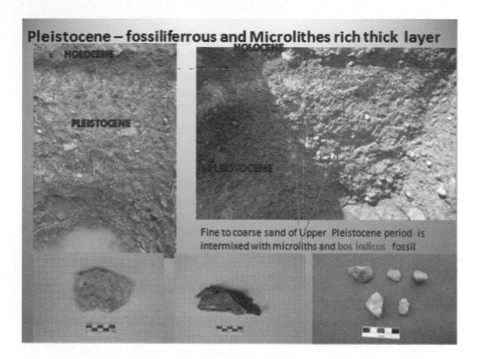

PLATE 14.1   Shows the picture of different fossil-bearing Quaternary layers.

DETAILS OF TOOLS AND FAUNAL REMAINS OBSERVED IN DIFFERENT STRATAS

a. Mesolithic tools:

A large number of tools were recovered from the right bank side of river surface as well as from the stratigraphical succession of Quaternary deposit (Plate 14.1).

b. Faunal remains:

Layer F: An ischium portion of pelvic girdle of *Equid* was collected through this layer.

Layer E: A second phalanx of young cattle *Bos namadicus* was collected from this layer.

Layer D: A shaft fragment of long bone of large mammals (such as cattle, buffalo, *Boselaphus, tragocamelus*) was recovered from this layer.

Layer C: Carpel of unidentified mammal (camel or elephant) was collected.

Layer B: Absent.

Layer A:

1. In this layer, huge two fossils of (vertebra) *Bos namadicus* (cattle) were recovered.

2. A huge number of fragments of *Lamellibranchs* shell mostly in soft and very little hard forms were collected.

Discussion Based on Taphonomy

The above-mentioned vertebrate assemblage was collected from the Quaternary trench in two successive seasons 2013–2014 and 2014–2015 of Quaternary excavation. The assemblages of an ischium portion of pelvic girdle of equid; phalanx of young cattle *Bos namadicus;* a shaft fragment of long bone of large mammals (viz., cattle, buffalo, *Boselaphus, tragocamelus);* carpel of unidentified mammal (camel or elephant); huge two fossil of (vertebrae) *Bos namadicus* (cattle); and a, huge number of fragments of *Lamellibranchs* shell were collected within the different taphonomic modes and were well preserved (Sabale, 2012) (Plates 14.1–14.3). The presence of remains of these animals acts as a local source. The mollusks and shell fragments collected from each layer suggest that most of the layers are formed by flood deposition material and the animals lived in lacustrine habitats. The excellent preservation of several big animal fossil specimens in fluvial deposits at the river bank section in upper reaches of Sina River at this locality also indicates a short transport followed by a rapid burial at the local site and suggests that this could also live in fluvial environments. The presence of carpel of unidentified mammal (camel or elephant) is rather unusual. However, the presence of shell and mollusk is abundant in freshwater environments during the Cretaceous.

Ischium portion of Pelvic girdle Equid. Second phalanx of young cattle Boss Namadicus.

Shaft fragment of long bone of large sized mammals (such as Cattle, Buffalow, Boss Elephus stragocamelous).

PLATE 14.2   Details of the fossils collected from Locality-I.

Carpel of unidentified mammal (Camel / elephant).

Shell (lamellibranches).

(a)   (b)

Vertebra of Boss Namadicus (Cattle).

PLATE 14.3   Shows the picture of fossils recovered from the Pleistocene deposit at Locality-I.

The fossil assemblage's surface morphology shows sharp edges and faces at places, which clearly indicates the less transportation of the fossil assemblages from their source area.

Most of vertebrate remains are very well preserved in the above-mentioned fossil-bearing layers. The exceptional preservation of ribs, associated with impressions in the lowermost layer, reflects a rapid burial in fine-grained sediments after a short transport in the alluvial plain, which are deposited under low-energy flood deposits. However, most of the fossils (except the above-mentioned) are deposited with coarse-grained sediment under high-energy depositional conditions.

The taxonomic diversity of the above-mentioned bone beds clearly suggests distinct accumulations of vertebrate remains belonging to individuals that died in different places and/or at different times. Such remains of such animals were then transported by river streams from their different habitats, finally deposited, and gathered in the same area in a channel (Cincotta et al., 2015).

It appears that the big bones lowermost bed of *Bos namadicus* presents a nonrandom orientation pattern with a major north–south trend for the elongate elements (Plate 14.3). In this, pebbly–cobble–sandy layer elements are heterogeneous with various sizes and shape ranges. The orientation pattern with heterogeneous elements probably results from larger elements acting as obstacles for the other smaller elements, while in case of upper layers in which sediments are deposited in a haphazard position, there is no clear preferential orientation that probably results from low-energy depositional conditions. During the deposition of such beds, currents were probably too weak to orient all the elements in the same direction. In other case, by the sudden deposition irregular orientation of the elements in the layers could also originate in the abundance of elements. This has been demonstrated by the bones transported in shallow water deposited in a parallel orientation to the flow, suggesting nearly a north–south trend of the currents that transported these bones.

Organic particles are typically found in the finer sediments because of their low densities. Thus, organic matter reflects a low-energy depositional environment far from the river stream.

In the sediment layers, numerous fragments of charcoal might indicate the presence of wildfires or other activities in the concerned time period. Moreover, wildfires might intensify landslides or firing activities and thus influence sedimentation. On the other hand, charcoal might also be the result of plant debris maturation. The scarcity of pollen grains in the sediment is more likely the result of the poor preservation in the basic soils and the distance from the vegetation sources other than their nonpreservation.

The vertebrate assemblage discovered at Math Pimpri Quaternary trench at Locality-I clearly represents a mixture of aquatic and terrestrial animals that lived in different habitats. Transported vertebrate species appear to be well preserved in over bank deposits, due to rather short transport in a low-energy fluvial system. On the basis of their morphology, the fossil bodies clearly indicate that there is very little transport inside their life environment. While some freshwater shell–mollusk transported little more distance inside the fluvial realm, the bones of big animals in their vicinity and their remains were transported.

CONCLUSION

The sedimentological record observed in upper reaches of Sina, at Math Pimpri site, represents a vertical succession of several facies that belong to the channel and floodplain deposits in the lower–middle and upper parts of a fluvial depositional system. The lithostratigraphic section shows the evidence for both fluvial and dry environments. There are 37 sedimentological sequences recorded at Locality-I, which are dominated by overbank deposits, indicating the successive flooding episodes. The upper part of each layer shows slightly erosive surface, while the middle and lower portions of sequences are characterized by depositional episodes. The relatively fine-grained sediments and amalgamation of channel-fills suggest a river system surrounded by a floodplain. However, at places, the calcrete layers in between the flood deposits layer indicate drought situation due to very less rainfall in the upper catchment areas. Such calcium carbonates in calcrete layers cement, spread, and capture the lose sediments in the upper and lower contact layers. On the basis of the results of the present work, it is clear that the river flowing in severe chronic drought-prone, semi-arid region of Maharashtra is also rich in faunal records of a variety of species, which occurs in other comparatively perennial rivers like Bhima. Therefore, one can conclude that the paleoenvironmental conditions in Sina were not that much different than those in Bhima even it is extremely dry today. Detailed investigations are still required to understand the biodiversity of the basin in Quaternary Era.

ACKNOWLEDGMENT

I thank the authorities of Deccan College Post Graduate and Research Institute, Pune, especially Prof. Vasant Shinde (Director) and Dr. G.B. Deglurkar (President), for their guidance and encouragement during the work. I am grateful to the colleagues who visited the site and offered suggestions. I am also thankful to villagers for their help during the progress of excavation.

REFERENCES

A. Cincotta, J. Yans, P. Godefroit, G. Garcia, J. Dejax, M. Benammi, S. Amico, and X. Valentin. 2015. Integrated paleoenvironmental reconstruction and taphonomy of a unique upper cretaceous vertebrate-bearing locality (Velaux, Southeastern France). *PLOS One* 10: e0134231.

K. Shivaji, S. Mishra, and P. D. Sabale. 2014. Some acheulian localities in the Sina basin. *Bulletin of the Deccan College* 74: 163–168.

P. D. Sabale. 2012. Geoarchaeological investigations in middle reaches of Bhima. An unpublished report of major Research Project submitted to University Grant Commission, New Delhi.

P. D. Sabale, J. D. Deshmukh, S. D. Kshirsagar, D.M. Pokharkar, K. G. Sarwade, P. V. Kamble, and Nagaraja Rao. 2015. Preliminary report on quaternary geoarchaeological excavation at Math Pimpri, District Ahmednagar. *Bulletin of the Deccan College Research Institute* 75: 25–36.

P. D. Sabale and S. D. Kshirsagar. 2014. A preliminary report on geoarchaeological explorations in upper reaches of Sina river basin: A case study of Ahmednagar, Bid, Osmanabad and Solapur districts of M.S. *Bulletin of the Deccan College* 7: 127–138.

S. D. Kshirsagar, G. L. Badam, and P. D. Sable. 2016. A preliminary note on the quaternary geology and middle –late Pleistocene fauna collected from various tributaries of the Bhima valley, Maharashtra. *Journal of the Palaeontological Society of India* 61(1): 123–132.

V. Mushrif-Tripathy and P. D. Sabale. 2015. Note on human skeletal finds belonging to modern time from Math Pimpri. *Bulletin of the Deccan College* 74: 153–158.

Glossary

Accelerator mass spectrometry (AMS)   It is used primarily for the quick and accurate measurement of isobaric isotope ratios and rare radioactive isotopes, particularly radiocarbon.

An algal bloom or algae bloom   It is a rapid increase or accumulation in the algae population in freshwater or marine water systems, and is often recognized by the water's discoloration from their pigments.

Archaeological findings   It is the body of physical (not written) evidence about the past. Archaeological theory is used to interpret archaeological record for a better understanding of human cultures. The archaeological record can consist of the earliest ancient findings as well as contemporary artifacts.

Atlantic Multidecadal Oscillation (AMO)   It has been identified as a coherent mode of natural variability occurring in the North Atlantic Ocean for an estimated 60–80 years. It is based upon the average anomalies of sea surface temperatures (SST) in the North Atlantic Basin, typically over 0–80 N.

Benthic foraminifera   These are single-celled organisms similar to amoeboid organisms in cell structure. Benthic foraminifera occupies a wide range of marine environments, from brackish estuaries to the deep ocean basins, and occurs at all latitudes.

Bottom surface sediments (BSS)   Bottom sediment heat exchangers consist of pipes laid on or buried in the surface water body's bottom sediment. Because these pipes can contact the sediment layer, the heat exchangers can absorb or reject heat directly to/from the sediment layer and the surface water body.

Bottom water temperature (BWT)   The Antarctic bottom water (AABW) is a type of water mass in the Southern Ocean surrounding Antarctica with temperatures ranging from −0.8°C to 2°C (35°F), and salinities from 34.6 to 34.7 psu. This is due to the oxidation of deteriorating organic content in the rest of the deep oceans.

Braided river system   A braided river is a network of tiny channels separated by islands that are often not fixed. In other words, the river channels wander across a flat area. Braided rivers are found in places where the river carries a lot of sediment and slows down and spreads out.

Calcium carbonate   It is a chemical compound with the formula $CaCO_3$. It is a common substance found in rocks as the minerals calcite and aragonite (most notably as limestone, a sedimentary rock consisting mainly of calcite). It is the main component of eggshells, snail shells, seashells and pearls.

Cenozoic Era   It is the third of the significant eras of Earth's history, beginning about 66 million years ago and extending to the present. It was the interval of time during which the continents assumed their modern configuration and geographic positions and during which Earth's flora and fauna evolved toward those of the present.

Coastal variability   Along active margins, uplifted rocky coasts dominate. Emergent coasts result from local tectonic or uplift or a drop in eustatic sea level. Emergent coasts are characterized by rocky shores with sea cliffs and raised wave-cut benches (marine terraces).

Coral fragments   Corals are modular organisms that have the potential to undergo colony fragmentation regularly. A coral fragment is defined as "a live portion of the colony that has become physically separated, due to the breakage of the skeleton, from the rest of the colony."

Coral sand   It is a collection of sand of particles originating in tropical and subtropical marine environments from bioerosion of limestone skeletal material of marine organisms. Because it is composed of limestone, coral sand is acid soluble.

Cosmogenic nuclides (or cosmogenic isotopes)   These are rare nuclides (isotopes) created when a high-energy cosmic ray interacts with the nucleus of an *in situ* solar system atom, causing nucleons (protons and neutrons) to be expelled from the bit (see cosmic ray spallation). These nuclides are produced within Earth materials such as rocks or soil, the Earth's atmosphere, and extraterrestrial items such as meteorites.

Cylinder-shaped stalactites   Dripstone is calcium carbonate in the form of stalactites or stalagmites. Stalactites are pointed pendants hanging from the cave ceiling, from which they grow. Soda straws are fragile but long stalactites with an elongated cylindrical shape rather than the usual more conical shape of stalactites.

Dansgaard–Oeschger cycles   Dansgaard–Oeschger event, also called the D-O event, any of several dramatic but fleeting global climatic swings characterized by a period of abrupt warming followed by a period of slow cooling during the last ice age.

Deep-Sea Drilling Project (DSDP)   It was an ocean drilling project operated from 1968 to 1983. DSDP provided crucial data to support the seafloor spreading hypothesis and helped prove the theory of plate tectonics.

East Asian Monsoon (EAM)   It is a "sea breeze monsoon," the most common type. They form because land and sea heat up at different rates, so high pressure forms over the ocean and low pressure over land, resulting in the wind blowing onshore in the summer.

Ecological niche   It is a term for a species' position within an ecosystem, describing the range of conditions necessary for its persistence and its ecological role in the ecosystem.

Ecosystem stresses   Ecological (or biological) stress occurs when a physical factor harms an ecosystem or its biotic components. In living organisms, this may result in risks to survival or restrictions in growth or reproduction.

El Nino–Southern Oscillation (ENSO)   ENSO is also known as a periodic fluctuation in sea surface temperature (El Niño) and the overlying atmosphere's air pressure (Southern Oscillation) across the equatorial Pacific Ocean.

En glacial lake   It is defined as a water mass existing in a sufficient amount and extending with a free surface in, under, beside, and/or in front of a glacier and originating from glacier activities and/or retreating processes of a glacier.

Enhancing Coastal and Ocean Resource Efficiency (ENCORE) ENCORE aims to strengthen integrated coastal zone management in all coastal States and Union Territories of India. The project seeks to assist the Government of India (GoI) in enhancing coastal resource efficiency and resilience by building collective capacity (including communities and decentralized governance) for adopting and implementing integrated coastal management approaches. Recognizing Integrated Coastal Zone Management (ICZM) as a continuous process rather than a one-off investment action, ENCORE will build upon and draw from the experience of the ongoing World Bank-supported Integrated Coastal Zone Management Project (ICZMP), including the linkages between coastal conservation, climate resilience, and poverty reduction.

Equatorial Counter Current (ECC) Equatorial counter current, a current phenomenon noted near the equator, an eastward flow of oceanic water in opposition to and flanked by the low westward currents of the Atlantic, Pacific, and Indian Oceans.

Equatorial Pacific The equatorial Pacific is an exception. This reversal drives warm surface water from the western Pacific eastward, leading to the development of hot sea surface temperatures east of the international dateline.

Estuaries The estuary where fresh and saltwater mix. Estuaries and their surrounding wetlands are bodies of water usually found where rivers meet the sea. Estuaries are home to unique plant and animal communities adapted to brackish water, a mixture of freshwater draining from the land and salty seawater.

Euro-Vector Elemental Analyzer Euro-Vector Elemental Analyzer provides the highest level of analysis performances in new energetic tasks such as selecting cleaner combustibles and transforming hazardous waste into safe energy.

Eutrophic lakes A eutrophic condition describes a situation where a water body has lost so much of its dissolved oxygen that everyday aquatic life begins to die off. Eutrophic conditions form when a water body is "fed" too many nutrients, especially phosphorus and nitrogen.

Flowstones Flowstone, mineral deposit found in "solution" caves in limestone. Flowing water films that move along floors or down positive-sloping walls build up layers of calcium carbonate (calcite), gypsum, or other cave minerals. These minerals are dissolved in the water and deposited when the water loses its dissolved carbon dioxide and its carrying ability. Flowstone is usually white or translucent but may be stained with various colors by minerals dissolved in the water.

Fluvial and dry environments Fluvial environments are a type of sedimentary environment describing where fluvial landforms (geomorphology) and fluvial deposits (facies) are created, modified, destroyed, and/or preserved through erosion, transport, and deposition of sediment.

Fluvial geomorphic studies   It is the study of river process and form. Water flowing within a channel transfers sediment in solution, suspension, and contact
   with the bed. Fluvial geomorphology focuses on rivers' physical characteristics, but biota interactions (including humans) and biological processes
   are vital.

Fluvial morphology   It is a subdiscipline of geomorphology that investigates how
   flowing water shapes and modifies the Earth's surface through erosional and
   depositional processes. The origin of fluvial comes from the Latin fluvialis,
   from fluvius (river), and its first known usage occurred in the 14th century.

Foraminiferal species   These are single-celled protozoans commonly found in
   marine environments (some are much bigger). Despite being single-celled,
   microscopic organisms, Foraminifera species are characterized by shells
   known as tests.

Glacial Isostatic Adjustment (GIA)   It describes the adjustment process of the
   Earth to an equilibrium state when loaded by ice sheets.

Glacial Lake Outburst Floods (GLOF)   It is a type of outburst flood caused by a
   dam's failure containing a glacial lake. An event similar to a GLOF, where
   a body of water held by a glacier melts or overflows the glacier, is called a
   jökulhlaup. The dam can consist of glacier ice or a terminal moraine.

Greenhouse gas (GHG) emission   Greenhouse gas, any gas that has the property
   of absorbing infrared radiation (net heat energy) emitted from the Earth's
   surface and reradiating it back to the Earth's surface, contributes to the
   greenhouse effect. Carbon dioxide, methane, and water vapor are the most
   important greenhouse gases.

Greenland ice cores   It was a multinational European research project organized
   through the European Science Foundation. GRIP successfully drilled a
   3,029-m ice core to the bed of the Greenland ice sheet at Summit, Central
   Greenland, from 1989 to 1992 at 72°34.74′N 37°33.92′W.

Ground Penetrating Radar (GPR)   It is a geophysical locating method that uses
   radio waves to capture images below the ground's surface in a minimally
   invasive way. The massive advantage of GPR is that it allows crews to pin-
   point the location of underground utilities without disturbing the ground.

Gulf of Mannar Biosphere Reserve (GOMBR)   It is located in the southeastern tip
   of Tamil Nadu, extending from Rameswaram in the north to Kanyakumari
   in the south. The extent of GOMBR is 10,500 km², with the core area covering 560 km, having 3,600 species of fauna and flora.

Holocene Climatic Optimum   It was a warm period during roughly the interval
   9,000–5,000 years BP, with a thermal maximum of around 8,000 years BP.
   Many other names have also been known, such as Altithermal, Climatic
   Optimum, Holocene Megathermal, Holocene Optimum, Holocene Thermal
   Maximum, Hypsithermal Mid-Holocene Warm Period.

Holocene layers   The Holocene is the name given to the last 11,700 years of the
   Earth's history since the end of the last major glacial epoch, or "ice age."
   Since then, there have been small-scale climate shifts, notably the "Little
   Ice Age" between about 1200 and 1700 AD, but in general, the Holocene
   has been a relatively warm period in between ice ages.

Holocene pale monsoon   The Holocene period has been classified further based on the globally observed abrupt climatic events at 8.2 and 4.2 ka. The 8.2 ka global cooling events have been recorded from northern Indian Ocean marine archives, but limited records from the continental archives of the Indian land-mass have demonstrated the 8.2 ka event. Simultaneously, the 4.2 ka dry climate has been endorsed by both marines and continental records, and agrees with the global studies. During the "Little Ice Age" (LIA), in the Indian subcontinent, wet conditions prevailed in the northern, central, and western regions. At the same time, a dry climate existed over the greater part of peninsular India. The present review offers an account of ISM signatures and possible mechanisms associated with the monsoon variability in the Indian subcontinent and the northern Indian Ocean during the Holocene period.

Holocene period   The Holocene Epoch is the current period of geologic time. Another term used is the Anthropocene Epoch, because its primary characteristic is the global changes caused by human activity. This term can be misleading, though; modern humans were already well established long before the epoch began.

Hydrostatic deformation   Hydrostatic deformation of isotactic polypropylene (i-PP) of an intermediate weight average molecular weight (2.9×10^5) has been achieved by uniaxial compression between two axially aligned circular cylinders at the desired draw temperature (TDR). Independent measurements of load (i.e., stress) and displacement (i.e., strain) are made during deformation. A systematic variation in TDR (from 30°C to 130°C) at a ram speed of 0.25 in./min has been studied. The stress-strain data corrected for machine compliance has been compared to a theoretical model. This model assumes a rigid-plastic behavior with a hydrostatic pressure effect.

Ice-dammed glacial lakes   These are common feature of glaciated mountain ranges. They form wherever glacial ice blocks the drainage of rivers or meltwater. This includes a glacier block, a trunk, or tributary valley, and a glacier fills an over-deepened valley created by glacial erosion.

Indian Ocean Dipole (IOD)   It is also known as the Indian Niño, is an irregular oscillation of sea surface temperatures in which the western Indian Ocean becomes alternately warmer (positive phase) and then colder (negative phase) than the eastern part of the ocean.

Indian Summer Monsoon (ISM)   The Indian summer monsoon rainfall (ISMR) or Southwest monsoon rainfall during June–September is a component of the Asian monsoon system, which accounts for 70%–90% of India's annual precipitation. The ISMR exhibits high temporal as well as spatial variations.

Integrated Ocean Drilling Program (IODP)   It was an international marine research program. The program used heavy drilling equipment-mounted aboard ships to monitor and sample sub-seafloor environments.

Intergovernmental Panel on Climate Change (IPCC)   It is the United Nations body for assessing the science related to climate change.

International Commission on Stratigraphy (ICS)   It is the largest and oldest constituent scientific body in the International Union of Geological Sciences (IUGS).

International Union of Geological Science (IUGS)   It was founded in 1961, with 121 national members representing over a million geoscientists, is one of the World's largest scientific organizations. It encourages international co-operation and participation in the Earth sciences with human welfare, and is a member of the International Science Council (ISC).

Intertropical Convergence Zone   The region circles the Earth near the equator, where the Northern and Southern Hemispheres' trade winds come together. The equator's intense sun and warm water heat the air in the ITCZ, raising its humidity and making it buoyant.

Jurassic Period   It is the second of three periods of the Mesozoic Era. Extending from 201.3 million to 145 million years ago, it immediately followed the Triassic Period (251.9–201.3 million years ago) and was succeeded by the Cretaceous Period (145–66 million years ago). The Morrison Formation of the United States and the Solnhofen Limestone of Germany, both famous for their exceptionally well-preserved fossils, are geologic features formed during Jurassic times.

Lacustrine sections   A lacustrine plain or lake plain is a plain formed due to the past existence and its accompanying sediment accumulation. Lacustrine plains can be created through one of three central mechanisms: glacial drainage, differential uplift, and inland lake creation and drainage.

Last glacial maximum (LGM)   It, also referred to as the Late Glacial Maximum, was the most recent time during the Last Glacial Period that ice sheets were at their greatest extent. The LGM is referred to in Britain as the Dimlington Stadial, dated between 31,000 and 16,000 years.

Lithostratigraphic   It is defined as "the description and systematic organization of rocks into distinctive named units based upon the lithological character of the rocks and their stratigraphic relations."

Marine Isotopic Stage (MIS)   It is marine oxygen-isotope stages, or oxygen iso-tope stages (OIS), are alternating warm and cool periods in the Earth's paleoclimate, deduced from oxygen isotope data reflecting changes in temperature derived from data from deep-sea core samples.

Medieval Climate Anomaly (MCA)   It is also known as the Medieval Climate Optimum, or Medieval Climatic Anomaly, a time of warm climate in the North Atlantic region lasting from c. 950 to c. 1250. It was likely related to warming elsewhere, while some other areas were colder, such as the tropi-cal Pacific.

Mesotrophic lakes   These are standing water bodies with a surface area greater than 1 ha, moderate alkalinity levels and nutrients. Although they may look similar to other standing water types, this is an unusual loch type in Britain. The sites have clear water and support a wide range of species.

Micropaleontology   Micropaleontology involves studying organisms so tiny that they can be observed only with the aid of a microscope. The size range of microscopic fossils, however, is immense. In most cases, the term "micropa-leontology" connotes that aspect of paleontology devoted to the Ostracoda.

Mid-oceanic ridges   A mid-ocean ridge (MOR) is a seafloor mountain system formed by plate tectonics. It typically has a depth of ~2,600 m (8,500 ft) and rises about 2 km above the ocean basin's deepest portion. This feature is where seafloor spreading takes place along a divergent plate boundary.

Milankovitch cycles   It includes the shape of the Earth's orbit, known as eccentricity; the angle earth's axis is tilted to Earth's orbital plane, known as obliquity . The direction Earth's axis of rotation is pointed, known as precession.

Millennial-scale changes   Traditionally, it has been thought that millennial-scale changes during the last glacial interval are driven either by instabilities in the Northern Hemisphere ice sheets or feedback related to Atlantic thermohaline circulation, driven by strange input of meltwater into the North Atlantic. Despite extensive efforts, however, tracking all the meltwater sources and mechanisms for thermohaline oscillations remains controversial. Other ideas for the origin of millennial-scale climate changes may involve rhythmic solar forcing or internal resonant fluxes of the coupled ocean–atmosphere system.

Miocene period   The late Miocene was a time of global drying and cooling. As ice rapidly accumulated at the poles, sea levels fell, rainfall decreased, and rainforests retreated. Many plant and animal groups died out and other forms, better adapted to a drying world, took their place.

Moderate Resolution Imaging Spectroradiometer (MODIS)   It is a key instrument aboard the Terra (originally known as EOS AM-1) and Aqua (originally known as EOS PM-1) satellites. Terra's orbit around the Earth is timed so that it passes from north to south across the equator in the morning, while Aqua passes south to north over the equator in the afternoon. Terra MODIS and Aqua MODIS view the entire Earth's surface every 1–2 days, acquiring data in 36 spectral bands or groups of wavelengths (see MODIS Technical Specifications). These data will improve our understanding of global dynamics and processes occurring on the land, the oceans, and the lower atmosphere. MODIS is playing a vital role in developing validated, global, interactive Earth system models able to predict global change accurately enough to assist policymakers in making sound decisions concerning the protection of our environment.

Moist deciduous forest   Moist deciduous forests are a mixture of trees and grasses. These forests are found in moderate rainfall areas of 100–200 cm per annum, a mean annual temperature of about 27°C, and an average annual relative humidity of 60%–75%.

Moonmilk cave deposit   Moonmilk (sometimes called mondmilch, also known as montmilch or as cave milk), is a white, creamy substance found inside limestone, dolomite, and possibly other caves.

Moraine-dammed glacial lakes   A moraine-dammed lake occurs when the terminal moraine has prevented some meltwater from leaving the valley. Its most common shape is that of a long ribbon (ribbon lake).

Morphometric analyses   Morphometric analysis, quantitative description, and analysis of landforms as practiced in geomorphology may be applied to a particular kind of landform or drainage basins and large regions. Formulas for right circular cones have been fitted to alluvial fans' configurations, logarithmic spirals have been used to describe specific shapes of beaches, and drumlins, spoon-shaped glacial landforms, have been found to accord to the form of the lemniscate curve.

North Atlantic Oscillation (NAO) It is a weather phenomenon over the North Atlantic Ocean of fluctuations in atmospheric pressure at sea level (SLP) between the Icelandic Low and the Azores High. The NAO was discovered through several studies in the late 19th and early 20th centuries.

North Equatorial Current (NEC) North Equatorial Current (NEC) is represented by the Mariano Global Surface Velocity Analysis (MGSVA). The NEC is the broad westward flow that is the southern component of the North Atlantic subtropical gyre. The NEC is found in the North Atlantic from about 7°N to about 20°N.

Northeast monsoon (NEM) India receives rainfall during two seasons. About 75% of the country's annual rainfall is obtained from the Southwest monsoon between June and September. The Northeast monsoon, on the other hand, occurs from October to December and is a comparatively small-scale monsoon, which is confined to the Southern peninsula.

Also called the winter monsoon, the Northeast monsoon's rainfall is vital for Tamil Nadu, Puducherry, Karaikal, Yanam, coastal Andhra Pradesh, Kerala, north interior Karnataka, Mahe, and Lakshadweep.

Some South Asian countries such as Maldives, Sri Lanka, and Myanmar, too, record rainfall from October to December.

Ocean Drilling Program (ODP) It was funded by the US National Science Foundation and 22 international partners (JOIDES) to conduct primary research into the ocean basins' history and the crust's general nature beneath the ocean floor using the scientific drillship JOIDES Resolution.

Oligotrophic lakes Limnologists use the term "oligotrophic" to describe lakes with low primary productivity due to nutrient deficiency. Due to their low algal production, these lakes have clear waters with high drinking-water quality.

Orbital-scale variations The Earth's orbit varies between nearly circular and mildly elliptical (its eccentricity varies). When the rotation is more elongated, there is more variation in the distance between the Earth and the sun and solar radiation at different times in the year.

Organic carbon (OC) Soil organic carbon is a measurable component of soil organic matter. Organic matter contributes to nutrient retention and turnover, soil structure, moisture retention and availability, degradation of pollutants, and carbon sequestration.

Organic matter Organic compounds have come from the remains of organisms such as plants and animals and their waste products in the environment. Organic molecules can also be made by chemical reactions that don't involve life.

Ostracod carapace One of the defining characteristics of ostracods is their carapace or shell, which, when closed, covers the noncalcified body parts and appendages. The carapace originates from the head region and consists of two valves hinged along the dorsal margin.

Ostracoda   These are a class of the Crustacea (class Ostracoda), sometimes known as seed shrimp. They are small crustaceans, typically around 1 mm (0.039 in) in size, but varying from 0.2 to 30 mm (0.008–1.181 in) in the case of Gigantocypris.

Pacific Decadal Oscillation dynamics   It is a robust, recurring pattern of ocean–atmosphere climate variability centered over the mid-latitude Pacific Basin. The PDO is detected as warm or cool surface waters in the Pacific Ocean, north of 20°N. This PDO index is the standardized principal component time series.

Palynomorphs   Palynomorphs are broadly defined as organic-walled microfossils between 5 and 500 μm in size. Palynomorphs may be composed of organic materials such as chitin, pseudochitin, and sporopollenin.

Peninsular   Use the adjective peninsular to describe a near-island that is connected to the mainland. A peninsula is a piece of land that juts out into the water, nearly an island. Something that's peninsular looks like a peninsula or is a geographical area with a lot of peninsulas.

Pleistocene deposit   About 2 million years ago, continental glaciers moved generally southward across North America, covering eastern South Dakota several times. As each ice sheet advanced, it transported large volumes of rock debris frozen into the lower layers of ice.

Pleistocene ice sheets   The Quaternary glaciation, also known as the Pleistocene glaciation, is an alternating series of glacial and interglacial periods during the Quaternary period that began 2.58 Ma (million years ago) and is ongoing. During glacial periods, they expanded, and during interglacial periods, they contracted.

Pollen analysis   Pollen analysis is a scientific method that can reveal evidence of past ecological and climate changes: it combines the principles of stratigraphy with observations of actual (modern) pollen–vegetation relationships to reconstruct the terrestrial vegetation of the past.

Post-Archean Australian Shale   The composition of post-Archaean terrigenous clastic sedimentary rocks is very uniform for several insoluble trace elements (e.g., REE, Th, Sc). It is thought to reflect the composition of the exposed upper continental crust.

Potential flood volume (PFV)   The potential flood volume (PFV) for each lake was assessed by calculating the maximum PFV of self-destructive and dynamic failures. In case of operational failure, the PFV was estimated based on the assumption that the water displaced will be equal to the volume of mass entering the lake.

Precambrian geochemistry   The Precambrian section is divided clearly into two carbonate rock associations. The Archean–Early Proterozoic stage was dominated by calcium and iron carbonates. Such differentiation of carbonate rock types suggests relatively acid and alkaline geochemical settings at the first and second stages.

Primordial radionuclides   These are residues from the Big Bang, from cosmogenic sources, and from ancient supernova explosions before forming the solar system. Bismuth, thorium, uranium, and plutonium are primordial radionuclides because they have half-lives long enough to be still found on the Earth.

Relatively undisturbed-relict sediment   Relict sediments are remnants from an earlier environment and are now in disequilibrium. Approximately 50% of the present continental shelves are covered by relict sediments deposited during lower sea levels in the Pleistocene.

River runoff   It is the amount of water entering local rivers, and the discharge is the accumulation of all upstream runoff. In most regions, between 40% and 70% of the precipitation flow flows into the river network.

Scanning electron microscopy (SEM)   A scanning electron microscope (SEM) is a type of electron microscope that produces images of a sample by scanning the surface with a focused beam of electrons. The electrons interact with atoms in the model, producing various signals that contain information about the surface topography and composition of the sample. The electron beam is scanned in a raster scan pattern, and the position of the shaft is combined with the intensity of the detected signal to produce an image.

Sea level rise (SLR)   It is a major global concern and a pressure for several goals of the Ocean Health Index. Its ultimate cause is the rise in global temperature caused by the release of carbon dioxide and other heat-trapping gases resulting from fossil fuel combustion.

Sea surface salinity (SSS)   It plays a fundamental role in the density-driven global ocean circulation, the water cycle, and climate. The CCI+ phase of the sea surface salinity project will involve fully exploiting the ESA/Earth explorer SMOS mission complemented with SMAP and AQUARIUS satellite missions.

Stream length (SL) indices   Among the geomorphic indices, the stream length-gradient (SL) index represents a practical tool to highlight anomalous changes in river gradients.

Soda straws   A soda straw (or simply straw) is a speleothem in the form of a hollow mineral cylindrical tube. They are also known as tubular stalactites. Soda straws grow in places where water leaches slowly through cracks in the rock, such as on caves' roofs.

South Equatorial Current (SEC)   It is a broad, westward-flowing current extending from the surface to a nominal depth of 100 m. Its northern boundary is usually near 4°N, while the southern border is generally found between 15° and 25°S, depending primarily on longitudinal location and the time of the year.

Southern Annular Mode (SAM)   It is also known as the Antarctic Oscillation (AAO), which describes the north–south movement of the westerly wind belt that circles Antarctica, dominating the middle to higher latitudes of the Southern Hemisphere.

Southwest Monsoon (SWM)   The term "monsoon" was first used in English in British India, including present-day India, Bangladesh, and Pakistan. The Southwest Monsoon winds are called "Nairutya Maarut" in India. Southwest Monsoon in India is conceived as a complex phenomenon. A cautious approach is generally taken before announcing the onset or withdrawal of monsoon in India.

July and August are generally the active monsoon months for the entire country. June is the onset month that witnesses an outburst of rain, while September is the withdrawal month, receiving more sporadic rain.

Speleothem Isotopes Synthesis and Analysis (SISAL)   SISAL (Speleothem Isotope Synthesis and Analysis) is an international working group for the Past Global Changes (PAGES) project. The working group aims to provide a comprehensive compilation of speleothem isotope records for climate reconstruction and model evaluation.

Speleothems   Speleothems or cave formations are secondary mineral deposits formed in a cave. Speleothems typically include limestone or solutional dolomite caves.

Stalactites   A stalactite is an icicle-shaped formation that hangs from the ceiling of a cave and is produced by the precipitation of minerals from water dripping through the cave ceiling. Most stalactites have pointed tips.

Stalagmites   A stalagmite is an upward-growing mound of mineral deposits precipitated from water dripping onto a cave floor. Most stalagmites have rounded or flattened tips. There are many other types of mineral formations found in caves.

Subglacial lake   It is a lake found under a glacier, typically beneath an ice cap or ice sheet. As ecosystems isolated from the Earth's atmosphere, subglacial lakes are influenced by interactions between ice, water, sediments, and organisms.

Submarine Mass Failures (SMF)   These are a potential source of dangerous tsunamis. While the link between seismic events and the magnitude of tsunami waves has been extensively studied and corresponding approaches are included in numerical tsunami warning models, the basic implementation of SMF-generated waves is subject to ongoing research.

Supraglacial lake   A supraglacial lake is any pond of liquid water on the top of a glacier. Although these pools are temporary, they may reach kilometers in diameter and be several meters deep. They may last for months or even decades at a time but can empty in hours.

Swampy and marshy habitat   Swamps and marshes are specific types of wetlands that form along waterbodies containing rich, hydric soils. Marshes are wetlands, continually or frequently flooded by nearby running bodies of water, that are dominated by emergent soft-stem vegetation and herbaceous plants.

Taxonomic diversity   It can be defined as the average taxonomic path between randomly chosen individuals. It takes into consideration taxonomic differences and heterogeneity (species richness and evenness). Taxonomic distinctness can be defined as the average taxonomic path between two individuals from different species.

Tertiary age   Tertiary period, a former official interval of geologic time lasting from approximately 66 to 2.6 million years ago. It is the traditional name for the first of two periods in the Cenozoic Era (66 million years ago to the present); the second is the Quaternary Period (2.6 million years ago to the present).

Thermal expansion   Thermal expansion is the general increase in the volume of a material as its temperature is increased. It is usually expressed as a fractional change in length or volume per unit temperature change; a linear expansion coefficient is generally employed in describing the expansion of a solid, while a volume expansion coefficient is more beneficial for a liquid or a gas.

Tropical zone   The tropics are the Earth's region near the equator and between the Tropic of Cancer in the Northern Hemisphere and the Tropic of Capricorn in the Southern Hemisphere. This region is also referred to as the tropical zone and the torrid zone. The word "tropical" specifically means places near the equator.

Tsunami   It is a series of waves caused by earthquakes or undersea volcanic eruptions. Tsunamis are giant waves caused by earthquakes or volcanic eruptions under the sea. Out in the depths of the ocean, tsunami waves do not dramatically increase in height.

West Indian Coastal Current (WICC)   The West India Coastal Current (WICC) flows northward during November–February and southward during April–September. In terms of these waves, the West India Coastal Current is a superposition of annual and semiannual coastally trapped Kelvin waves.

Younger Dryas (YD)   The Younger Dryas (around 12,900–11,700 years BP) returned to glacial conditions after the Late Glacial Interstadial, which temporarily reversed the gradual climatic warming after the Last Glacial Maximum (LGM) started receding around 20,000 BP. The current theory is that the Younger Dryas was caused by significant reduction or shutdown of the North Atlantic "Conveyor," which circulates warm tropical waters northward to a sudden influx of fresh water from Lake Agassiz deglaciation in North America.

Index